ENDANGERED ANIMALS

A Reference Guide to Conflicting Issues

Edited by
Richard P. Reading and Brian Miller

D1368585

Greenwood Press
Westport, Connecticut • London

Library of Congress Cataloging–in–Publication Data

Endangered animals : a reference guide to conflicting issues / edited by Richard P.
Reading and Brian Miller.
    p. cm.
    Includes bibliographical references and index.
    ISBN 0–313–30816–0 (alk. paper)
    1. Endangered species. 2. Conservation biology. 3. Rare vertebrates. 4. Wildlife
conservation. I. Reading, Richard P. II. Miller, Brian, 1948–
QL82.E55 2000
333.95'42—dc21   99–049149

British Library Cataloguing in Publication Data is available.

Library of Congress Catalog Card Number: 99–049149
ISBN: 0–313–30816–0

First published in 2000

Greenwood Press, 88 Post Road West, Westport, CT 06881
An imprint of Greenwood Publishing Group, Inc.
www.greenwood.com

Printed in the United States of America

The paper used in this book complies with the
Permanent Paper Standard issued by the National
Information Standards Organization (Z39.48–1984).

10 9 8 7 6 5 4 3 2 1

In memory of our mothers,
who always encouraged us to follow our passions

Contents

SYNTHESIS

Acknowledgments

Editing a book is no easy task. And when that book contains 50 contributions from 70 authors representing 16 different countries, the task becomes even more formidable. Looking back, perhaps we were crazy when we agreed to be editors! Was it a blessing or a curse when Ben Beck recommended to Greenwood Press's Emily Birch that she contact us? On some days it felt like the former, and on others we were sure it was the latter! Thankfully, we had a lot of help and support in our efforts. It was a joy to work with Emily Birch. She was supportive, helpful, and understanding when we inevitably missed our deadlines. We also received the assistance of several staff members at the Denver Zoo, including Debbie Perego, Belinda Foreman, Molly Pound, Nicole Carter, and Pat Moredock. One of our interns, Ethan Borg, was invaluable in helping to keep this sometimes daunting task organized and manageable. Finally, the Denver Zoological Foundation backed our efforts as we worked on the manuscript, and the support of our bosses, Clayton Freiheit and Brian Kleppinger, was never in doubt.

Special thanks to all of the contributors for volunteering their time and energy to this book. We remain impressed by the quality of their work, their dedication to conservation, and their patience with us. Tim Clark deserves special recognition for his assistance and insight. None of the contributors to this volume received, or will receive, any compensation for their effort (other than a copy of the finished work). Instead, all proceeds will support the conservation and research efforts of the Denver Zoo's Department of Conservation Biology.

Books, whether you write or edit them, take time. A lot of time. This often translates to time away from family and friends. As such, our families and friends deserve special recognition. We wish to thank Carina, Mary Aseneth, Annabella Isabel, and Orville Miller, and Ron and Mary Reading for giving us their love and encouraging us to pursue our passions.

Introduction

Brian Miller and Richard P. Reading

Our planet is losing its diversity of life at a rate unparalleled in recent times (see Wilson 1988). As human population and needs (real or perceived) expand, the globe becomes scarred by a deadly scythe. The habitats we increasingly harvest, such as tropical forests and wetlands, are often the crucibles of biotic richness. Is this loss simply fate? Are *Homo sapiens* following some specific manifest destiny? Should people just accept the trend and go about their daily business? After all, don't people simply represent one species on the planet, all of which are struggling for survival?

Many adopt that attitude. It is certainly the easy path. After all, no matter how apocalyptic the outcome, the process is so slow (at least on human time scales) that it is nearly imperceptible. Also, each generation of increasingly urbanized populations throughout the world moves farther from nature. A few people may notice that the howl of a wolf no longer floats over the hills or that the springtime song of their favorite prairie bird rings less frequently than in their youth. But by and large, many lost life forms are too distant and obscure to be missed, and in thousands to millions of cases the forms may be gone before they are even known to science (Wilson 1992).

Yet there is a myriad of people who find this trend unacceptable. The stories of some of those people are encapsulated in this book. Issues that surround the dramatic declines of species are complex, often conflict-laden, and not easy to reverse. However, one can learn from past practices, improve performance, and avoid the problems common to endangered species conservation. To that end, this volume provides 49 case studies of subspecies, species, or groups of species that have been pushed to the brink of extinction. The contributing authors have dedicated an incredible amount of time and effort toward preserving the organisms about which they write, and they describe the controversies and complexities of each struggle. They do not want to be part of a modern extinction spasm, in which a large number of species go extinct in a relatively short period of time.

EXTINCTION: PAST AND PRESENT

Extinction can be a natural process. Indeed, the vast majority of species that ever existed are now extinct. In light of this, some people have justified the growing global impoverishment of species by arguing that because extinction is a natural process, the extinction of modern species is simply a continuation of a normal phenomenon. Yet historically, typical, or background extinction rates usually have been lower than the rate at which new species evolved (except during extinction spasms). But because modern extinction rates are roughly 1,000 times higher than background extinction rates (Myers 1988; Soulé 1996), today only a handful of people deny that the planet is in, or rapidly approaching, an extinction spasm.

Lately, a new rationale to exonerate human actions holds that extinction spasms have occurred periodically throughout geologic time and that nevertheless the planet has always recovered (albeit over periods of 10 to 30 million years). Some people claim it is important only to recognize that there are periodic mass extinctions, not to know their cause (Rosenzweig 1995). However, this argument may be biologically faulty, as the subsequent discussion reveals.

There have been five historic extinction spasms, the most recent occurring at the end of the Cretaceous when dinosaurs disappeared. Although current extinctions may not have yet reached the extent of those of the Permian Period 250,000,000 years ago, the present extinction spasm has a unique aspect. The causal agent is not an external perturbation such as volcanic eruption or collision with an asteroid. The agent is one of nature's own creations, *Homo sapiens* (Soulé 1996).

Human influence can be seen from the time humans arrived in Australia, North America, Europe, South America, and several island archipelagos (Ward 1997). In each of these regions, the arrival of people was quickly followed by a loss of 75% to 85% of the mammalian megafauna, or species larger than 44 kg (Primack 1998). The rate of loss accelerated as humans increased in number and acquired the technological capacity to alter habitat and directly exploit a wider array of animal and plant resources. It has been estimated that at present population levels, humans use 40% of the primary productivity of plants (Vitousek et al. 1986). If the human population doubles in the next half-century, that will leave little productivity for other forms of life.

In the United States alone there are now well over 1,000 species federally protected (Noss et al. 1997), and there may be another 5,000 species threatened but not included on federal lists (U.S. General Accounting Office 1992). Moreover the number of plants and animals formally listed as threatened or endangered continues to rise; only a small handful of species have been removed from the U.S. Endangered Species List because their numbers have recovered (U.S. General Accounting Office 1992; Murphy et al. 1994; Wilcove et al. 1996). In a recent analysis of the causes for these astounding numbers, 85% were linked to habitat loss, 49% to invasion of alien species, 24% to pollution, 17% to overexploitation, and 3% to disease[1] (Wilcove et al. 1998). The main causes of habitat destruction included agriculture, livestock grazing, mining, logging, infrastructure development, road construction, military activities, outdoor recreation, water development, urban and commercial development, and disruption of fire ecology (Wilcove et al. 1998).

Habitat changes reduce biotic integrity (i.e., ecosystem health), deplete native species, and greatly simplify the system and its habitats (e.g., crop agriculture). The process of habitat destruction is incremental. Each piece of habitat may not seem important individually, but there are cumulative effects. The process is more insidious than direct overexploitation. No one holds a "smoking gun." The native species simply vanish.

The effects of these changes can be predicted mathematically. Roughly, when 90% of the habitat is eliminated, 50% of the species will be lost (Wilson 1992). Selection of the lost species, however, is not random. The larger, wide-ranging species, such as large carnivores, suffer first. Because those groups often contribute to healthy ecosystem processes, a wave of secondary losses may follow their decline (see Terborgh et al. 1999). Animals that conflict with humans are also the victims of concerted eradication efforts. Species with a narrow geographic range, or species that were never common, are vulnerable as well. Species that are not effective dispersers are limited when their habitat is disrupted. Species with narrow niche requirements may see that niche disappear quickly. And species that live in colonies, or social groups, are often affected when numbers decline.

When habitat is fragmented, some species die as a direct result of lost resources. Other species survive in reduced numbers in the habitat fragments, but their vulnerability to extinction increases. Populations existing in fragments become susceptible to genetic disorders, demographic problems, environmental variability, and catastrophic events. Fragmented populations are especially vulnerable to deterministic events, such as susceptibility to poaching, as border areas become population sinks, where population death rates exceed birth rates. Woodroffe and Ginsberg (1998) reported that 74% of known mortality of large carnivores living in protected areas occurred at the boundary and was caused by humans. When deterministic factors (e.g., habitat destruction) reduce populations such that random genetic, demographic, and environmental forces are able to drive them to extinction, the condition is called an extinction vortex; it is usually extremely difficult to recover species entering such a whirlpool (Gilpin & Soulé 1986).

In short, human impacts have triggered the present extinction spasm. During the other five spasms the causal agent was a temporary disturbance. When the causal agent subsided, recovery proceeded. Some people have speculated that human activities may ultimately result in the elimination of

Homo sapiens, and if that happens, recovery over geologic time may not be very different from the other five spasms. However, the causal agent of this current extinction episode may well endure. After all, humans are the ultimate weed species (Quammen 1998). We have shown an incredible ability to invade, change, and inhabit every habitat type on the planet. Sheer numbers of humans alone can overwhelm our co-species, and technology heightens our capabilities. The only other species that will likely survive our impact will be those adapted to living with humans—in short, the species that can utilize the pauperized systems left in our wake. Because evolution, and recovery from an extinction spasm, rely on vacant niches, the human species will do more than just eliminate other species. By simplifying habitat, humans will also limit other species' capacity to recover from that blow (Soulé 1996). Thus it is important to identify and understand the cause of an extinction spasm.

CONSERVATION: SLOWING, HALTING, AND REVERSING THE EXTINCTION SPASM

Conservation biology is an applied, goal-oriented discipline that seeks to stop the current extinction spasm and recover the Earth's natural systems so that natural selection and evolution can continue. As such, conservation biology is value laden, based on the premises that biological diversity and evolution are good and that untimely extinction is bad. Conservationists possess expertise in a wide variety of fields but are united in their desire to stem the loss of global biodiversity at the hands of humanity.

Despite growing recognition of the current extinction crisis and the importance of biological diversity, there is a continuing loss of ground in the effort to conserve the amazing array of life inhabiting our planet. Efforts to conserve life forms appear to be less effective than they could be—and certainly less effective than they must be if they are to prevent future extinctions. Even in the United States—where relative to other countries strong legislation has been in place for decades, funding and other resources are plentiful, and public support remains high—success rates for conservation and recovery programs remain low (Murphy et al. 1994; Clark 1996). Clearly, if we hope to maximize the biodiversity remaining on Earth, we must improve our approaches to conservation.

Understanding the causes of extinction and endangerment for individual species is crucial, as this is the first step toward developing more effective conservation strategies. Without a good definition of the problem, effective solutions will remain elusive (Clark [this volume]). Although the proximate causes of species endangerment (e.g., habitat loss, overexploitation) are often relatively obvious and easy to understand, the ultimate or underlying factors responsible for those proximate causes are usually far more complex and difficult to address. For that reason, many conservation strategies merely treat the symptoms of decline, not the underlying "disease." The ultimate reasons for species decline are primarily social, political, and economic. They relate to the values humans place on nature (Kellert 1996). Environmental problems are intertwined with issues of power and authority (nation, state, organization, or individual), attitudes and beliefs (fear of predators or animal rights), development (natural resources extraction or urban sprawl), economy (global trade, poverty, personal consumption, and national debt), and more. These factors generally lead to conflict surrounding endangered species conservation efforts.

Conservationists are increasingly recognizing that the ultimate causes of extinction are primarily socioeconomic and political, yet biological approaches to recovery continue to dominate (Clark & Wallace 1998). However, more inclusive, interdisciplinary approaches to conservation may offer better prospects for managing problems and conflict. Such an approach is outlined at the end of this volume by Tim Clark and elsewhere in far more detail (e.g., Lasswell 1971; Lasswell & McDougal 1992; Gunderson et al. 1995; Clark 1996; Clark et al. in press).

The extinction crisis is a huge and complex problem. For that reason it is important to think proactively and on large temporal and geographic scales, for example, over evolutionary time periods and across entire landscapes (Soulé & Noss 1998). Although the challenge is formidable, it cannot be avoided. The consequences of our present path are unfolding rapidly, and the longer we avoid them the worse they will become.

CASE STUDIES: LEARNING FROM THE PAST

Case studies provide in-depth analyses rich with lessons for learning and for improving performance. The 49 case studies presented in this book demonstrate the causes and context of endangerment for a wide variety of animals, mostly vertebrates. A diversity of case material is included from a range of perspectives. Although there is a predominance of U.S. authors and case studies, the book includes contributors from 16 nations and addresses species inhabiting over 20 countries (and several oceans) and a wide variety of ecological systems on every continent except Antarctica. The case studies are dominated by mammals ($N = 27$), but there are also entries on birds ($N = 12$), reptiles ($N = 5$), fish ($N = 3$), an amphibian, and an insect.

Each entry starts with brief information on taxonomy, status, threats, habitat, distribution, and physical description. Then the text covers the species' natural history, conflicting issues, and prognosis. Because of the importance of social, economic, and political considerations to improved conservation, the contributors were asked to focus on the *nonbiological* conflicts surrounding species recovery programs. Although problems vary from case to case, there are an incredible number of similarities regardless of taxa and region. Much of this information is summarized in the conclusion. These comparisons provide opportunities to improve our approach to endangered species conservation. Tim Clark's chapter concerning the Yellowstone Grizzly Bear case provides an intellectual framework for doing just that.

The entries are about 2,500 words each. If the reader requires more detail, the reference lists provide access to the primary literature. For ease of use, references have been grouped together at the end of the book. A glossary is provided to define terms associated with the study of endangered animals.

This volume provides useful examples of what people can do to change the present situation. Time is short. Even if one could stop the decline today, the recent effects will still impact the planet into the future. The conservation effort will be long, and in some cases we may not live to see the fruits of these labors. Yet the stories in this book are an inspiration for people who do not want to accept the status quo. If you agree with them, do not be timid. There is a struggling species that can use your help.

NOTE

1. The percentages exceed 100% because some of the factors overlap (e.g., habitat changes make it easier for alien species to invade).

African Wild Dog

M.G.L. Mills

Common Name: African wild dog, Cape hunting dog
Scientific Name: *Lycaon pictus*
Order: Carnivora
Family: Canidae
Status: Endangered under the World Conservation Union's (IUCN) Red List categories.
Threats: Habitat destruction and fragmentation; human intervention by shooting, snaring, poisoning, and road kills; diseases such as rabies and canine distemper.
Habitat: Wide habitat tolerance from open plains to moderately dense bush and low montane forests; reaches highest densities in woodland savannah and broken, hilly country.
Distribution: Formerly distributed over most of Africa south of the Sahara, except for the rain forests and extreme deserts. Today most are found in northern Botswana, northeastern Namibia, western and northern Zimbabwe, and southern Tanzania. Viable populations also occur in the Kruger National Park, South Africa, and probably in the Kafue National Park and Luanga Valley, Zambia. In East Africa, Kenya and Ethiopia have small populations that may be viable, as do Senegal and Cameroon in West Africa.

DESCRIPTION

The African wild dog is a medium-sized member of the Canidae family standing 0.6–0.75 m at the shoulder and weighing 20–30 kg. Males are slightly larger than females, and southern African dogs are larger than eastern African dogs. The wild dog is long in the leg and has a variable and patterned yellow, brown, black, and white coat, large round ears, a dark muzzle, and a white-tipped tail. It has four claws on each foot (no dewclaw), 12–14 mammae or nipples, and a dental formula of 3/3 incisors, 1/1 canines, 4/4 premolars, and 2/3 molars, which are sharp and adapted to slicing meat (flesh).

NATURAL HISTORY

The wild dog represents a unique lineage within the wolflike canids and is the only living member of the genus *Lycaon*. Although no subspecies are recognized, the genetic differences between southern, eastern, and western African wild dog populations suggest that all these populations must be conserved to preserve the genetic diversity of the species (Girman & Wayne 1997).

The wild dog is an efficient hunter. It preys mainly on small to medium-sized ungulates, particularly impala *(Aepyceros melampus)* in southern Africa and Thomson's gazelle *(Gazella thomsonii)* and blue wildebeest *(Connochaetes taurinus)* in east Africa. It may occasionally become a severe problem for small livestock, but it rarely preys on cattle (Fuller & Kat 1990; Rasmussen 1996).

Wild dogs are highly social animals, spending most of their time in the company of others. A pack may be as small as a pair or as large as 49 (personal observations) but usually comprises 4 to 8 adults, 2 to 6 yearlings, and 5 to 11 pups (Fuller et al. 1992; Creel & Creel 1995; Mills & Gorman 1997). The number of dogs in a pack fluctuates markedly over time due to emigration and mortality (Frame et al. 1979. Mills 1993, Burrows 1995, McNutt 1996). A pack is formed when a single sex group leaves the natal pack and joins up with another group of dogs lacking adults of their sex, or occasionally when a large pack splits into two. In a pack, the adult males are relatives and so are the adult females, but the adult males and females are unrelated (McNutt 1996; Girman et al. 1997).

The wild dog is a seasonal breeder. In the Serengeti most litters are born in the latter part of the wet season (March–June; Schaller 1972), whereas in southern Africa they are born in the midwinter/dry season (May–June; Mills 1993). Usually only the dominate or alpha male and female breed, but subordinate animals do sire some young (Girman et al. 1997). Litter size is usually 10 to 12, but litters of 21 have been recorded (Fuller et al. 1992). The denning period lasts 3 months. All pack members help raise the pups by regurgitating meat to feed them and taking turns guarding the den when the pack is hunting (Frame et al. 1979; Malcolm & Marten 1982).

Outside the breeding season, wild dogs have vast home ranges of 350–2,400 km² depending on the area (Fuller & Kat 1990; Fuller et al. 1992; Creel & Creel 1996; Mills & Gorman 1997). Although there is a measure of overlap between the home ranges of neighboring packs, they rarely meet. Because of their extensive home ranges, wild dogs live at low densities of 0.7 to 4.6 adults/100 km² (Fuller & Kat 1990; Creel & Creel 1996; Mills & Gorman 1997).

CONFLICTING ISSUES

˙ Over the last 30 to 50 years, the wild dog is believed to have become extinct in 25 of the 39 countries in which it was formerly recorded. There are perhaps 3,000 to 5,500 individuals in 600 to 1,000 packs surviving in the wild (Fanshawe et al. 1997). Countries where the wild dog has been driven to extinction are characterized by high human densities where habitat fragmentation, persecution, and loss of prey have occurred on a large scale.

The wild dog has suffered from extremely negative attitudes on the part of humans; for example, Stevenson-Hamilton (1947) described it as "a terrible foe to game with a wasteful method of hunting." Consequently, stock farmers have held a concerted campaign to exterminate it. Until the mid-1960s even game wardens were controlling wild dog numbers in conservation areas to "protect" prey populations (Bere 1955; Atwell 1959).

Food does not seem to be important in limiting wild dog numbers in undisturbed ecosystems. The most important factors appear to be predation by lions *(Panthera leo)* and stealing of food (kleptoparasitism) by spotted hyaenas *(Crocuta crocuta)* (van Heerden et al. 1995; Creel & Creel 1996; Mills & Gorman 1997). Extremely high rates of daily energy expenditure make wild dogs particularly susceptible to the latter form of competition (Gorman et al. 1998).

Human activities are important in limiting wild dog numbers (Woodroffe & Ginsberg 1997b). In Hwange National Park, Zimbabwe, more than half the recorded wild dog mortality was from road kills along a highway that runs along the boundary of the park (Woodroffe & Ginsberg 1997b). Similar cases of mortality have been recorded in protected areas in southern Tanzania and Zambia (Woodroffe & Ginsberg 1997b).

Snaring, poisoning, and shooting are important mortality factors around protected areas (van Heerden et al. 1995; Woodroffe & Ginsberg 1997b). Snaring is mainly secondary, in that the target species are antelope. Commercial game ranchers have varying attitudes toward wild dogs; some are tolerant, even regarding them as an asset, but most see them as a threat to their livelihood as they kill animals that could be sold or hunted. However, these arguments are simplistic and do not consider broader ecological principles. Although it is illegal to kill a dog in game ranching areas in South Africa without permission from the conservation authorities, farmers often take the law into their own hands and are seldom prosecuted for doing so.

The wild dog is susceptible to rabies, and this may be disastrous for small populations. Indeed, in the Serengeti, rabies has been recorded in wild dogs (Gascoyne et al. 1993; Kat et al. 1995) and may have exterminated the small population there (Burrows 1992). Reintroduced packs in Etosha National Park, Namibia (Scheepers & Venzke 1995), and Madikwe Game Reserve, South Africa (Hofmeyr personal communication), were affected by rabies as well. Canine distemper virus may also adversely affect wild dog populations; 10 died in a pack in northern Botswana (Alexander et al. 1996), and circumstantial evidence exists for other packs in this area and in the Serengeti ecosystem contracting the disease (McNutt personal communication; Schaller 1972; Alexander & Appel 1994).

FUTURE AND PROGNOSIS

Maintaining the integrity of large conservation areas of over 10,000 km^2 is critical for wild dog conservation (Woodroffe & Ginsberg 1997a). It should be the driving force behind future strategies for wild dog management.

In certain areas, managing disease—particularly rabies, but also canine distemper—must be addressed. In small, introduced populations, vaccination against rabies is feasible; however, an effective and safe vaccination against distemper for wild dogs is not available. In large areas, vaccination is impossible and usually unnecessary. As domestic dogs are often likely transmitters of rabies (and canine distemper) to wild dog populations, contact between these species should be minimized. Where rabies and canine distemper are present in domestic dogs, a vaccination campaign of domestic dogs could be of benefit to both the owners of domestic dogs and the wild dogs.

A strategy to improve the conservation status of the wild dog in southern Africa was discussed at a workshop in October 1997 (Mills et al. 1998). Establishment of a second wild dog population in South Africa comprising a minimum of nine packs distributed in several suitable small reserves and managed as a metapopulation (population of smaller subpopulations) was identified as a priority. Furthermore, it was recommended that the possibility of conserving the wild dog in the growing number of game ranching areas in southern Africa be investigated. This would involve research not only into the biological implications of such an effort, but also into the potential economic benefits that might accrue through enhanced ecological-based tourism or ecotourism. Education campaigns among game ranchers would be important as well. A Wild Dog Action Group comprising all important stakeholders to accomplish the aims of the workshop has been formed.

Altai Argali Sheep

Sukhiin Amgalanbaatar and Richard P. Reading

Common Name: Altai argali sheep
Scientific Name: *Ovis ammon ammon*
Order: Artiodactyla (even-toed ungulates)
Family: Bovidae (antelope, cattle, sheep, goats, and their relatives)
Status: Threatened in the Mongolia Red Book; Endangered in Russia and the People's Republic of China; listed in Appendix II of the Convention on International Trade of Endangered Species; Threatened on the U.S. Endangered Species List; Vulnerable on the 1996 IUCN Red List of Threatened Animals (Baillie & Groombridge 1996 ; Schackleton 1997; Shiirevdamba et al. 1997).
Threats: Primarily poaching and competition with domestic livestock for forage and habitat.
Habitat: Altai argali inhabit cold, arid grasslands of mountains, intermountain valleys, and rocky outcrops in Central Asia. They prefer rolling hills, plateaus, step hills, and gentle slopes over rugged mountainous terrain.
Distribution: Altai argali live in the Altai Mountains and adjacent regions of China, Mongolia, and Russia; they also inhabit mountainous regions and rocky outcrops of northern, central, and western Mongolia and southern Tuva, Russia (Schackleton 1997).

DESCRIPTION

Altai subspecies of argali are the world's largest sheep, with some males growing to 127 cm at the shoulder and weighing over 180 kg (Valdez 1982; Schaller 1977, 1998). Rams have huge, curled horns that spiral outward and reach over 150 cm (record length = 169 cm) (Valdez 1982). Highly sexually dimorphic, female argali ewes are only 60% to 70% the size of rams, and their horns are seldom over 50 cm in length. Argali have relatively long, thin legs and compact bodies built for running speed. The sheep are light brown to brownish grey on their upper parts, with white faces, abdomens, inner legs, and conspicuous rump patches (Schaller 1977). The anterior portion of the body is darker than the back (Valdez 1982). Rams grow neck ruffs and dorsal crests in the winter and have more distinct rump patches than ewes do (Schaller 1977).

NATURAL HISTORY

Altai argali inhabit the Altai Mountains of China, Russia, and Mongolia and nearby mountains and rocky outcrop regions of northern and central

Mongolia (Schackleton 1997). Argali prefer less rugged and precipitous terrain, living instead in foothills, high plateaus, intermountain valleys, gentle slopes, and rolling steppes (Reading, Amgalanbaatar, & Mix 1998; Schaller 1998). Within these regions, argali feed on graminoids, forbs, legumes, and some shrubs (Schaller 1998). The most significant source of mortality for argali in Mongolia is human hunting and poaching. Other important predators include snow leopards (*Uncia uncia*), gray wolves (*Canis lupus*), large raptors, and lynx (*Lynx lynx*).

The Altai argali rut begins in late October or early November, with actual breeding starting in late November to early December. Argali give birth the following late April to early May after an approximately 150-day gestation. Ewes give birth to one, or occasionally two, lambs (Schaller 1977, 1998). Females can reproduce in their second year, although pregnancy rates at that age are relatively low (Schaller 1977). Males are usually prevented from breeding by larger, more dominant rams until they are much older. Few males live beyond 10 years of age (Schaller 1977).

Argali form flexible herd structures, with the only stable bond being that between a female and her lamb, or occasionally her yearling (Schaller 1977). There appears to be no territoriality, although rams defend ewes during the rut (Schaller 1977). Reported mean group size varies considerably, from 2.5 to 39.2, as do lamb : female and adult male : female ratios (11.0–68.5 : 100 and 52.6–92.5 : 100, respectively) (Reading et al. 1997). Groups may be all males (including single males), females with lambs and yearlings of both sexes, and mixed (Schaller 1977; Reading et al. 1997). Generally, rams and ewes (with young) separate following the rut.

CONFLICTING ISSUES

Currently, little is known about the ecology and status of Altai argali. Although Altai argali appear to be declining, population estimates are difficult to make. Most people agree that the subspecies is experiencing marked population declines and fragmentation (although local and foreign trophy-hunting organizations contend otherwise). Estimates vary considerably, with some researchers expressing concern for the status of the subspecies and others suggesting that argali are relatively widespread and less threatened (see review in Luschekina 1994).

Argali population decline appears to be a result of direct mortality (mostly from poaching) and competition with domestic livestock (Reading et al. 1977; Reading, Amgalanbaatar, & Mix 1998). Although argali have been protected from general hunting in Mongolia since 1953 (the species still may be hunted by foreign hunters with a special permit), poaching continues to be an important source of mortality (Luschekina 1994). Lax law enforcement permitted expansion of poaching activity in the wake of the major political and economic changes that have accompanied Mongolia's transformation to a democratic, free market system (MNE 1996). The subsequent weakening of central authority, a severe economic crisis, an increase in the human population growth rate, and the growing importance of meat in the Mongolian diet have all combined to increase the amount of illegal hunting (MNE 1996; Mallon et al. 1997). Indeed, many local people readily admit to shooting argali for meat (Reading, Amgalanbaatar, & Mix 1998).

Argali also suffer from competition with domestic livestock for water and forage (Luschekina 1994). Livestock numbers have increased dramatically over the past few years following privatization of herd ownership (Reading, Amgalanbaatar, & Mix 1998). As the nation's human and livestock numbers increase, herders are expanding grazing into more marginal pastures, resulting in increased competition with wild ungulates and displacement of argali from their former population strongholds (Mallon et al. 1997). The resulting overgrazing and displacement by livestock has substantially reduced and degraded argali habitat (Luschekina 1994). The situation is particularly pronounced in western Mongolia, where the human population has expanded greatly in recent years and livestock grazing has pushed deep into the high mountain regions and even into protected areas (Mallon et al. 1997; Reading, Amgalanbaatar, & Mix 1998).

The primary conflicts surrounding argali conservation in Mongolia revolve around trophy hunting and domestic livestock grazing (Luschekina 1994; Mallon et al. 1997; Reading, Amgalanbaatar, & Mix 1999). Poaching, although a significant problem, is the cause of much less conflict but may grow in importance as law enforcement efforts improve.

The Altai argali is the world's largest sheep. Because of its size and impressive horns, it is greatly sought by trophy hunters. In fact, foreign sports hunters paid over U.S. $20 million to harvest 1,630 rams from 1967 to 1989 (Luschekina 1994). However, trophy hunting of argali is a contentious issue both locally and internationally (Reading, Amgalanbaatar, & Mix 1998). Most local people oppose trophy hunting, especially by foreign hunters, which they blame for argali population declines. This blame is probably misplaced. Trophy hunting may have negative impacts on selected populations; however, the 20–30 animals trophy hunters harvest each year is a small fraction of the number poached by local people and displaced by their livestock.

Internationally, the situation has pitted hunting organizations against conservation organizations. The European Union had banned importation of argali from Mongolia, but pressure from hunters recently caused the ban to be lifted. The United States provided Mongolian argali with a Threatened status but has been issuing permits recently for importation by trophy hunters. A lawsuit challenging the legality of U.S. permit issuance is pending.

The United States provided Threatened status to Mongolian argali because the species' status is not clear and because U.S. authorities require, among other things, that hunted species be actively managed and that

money generated from hunting fees be used for the conservation management of that species (Reading, Amgalanbaatar, & Mix 1999). Conservationists claim that neither case holds in Mongolia. Indeed, under the
Mongolian Hunting Law of 1995 none of the revenue generated from argali
hunting goes directly to conservation or management. Hunting fees are
instead divided among the federal government's general funds (70%), the
local Sum (or county) government (20%), and the hunting organization
(10%) (Reading, Amgalanbaatar, & Mix 1999). The government does not
actively manage argali, and very little government-sponsored conservation
or management activity has been undertaken on behalf of argali over the
past several years (Mallon et al. 1997; Reading et al. 1997; Reading, Amgalanbaatar, & Mix 1998). Hunting organizations have provided modest
support for argali surveys and conservation activities. The results of this work
suggest that Altai argali are declining (Valdez & Frisina 1993).

Trophy hunters and the organizations that represent them argue that trophy hunting can provide an important source of income for conservation as
well as for local communities. However, this is only true if at least some of
the money goes to the local communities and to the conservation management and research of the hunted species. Trophy hunting in the absence of
a well-managed population could have negative impacts on local argali populations, conservation of the species, and future hunting opportunities. Alternatively, adequate conservation management would ensure survival of the
species, thereby benefiting the species, the hunters, the government of Mongolia (through the revenue generated), and the ecology of the region.

Livestock numbers have increased rapidly in Mongolia since the privatization of livestock herds that accompanied the transition from a communist,
centrally planned society to a democratic, free market system (MNE 1996;
Reading, Amgalanbaatar, & Mix 1998). Local nomadic herders have moved
quickly to improve their standards of living by enlarging and diversifying
their herds. Many of these people live a marginal existence, barely able to
feed and clothe their families. Most are also concerned about conserving
nature and wildlife, which they view as part of their cultural heritage.

Local herders argue that efforts to curtail grazing within protected areas,
some newly created, are unnecessary. Herders suggest that foreign hunters
are actually causing argali population declines and that they, the native herders, are able to co-exist with argali. This is probably not true (see above).
Individuals who admit to poaching argali suggest that their actions have
little impact on the overall argali population. Fewer individuals argue that
argali conservation should be secondary to human development and resource use by people.

FUTURE AND PROGNOSIS

More active argali conservation and management are necessary to halt and
reverse the current population decline and fragmentation. Perhaps the

greatest challenges to Altai argali conservation are poaching and competition with domestic livestock. Yet Mongolia has greatly expanded its protected area system since 1991, and argali currently inhabit or recently inhabited over 20 protected areas there. However, both poaching and overgrazing are prevalent throughout most of these protected areas (Mallon et al. 1997).

There are no easy solutions to argali conservation in the face of increased grazing pressure. By law, most protected areas in Mongolia permit limited grazing within at least a portion of their boundaries. However, the Mongolian Protected Areas Law of 1994 provides little guidance for doing so (MNE 1996). Currently, protected area managers, conservationists, and local herders are struggling to zone protected areas and devise management plans that are satisfactory to all stakeholders.

Poaching also presents a challenge to wildlife managers in Mongolia, especially given the size of the country and the limited numbers of rangers and other wildlife enforcement officials. Although an increase in the number of enforcement personnel would help, a more comprehensive strategy that couples enforcement with an education campaign would likely be more effective. Such a program should focus on Mongolia's long tradition of nature conservation while simultaneously making people aware of the rapid, recent decline of argali and the role of poaching in this decline. Making people aware of the inconsistency between their behaviors (poaching) and their values (nature conservation) might improve the situation.

Conflicts over trophy hunting are likely to continue in the near future (Reading, Amgalanbaatar, & Mix 1999). Trophy hunting should be permitted only if argali populations are more carefully managed and deemed capable of sustaining such harvests. Management in this regard would require systematic, rigorous, and comprehensive surveys for more accurate estimates of numbers and distribution. At least a substantial portion of money generated from sports hunting should be directed toward management of argali and their habitat. Given the high fees being charged to hunt Altai argali, it should be possible to increase fees marginally (from a percentage basis) but still generate significant revenue for conservation. Funds should go to increasing ranger staff, equipment, and training and to conducting more rigorous and regular argali surveys and research (see also Mallon et al. 1997). Without more active conservation management measures, Mongolia risks further declines in argali numbers and distribution, including the imminent loss of several populations.

American Burying Beetle

Erin Muths and Michelle Pellissier Scott

Common name: American burying beetle
Scientific name: *Nicrophorus americanus*
Order: Coleoptera
Family: Silphidae
Status: Endangered under the Endangered Species Act (ESA), July 1989 (*Federal Register 54*(133):29652–5); Endangered on CITES.
Threats: Destruction of habitat and the associated decline in mammal and bird carcasses of appropriate size for the beetle's reproduction.
Habitat: One remnant population occurs on glacial moraine vegetated by maritime-grassland plants. Others occur in areas of forest-pasture ecotone and open pasture (USFWS 1991).
Distribution: *Nicrophorus americanus* was formerly distributed throughout the temperate, eastern regions of North America. The American burying beetle (ABB) is now restricted in the East to one population on Block Island, Rhode Island, and reintroduced populations on Penikese and Nantucket Islands, Massachusetts (Amaral et al. 1997). West of the Mississippi, *Nicrophorus americanus* occurs more frequently.

DESCRIPTION

Nicrophorus americanus is the largest (35–45 mm) burying beetle in North America (there are 15 extant species). It is comparable in size and habit to *N. germanicus* in Europe, which is rare or possibly extinct. Worldwide, there are about 85 species in the genus *Nicrophorus* (Ratcliffe 1996). Like most other burying beetles, the ABB is shiny black with orange spots on the wing covers (elytra), has large antennae with orange-red tips, and has orange-red "facial" markings located above powerful mouthparts.

NATURAL HISTORY

ABBs are recorded from 35 states and 3 Canadian provinces (Figure 1). The main eastern U.S. population, numbering approximately 1,000 individuals, resides on Block Island off the coast of Rhode Island; smaller, unquantified populations exist on Penikese and Nantucket Island, Massachusetts. The species is more common in the West.

The recovery plan (USFWS 1991) states that vegetational structures and soil types are not limiting for this beetle, although habitat parameters influ-

Figure 1. The historical (gray) and present (black) distributions of the American burying beetle, *Nicrophorus americanus.* **The Block Island, Rhode Island, population is represented by a single black dot just south of Cape Cod (modified from Lomolino et al. 1995).**

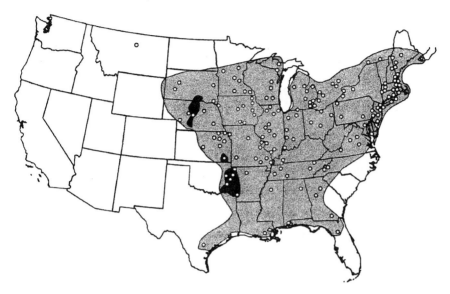

encing carcass availability and competitor presence are important. Other studies describe a bias toward forested sites with relatively deep soils or soils with added structure (Muths 1992; Lomolino et al. 1995; Lomolino & Creighton 1996).

ABBs are active from May to September. Adults are nocturnal, beginning activities just after dusk (Scott 1998). They are most active when night temperatures are above 15°C. Adults feed on available carrion and may consume live insects, but they depend on small vertebrate carcasses as a food source for their offspring. Optimum carcass size for reproduction of the beetle is between 100 and 250 g (Kozol et al. 1988) and is critical for rearing young. Carrion is distributed patchily in time and space, making carcasses a valuable but unpredictable resource to be defended from competitors.

Burying beetles are strong fliers and search widely for carrion, using their keen antennal chemical receptors. If a male discovers a suitable carcass before females arrive, he may broadcast pheromones to attract mates (Bartlett 1987; Eggert & Muller 1989). Beetles compete within each sex, and the largest male and female generally mate (Wilson & Fudge 1984, references in Scott 1998). Although a single beetle is capable of burying a carcass, usually a male-female pair cooperates. ABBs compete for carcasses with other burying beetles, but they are the largest species and can usurp a carcass from another species even after it has been buried (Kozol et al. 1988). They may bury the carcass where found or transport it several meters before burial. ABBs generally have the carcass concealed by morning and may eventually bury it 6–8 inches deep (Amaral personal communication). During burial the carcass is rolled into a ball, denuded of fur or feathers, and coated with secretions that retard decay and aid preservation (Pukowski 1933; Scott 1998). Beetle pairs mate during and after internment, and 36–48 hours later females lay eggs in the soil adjacent to the burial chamber. Eggs hatch 3 days later. Larvae make their way to the burial chamber, where they are fed regurgitated tissue by both parents. Larvae are dependent on parental re-gurgitation and die if they receive no parental care. Larval development is complete in 6 days, after which they disperse to pupate in the soil (Pukowski 1933; Kozol et al. 1988; Scott 1998). Both laboratory and field studies have found that the total brood mass, as well as the number of larvae reared, are correlated highly with the size of the carcass (Kozol et al. 1988).

CONFLICTING ISSUES

The American burying beetle was added to the Endangered Species List in 1989 after Wells et al. (1983) noted its range decline as one of the most drastic ever recorded for an insect. The pattern of the decline, inferred from specimen documentation, was from north to south and was nearly complete by 1923 (USFWS 1991). The reasons for decline remain unclear, but the pattern in the Northeast suggests strongly that DDT or other pesticides were not responsible because most *N. americanus* populations were gone in the Northeast 25 years before such compounds were applied broadly (USFWS 1991; Amaral et al. 1997).

Direct habitat loss is an improbable cause because *N. americanus* are found over a broad geographical range and in a variety of microhabitats. Anderson (1982) suggests that *N. americanus* was dependent mainly on primary deciduous forests, which are now reduced to less than 1% of their former area in the United States. However, short-term movement studies indicate that the beetles are generalists, at least in their search for carrion. Creighton and Schnell (1998) found that 71% of beetles recaptured were in habitat different from that at the time of their previous capture.

It is unlikely that *N. americanus* has been outcompeted by other burying beetles; they displace members of smaller species and can utilize larger car-casses. The proximate cause of the decline of the ABB may be its depen-dence for optimal reproduction on larger carrion than that used by other burying beetles. As this necessary resource has declined, the beetle's mam-malian competitors have increased (USFWS 1991; Holloway & Schnell 1997). Block Island may have served as a refuge for the ABB because of the absence of virtually all mammalian scavengers/predators and the success of ground nesting birds, which produce vulnerable fledglings when *N. americanus* is reproductive (Amaral et al. 1997; Raithel, Rhode Island Division of Fish and Wildlife, personal communication).

Relative to those of other endangered species, the ABB recovery program has rarely been controversial. However, three conflicts (one in Oklahoma and two in Arkansas) have been significant.

When the beetle was listed, the only known populations were on Block Island, Rhode Island, and in Latimer County, Oklahoma. A proposed highway project in Latimer County was routed through habitat occupied by the ABB. In response to a request by the Federal Highway Administration, the U.S. Fish and Wildlife Service (USFWS) reviewed the project and issued a biological opinion that determined construction of the highway might significantly impact and potentially jeopardize the species by adversely affecting one of the only known populations of burying beetles. The original, jeopardy biological opinion did not identify reasonable and prudent alternatives for the project and did not furnish a statement allowing incidental take (i.e., harm that is incidental to and not intended as part of an agency action, such as highway construction). The USFWS further suggested that the Oklahoma Department of Transportation (ODOT) consider alternate routes for the highway.

Surveys by the Oklahoma Natural Heritage Program found beetles at all alternate route sites, and as more surveys were conducted, more beetles were found. After new information became available, the USFWS re-initiated consultation with the ODOT and revised the biological opinion, including limited incidental take of the beetle during the construction and maintenance of the highway. The opinion stated that surveys and relocation of captured individuals be conducted prior to construction and in associated work areas and that construction be scheduled during periods of beetle inactivity. In the end, the construction of the highway was not halted or even delayed. The Ecological Services Office of the USFWS in Tulsa provided the biological opinion and worked closely with the ODOT from 1988 to 1992. The USFWS seldom re-initiates consultation after a biological opinion has been completed (Frazier, Oklahoma USFWS, personal communication). This and the fact that the conflict was generally nonadversarial make this case unusual.

A second issue with potential for controversy arose in 1993 when the city of Fort Smith, Arkansas, proposed a land-fill site using land transferred from Fort Chaffee. Because land was being transferred from a federal agency to the city, the National Environmental Policy Act (NEPA) was activated. In order to fulfill the conditions of the biological opinion issued by the USFWS, it was necessary for any ABBs found on the proposed site to be trapped and relocated. The condition was fulfilled, but few beetles were relocated. This situation and the associated costs were sensationalized and featured in the "That's Outrageous" section of the February 1994 issue of *Reader's Digest*. In this article, the authors erroneously stated that the trapping and removal of only seven beetles cost taxpayers over $78,000.

In this case, the public was presented with grossly incorrect statements. This points to the importance of strong public relations and the dissemination of accurate information on behalf of endangered species. For example, the National Audubon Society magazine and the Nature Conservancy of Rhode Island successfully promoted insect conservation by highlighting the beetle's "charming" parental behavior.

The third potentially contentious issue was precipitated by the formation of the Pond Creek Bottoms National Wildlife Refuge in Arizona. Some of the land was acquired through a land exchange between the Weyerhaeuser Corporation and the USFWS. Land managed by the Ouachita National Forest (ONF) was part of the trade and supported a known population of burying beetles. Before the exchange, the ONF consulted informally with the USFWS to determine if their management was jeopardizing the ABB. When it was determined that incidental take was harming the beetles, the ONF proceeded with a formal consultation. As a result, the Forest Service was required to survey for ABBs and relocate any individuals caught.

Because any land swap by a federal agency constitutes a federal action, the consequences of this swap were required to be examined. Furthermore, it was unclear whether the authority for incidental take, held by the Forest Service, would transfer to the Weyerhaeuser Corporation after the swap.

With incidental take in question, Weyerhaeuser elected to initiate a habitat conservation plan (HCP) to be carried out by the corporation. Federal law does not require an HCP, but Section 9 of the ESA prohibits "take" of endangered species. One can surmise that Weyerhaeuser initiated the HCP for three reasons. First, it wanted protection from the consequences of unauthorized, incidental take that would probably occur with its use and management of the forest resource. Second, it was interested in information on how its management practices might affect the ABB. Third, it wanted to be perceived as "doing the right thing" and cooperating with the USFWS and the ESA.

The HCP, produced by the Weyerhaeuser Corporation and written by a Weyerhaeuser biologist and an endangered species consultant, prescribed a different course of action from that initiated by the Forest Service. The HCP determined that habitat on Weyerhaeuser-managed lands provided a shifting mosaic habitat-type, one in which several habitat-types are present simultaneously. The HCP concluded that habitat necessary to ABBs, which are habitat generalists (Lomolino & Creighton 1996; Creighton & Schnell 1998), was available but that its location had shifted. Weyerhaeuser was required to fund long-term, USFWS-approved research to address the effects of forest management on the use of habitat by the ABB. Research topics and goals were not specified but came under discussion at annual meetings between the USFWS and Weyerhaeuser.

FUTURE AND PROGNOSIS

In general, real and potential conflicts and controversies have been well managed in recovery and protection efforts for the ABB. The recovery plan, completed in 1991, listed reintroduction as a priority. A pilot reintroduction on Penikese Island, Massachusetts, was initiated in 1990 after extensive surveys confirmed that the species was absent from the island. ABBs were present historically, the habitat was suitable, and the island supported carrion resources of suitable size. The island is owned by the Massachusetts Division of Fisheries and Wildlife (MDFW) and is managed as a wildlife sanctuary. Reintroductions occurred until 1993 with state support and without controversy. The population is monitored each summer and has persisted for five generations since the last release (Amaral et al. 1997). Those in charge are cautiously optimistic that the population is self-sustaining (Amaral personal communication).

Following the encouraging project at Penikese Island, a more ambitious reintroduction effort was initiated on Nantucket Island, Massachusetts, another historic locality for the beetle. Yearly releases of captive-reared beetles were conducted from 1994 through 1998 on private land held in conservation ownership. This recovery program was a cooperative effort among the MDFW, the Massachusetts Audubon Society (a large private landowner on the island), and the USFWS. Although this recovery program lacked overt controversy, Nantucket Island selectmen requested assurances that the "new" endangered species would not impede or restrict development of otherwise usable land (Amaral personal communication). Expanded survey efforts west of the Mississippi River have also identified remnant ABB populations, unknown at the time of listing in 1989. For example, beetles were found in south-central South Dakota in 1996 (Backlund & Marrone 1997).

The American burying beetle and its recovery program constitute an emerging success story. Surveys and research by state and federal agencies and private citizens have (1) increased knowledge of the ABB and its distribution, and (2) heightened awareness of biological diversity and the role of the beetle in the environment. Advances in knowledge through reintroductions, protection of habitat, and increased surveys, positive press, and well-managed conflicts have combined to give conservation biologists an excellent opportunity to recover the American burying beetle.

ACKNOWLEDGMENTS

We thank Mike Amaral, Doug Backlund, Robert Bastarache, Ken Frazier, Dale Guthrie, Marj Harney, Wally Jobman, Will McDearman, Nell McPhillips, and Gary Schnell.

Andean Condor

María Rosa Cuesta and Elides Aquiles Sulbarán

Common Name: Andean condor
Scientific Name: *Vultur gryphus*
Order: Falconiformes
Family: Cathartidae
Status: Listed as Endangered on Appendix I of CITES.
Threats: Loss of foraging habitat and indirect mortality from secondary poisoning of animals shot by human hunters.
Habitat: Andean condors inhabit open spaces free of tall vegetation. They are found from the high altitudes to barren coasts where thermal wind upliftings are common.
Distribution: Historically, the Andean condor was distributed along the Andean Cordillera, from Venezuela to Chile and Argentina, including the islands of Tierra del Fuego.

DESCRIPTION

Andean condors are the largest flying birds in the world, with a wingspan reaching around 3.3 m. Both sexes have naked heads and chocolate brown plumage while immature; as adults they are mostly black with broad, white patches on the back, or dorsa, of the wings. Males can be recognized from birth by a crest on the forehead and by their brown eyes as adults. Females have crestless heads and red eyes. Both sexes have white collars that, although already present on immature birds, are barely noticeable because at that age collars are the same color as the rest of the plumage. Average weight for females is 8 kg and for males, 11 kg.

NATURAL HISTORY

The Andean condor is the largest representative of the six species of South American vultures in the carrion-eating family known as Cathartidae. It is the only avian carrion-eating species in high-altitude habitats known in Venezuela and Colombia as *paramos* and as *puna* in the rest of South America. Other vultures rarely live at altitudes higher than 3,000 m. In lower lands they share food foraging with king vultures (*Sarcoramphus papa*), black vultures (*Coragyps atratus*), and turkey vultures (*Cathartes aura*) (Wallace & Temple 1987). Owing to its large size and long wings, the condor's use of habitats is limited to open areas; it is practically impossible for a condor to fly or take off within a heavily forested setting or to even find carrion under the forest canopy.

Condors do not build nests. Caves or rocky ledges situated in mostly inaccessible terrain are used to accommodate nesting and are delimited merely by sand or pebbles found at chosen spots. After a site is selected, both male and female press their bodies against the sandy substrate, shaping it into a rimmed depression.

The species is mainly diurnal (i.e., active during the day). Under normal conditions condors start flying around 8 A.M., although we have observed animals leaving overnight ledges as early as 6:45 A.M. Condors have become masters of gliding, using warm, uprising currents of air to gain altitude from which they can easily glide down or catch another thermal to rise even higher. If faced with danger, they beat their wings rather clumsily to gain height and begin gliding. Usually at 5 P.M. they begin returning to overnight sleeping sites, which are shared by large groups of condors.

The Andean condor has an extensive home range owing partially to the fact that it feeds on a resource that is unevenly and randomly distributed. Carrion is located visually, and the presence of large numbers of vultures and other carrion-feeding species facilitates a learning behavior to locate food by association. Condors, as do other vultures, start feeding on the softest parts of dead animals—such as the eyes, tongue, and udder—and progressively eat the rest of the carrion. Condors rarely start feeding as soon as they locate carrion. Once they sight a carcass, they limit themselves to flying over it or perching on an overlooking spot. One or two days pass before they come down to feed. Only the presence of other condors prompts them to come down sooner. Condors feed on any carcass they find. In areas with small populations of medium to large-size species, they make use of dead domestic animals, including cows, horses, and dogs.

Condors have one of the lowest reproductive potentials among birds. They attain sexual maturity after approximately 8 years of age and lay a single egg every other year. Both parents incubate the egg for 56 days and feed the chick, which remains dependent on its parents for over a year.

CONFLICTING ISSUES

The recovery of wild populations of Andean condors in South America is closely tied to efforts to save the California condor (*Gymnogyps californianus*) in the United States. That species, also a carrion feeder, faced extinction owing to the sudden drop of numbers throughout its distribution in California. A recovery program, based at the Los Angeles and San Diego zoos, was initiated in 1989 to get the species back to self-sustaining numbers and included the captive reproduction of Andean condors, a more abundant ecological equivalent (Toone & Risser 1987). Young, captive-bred Andean females were released in areas historically used by California condors to help investigators refine release techniques and field-monitoring equipment (Wallace 1987; Wallace & Temple 1989). At the end of the experiment, male and female Andeans were sent to Colombia to begin a population reestablishment effort in the species' native range (Rees 1989). Action was coordinated by personnel from Inderena (Colombia's Ministry of Environment), the San Diego Zoo and Wild Animal Park, the Los Angeles Zoo, and the U.S. Fish and Wildlife Service. Since then, more than 40 animals have been released within three national parks, almost doubling the wild population.

Emulating Colombia, Venezuela began an Andean Condor Recovery Program in 1992 to reinforce its own scarce wild population (Zonfrillo 1977). As in Colombia, the Venezuelan program had the cooperation of several U.S. zoological institutions, but mainly the Los Angeles Zoo and San Diego Zoo and Wild Animal Park. Condors in Venezuela inhabit one branch of the Andean Cordillera, which fractures from the main Colombian Cordillera entering Venezuela in a northeasterly direction. This mountain chain covers the territory of three states, Trujillo, Mérida, and Táchira. Mérida has the highest peaks in the country, with several reaching almost 5,000 m, and Bolivar Peak, the highest, reaching 5,007 m. Since its initiation, the Venezuela program has experienced several positive and negative outcomes.

Two groups of captive-born condors have been released within La Sierra de la Culata National Park in the western Venezuelan state of Mérida. Released animals made excellent progress in a very short time, ranging over 5,200 km² in 3 years and becoming completely food-independent 8 months after release.

People in general welcomed the recovery effort, which was financed by a relatively new regional bank. Two years later a financial crisis caused a dozen Venezuelan banking institutions to close, including the one that was supporting the condor project. From then on, turmoil troubled the project. Regular budgeted money was gone, forcing all but one member of the project to resign. Efforts to gain financial support from government agencies and large national corporations failed, primarily owing to the project's prior liaison with the (by then) ill-reputed bank. To enable the program to continue, the decision was made to create a new foundation completely free of political or banking relations. The renewed effort was able to fund the conservation work. The Cleveland Metropark's Zoo is currently covering the basic needs of the program and has promised to finance condor conservation for at least five years if the program continues to progress. This new support has permitted re-initiation of an educational campaign, and once more juvenile condors born in captivity are being brought to the country to be released within the condor's historic range.

Recovering a species that was almost extinct is not an easy task. The Venezuelan experience demonstrated that what is considered an important objective from a conservation point of view has little or no meaning to many other groups. As part of the program to augment Venezuela's condor's population, an intensive educational campaign was conducted in primarily rural areas where condors would fly once they were freed. Because of the birds' rarity, people had ceased thinking and talking about them. Even landmarks named after condors had been changed. Therefore a major task was to bring back the spirit of the condor. Staff visited more than 2,000 schools to give talks and introduce games as educational tools. More recently, live animals have been added to increase interest. Falcons, hawks, and king vultures, all confiscated from illegal traders, have proven especially useful for explaining the physiological characteristics of carrion eaters. Little by little, the main objective—to return the condor's name to the forefront of people's minds and mouths—has been accomplished. As a result, today dozens of small retail stores, taxi lines, hotels, and even radio stations are named after the condor.

In the state of Mérida, 47% (531,100 ha) of the territory is covered by national parks, some of which are shared with adjoining states. The Sierra Nevada National Park was created in 1952. The recent creation of the other three parks, Juan Pablo Peñaloza (1989), La Culata (1989), and Tapo Caparo (1993), resulted from the necessity to preserve mountain ecosystems within the Venezuelan Andean region. These parks contain very severe physiography, preserving the pristine quality of nature. They are extremely diverse and, most important, contain the source of water that supplies drinking water and electricity to the region's major cities and towns and to irrigation systems that make the area one of the most productive in the country. All this directly benefits the local human population.

The high proportion of land under the jurisdiction of the federal government's National Parks Institute (INPARQUES) has led to conflicts with state and local (municipal) governments, especially where rural communities lie within park boundaries. These regional and local governments are pushing for reforms that would allow them to assume administrative control over lands within national parks.

Even though regional planning permits legal permanence and social consolidation of rural communities within Mérida's national parks, active opposition to the parks arose from fears that the goal of the National Parks Institute is to evict farmers from their lands. This has created antienvironmental sentiments against any action intended to protect and consolidate the parks' frontiers. The situation has been detrimental to the condor recovery program owing in part to the fact that Bioandina, the private foundation administering program, has a working agreement with the National Parks Institute.

The Parks Institute signed agreements with the rural communities closest to the site chosen to develop the condor project's base camp and quarantine area for incoming animals. The program involved working with animals bred and raised in isolation to increase survival probabilities after release. As such, entrance to the area is restricted to people from the immediate area only. Rumors suggesting that INPARQUES was seizing houses and land circulated among the community, inducing local and state authorities to back public strikes demanding the removal of large tracts of land from the national parks system and protesting all INPARQUES environmental programs. Subsequently, four of the released condors were found dead. Two animals had been shot and two had died of unknown causes. Three years later, there was still no official account of the incident.

In contrast to the local situation, great acceptance of the condor project at international and national levels was proven by the thousands of letters of censure that poured in through regular mail and via fax following news of the four condors' death. A few months later, authorities named the Andean condor the state's environmental symbol.

Local authorities knew of the importance of the pioneering efforts to recover the Andean condor; nonetheless, they worked to discredit the National Parks Institute administration and revived false tales of condors kidnapping children and killing cattle. As in other cases, politicians used people's ignorance about environmental and conservation topics to achieve their own goals.

FUTURE AND PROGNOSIS

The status of the Andean condor in South America must be examined on a country-by-country basis. Venezuela and Colombia harbor the smallest numbers of wild animals, and both are carrying out programs to augment these populations. Colombia has doubled its wild populations through the use of reintroduced animals. These individuals are just now reaching sexual maturity, meaning that real success can only be measured after they start reproducing. Survival of a second generation for at least 8 to 9 years after the first successful breeding of the reintroduced animals will allow the program to be considered a success.

It is too early to predict the results of the Venezuelan experience. As of late 1998, a second phase of the program is beginning with new releases of animals born in zoos from America. The goal is to release at least five animals a year for the next 5 years by continuously monitoring and assessing the ability of released animals to become authentically independent. Because there are no historical accounts of how many animals populated the area, releases should continue until the animals' behaviors and survival rates indicate that a healthy, balanced population exists. Based on the territory covered by animals already released, we estimate that a maximum of 30 animals could represent the entire condor population of Venezuela. We hope that the population will eventually reconnect with the Colombian population, benefiting both.

Ecuador has a medium-sized population of condors in urgent need of conservation action. For the first time in that country there is a real effort to study the ecology of the species in the wild by using satellite transmitters.

Peru and Bolivia still have healthy populations, primarily owing to the fact that the condors there occupy vast areas with little human influence. Overall numbers should be monitored to facilitate an adequate response if needed.

Chilean and Argentinean populations are the most promising of all, with thousands of animals inhabiting a long stretch of Andean Cordillera. The Buenos Aires Zoo in Argentina successfully initiated a captive reproduction program for condors by combining the efforts of several zoos from around the country. They have raised seven animals, of which five have already been released in protected areas around the city of San Carlos de Bariloche. These juveniles, all marked with satellite transmitters, will provide excellent insight into the complex social structure they have joined. Fortunately, Chile and Argentina possess immense territories largely devoid of human presence owing to harsh climatic conditions. These countries harbor the best conservation hopes for the permanent presence of condors in South America.

Anegada Iguana

Numi Mitchell

Common Name: Anegada iguana
Scientific Name: *Cyclura pinguis*
Order: Sauria
Family: Iguanidae
Status: Listed on Appendix I of CITES; Critically Endangered under the IUCN.
Threats: Dog and cat predation; competition from feral livestock; habitat degradation and loss; poaching.
Habitat: Dry tropical forest and scrub; seasonally found in coastal strand; occasionally found in salt marsh or salt pond environments.
Distribution: Currently found on Anegada, Guana, and Necker Islands in the British Virgin Islands (BVI). Formerly widespread on the Puerto Rico Bank (Pregill 1981).

DESCRIPTION

The Anegada rock iguana is a relatively large and stout member of its genus. Males have been recorded with nose-cloaca lengths of 56 cm and may grow larger. These iguanas have dusty brown backs that are sometimes vertically barred with black. Their legs, sides, and dorsal spines are often a brilliant turquoise blue. Commonly the dorsal spines are quite small, especially on females. Juveniles are most colorful, patterned with a series of black chevrons crossing their green or blue-green dorsal surfaces. When Anegada iguanas are agitated, their eyes flush bright crimson.

NATURAL HISTORY

The range of *Cyclura pinguis* was reduced to Anegada when the Virgin Islands became densely settled. Anegada is an island made of old, reef-tract limestone and sand. It is honeycombed with holes, caves, and other rocky shelter sites—ideal living quarters, or escape retreats, for iguanas. Anegada's human population has always been relatively low and there are few dogs and cats—and to date, no mongooses that have decimated iguana populations on other islands to which they have been introduced. Jointly these qualities made Anegada the last reasonably safe environment for *C. pinguis*. On Anegada, iguana distribution remains closely tied to the more porous limestone habitats, although the iguana uses adjacent sandy areas for burrowing and nesting as well.

Both Guana and Necker are largely volcanic in origin and have few naturally occurring shelter sites. The iguanas there must dig their own burrows. As a result, on these small islands or cays, fewer burrows are used and animals are more arboreal (i.e., spend more time in trees). Iguanas are much more vulnerable on these privately owned islands, but non-native predators such as cats and dogs are either controlled or are not permitted there.

C. pinguis is predominantly herbivorous. Invertebrates such as centipedes *(Scolopendra)*, moth larvae *(Pseudosphinx tetrio)*, and scarab beetles are also eaten. Iguanas rely heavily on fruit in season. On Anegada, where the iguana's distribution entirely overlaps with feral ungulates, the diet consists mostly of plants that are not favored by livestock; many contain secondary compounds that are produced by the plants to deter grazing and browsing. One such plant, *Croton discolor*, has dramatically increased in abundance since livestock were released on the cay in the late 1960s. The flora of Guana is quite different. Iguanas there generally inhabit areas where sheep do not graze. Native plants eaten on Guana include large quantities of *Centrosema virginiana* (a tender-leaved pea), flowers of *Tabebuia heterophylla*, seed pods of *Capparis cynophallophora*, and leaves of *Stigmophyllon emarginatum* and *Capparis flexuosa*.

Guana and Necker Islands have low-quality iguana forage in most areas. Both islands have been affected by feral sheep, goats, and possibly other grazing ungulates. As with Anegada, the vegetation has shifted in composition to plants rich in secondary compounds that are either distasteful or toxic to livestock. Guana still has feral sheep and goats, although efforts are being made to reduce or eliminate them. Necker's livestock has been removed, but vegetation still resembles that of islands on which livestock grazing occurs.

Along with a decline in population and a change in diet since livestock release in 1968, social organization among iguanas now differs as well. In 1968, the average home range for iguanas on Anegada was less than 0.1 hectare, and male home ranges had free space between them. The sex ratio was 1 male : 1 female. Males and females appeared to be monogamous and lived in separate but closely adjacent burrows (Carey 1975). By 1988, home range size and spacing appeared to differ in two ways: (1) home range size averaged 100 times larger, and (2) male home ranges abutted and overlapped slightly (Mitchell 1999, in press). The number of females declined to a sex ratio of 2 males : 1 female. "Pairs" were no longer clearly definable because several males would enter the home range of the few females present. Males suspected of having a mate had a principal burrow near that of a female, used some of the same burrows as the female, and had noticeably smaller home ranges—presumably to guard the female against wandering bachelors (Mitchell 1999).

Females lay 12–16 eggs between May and June. Females inhabiting sandy areas nest in their principal burrows or in dunes; those in rocky areas travel to find a spot in which to dig. On Guana Island, iguanas nest in seagrape-dominated beach strand. Clutches hatch in August and September as the rainy season commences and vegetation becomes more lush. If conditions are favorable, hatchlings can mature to reproductive size (about 450 mm snout-vent length; Carey 1975) in 3 to 4 years.

CONFLICTING ISSUES

Anegada is no longer a safe refuge for iguanas. Since the late 1960s iguanas there have experienced a massive population decline, and their density in what used to be considered good habitat is almost 10 times lower than former levels. The drop in numbers is probably due to a number of causes: (1) feral dogs, first reported in 1994, are known to kill adult iguanas, (2) an exploding population of feral cats, which prey on juvenile and subadult iguanas, (3) human poachers trafficking exotic pets, and (4) feral livestock. The latter represent Anegada's biggest drawback—the island teems with sheep, goats, donkeys, and cattle that cyclically breed to carrying capacity (i.e., the population size that the island can support), then starve after stripping the landscape of all palatable vegetation. These ungulates compete with iguanas for food. As a result, the iguana's diet has shifted to plant species that livestock do not eat, mostly those with secondary compounds. Many of these plants are poorly digested and therefore of dubious nutritional value.

Attempts have been made to restore iguanas to parts of their former range. Between 1984 and 1986, eight iguanas from Anegada were relocated to Guana. In vegetatively diverse regions of Guana these individuals and their descendants are thriving and reproducing (Goodyear & Lazell 1994); some of the offspring have been relocated to Necker Island.

Prior to the 1960s, the human residents of Anegada—descendants of former slaves and pirates—relied for their livelihoods on a combination of farming, animal husbandry, and fishing. The island was neatly and effectively partitioned by a system of stone walls, painstakingly constructed by residents, that retained livestock, fenced agricultural crops, and defined ownership.

Residents maintain that ownership of the land was granted to them by Queen Victoria. In an 1885 ordinance pertaining to Anegada, the British Crown agreed to grant land with the proviso that landowners have their property boundaries surveyed. Not one Anegadian did so (Renwick 1987). In 1961, under new legislation known as the "Anegada Ordinance," the British Crown assumed administration of most of the island and, along with the government of the BVI, leased all but 607 hectares to a Canadian development firm in 1968. The company began to develop the cay by bulldozing the network of stone walls. Stock animals (goats, sheep, cattle, burros, and swine) escaped to the bush and began to freely range the cay.

In combination with a prolonged drought, crops were raided and failed. Anegadians turned to the sea for a living. The livestock release was also a turning point for iguanas, which now had mammalian herbivores as competitors. Ironically the development firm soon folded, abandoning warehouses, rock crushers, and excavating equipment. Some Anegadians claim they drove the firm off the cay. Shortly thereafter the government began a reassessment of land claims on Anegada, and some titles were granted (Lands Adjudication Act, 1970).

The idea of creating a national park on Anegada was endorsed by the BVI Executive Council in 1981 and proposed in 1986 in the BVI System Plan (Geoghegan et al. 1986). In 1987 the governor appointed a one-man commission that recommended an equitable division of lands on Anegada (Renwick 1987). A respected group of community leaders formed the Anegada Lands Committee to mediate and settle remaining landownership issues. In March 1993 they approved the Anegada National Park concept in principle.

Based on iguana survey work (Mitchell 1999) a joint proposal from the National Parks Trust and the Conservation Agency was submitted to the Town and Country Planning Department, which recommended establishment of three terrestrial conservation zones on Anegada (Goodyear & DeRavariere 1993). These regions did not include land for which titles were already held.

Town and Country Planning was at that time producing the Anegada Development Plan (Government of the British Virgin Islands 1993), in which the recommended conservation zones were closely adopted. In November 1993 the Development Plan was released to the public. With land claims still unresolved, the maps enraged residents, preventing dialogue when the chief planner arrived to formally introduce the plan. To date, disagreements are still rampant and consensus on ownership and property boundaries remains elusive. Therefore, the Anegada Development Plan and the National Park proposal have been tabled.

The residents of Anegada will not set aside land for iguanas until they have land for themselves. In fact, the prevailing sentiment on Anegada is to disallow or discourage help of any kind for the iguana until the government has given Anegadians land titles.

Several unresolved questions impede establishment of a national park, or form of sanctuary, to protect the Anegada iguana. First, who is entitled to claim land on Anegada: (1) everyone born on Anegada and their descendants (a number of former Anegadians, and first or second generation offspring, currently living in New York who have submitted land claims), (2) only those born on Anegada, (3) only those born and raised on Anegada, or (4) only those born, raised, and still living on Anegada? Second, should the iguana be granted land before the people's claims are settled? Third, who should decide whether the iguana gets a land allocation, and who chooses which land: (1) the British Crown (an appointed official), (2) the

BVI government (locally elected officials), (3) the citizens of Anegada (but see the first issue), or (4) some combination of 1, 2, and 3? Although there are only about 150 residents, Anegadians have swayed elections in the past. Elected BVI officials will not allocate land for iguanas because it would widely displease voters on Anegada.

Complicating these issues is the fact that the general community on Anegada has not been shown how or why a national park would benefit them. The potential benefits of ecotourism should be explained to generate enthusiasm and local support. In addition, although the National Park Trust ostensibly supports establishment of a national park on Anegada, it is rendered ineffective because, as a branch of the BVI government, ultimately it gets its directive from politicians. Finally, because of past interactions with the Crown and government officials, Anegadians are skeptical and suspicious of outsiders. They do not believe anyone has their best interests at heart.

FUTURE AND PROGNOSIS

From a conservation standpoint, the first order of business on Anegada is to protect, fence, and remove livestock from land for the iguana. It will probably be necessary to assist the recovery of native plant communities that provide iguana forage. Experiments are currently under way to determine whether livestock exclusion is sufficient to promote recovery of plant communities (a passive effort) or whether restoration is required. Conditions on Necker suggest that active management will be necessary, as the cay retains a plant community poor in species diversity more than 15 years after livestock removal.

Obtaining a land allocation for iguanas may be more difficult than managing it. To gain approval for an iguana sanctuary, the government should first attempt to satisfy the concerns of residents. It should set a target date for settling all private claims to land on Anegada. If agreements cannot be reached by this deadline, it is of paramount importance that the Crown assume responsibility for *C. pinguis* and, as holder of land-granting privilege as well as lands proposed for the national park, set aside a protected area for iguanas. No residents currently on Anegada claim areas in the proposed national park, except as historical grazing range. Most of the area is lowland or wetland and not buildable.

The Anegada Development Plan (Government of the British Virgin Islands 1993) did not discuss management strategies for proposed conservation zones. Some suggestions follow.

1. A national park should be established and managed to promote recovery and proliferation of the island's rare, indigenous, and endemic species, particularly the iguana. Focus should be on restoring and repairing native habitats that have been damaged by overgrazing. The area should be fenced. Nature trails and boardwalks could be developed, but buildings (ex-

cept shade shelters and observation platforms) should not be permitted. The park should be staffed with Anegadian wardens and interpretive personnel. Fishing and salt collection in ponds could continue in accordance with BVI regulations but must not interfere with the reproduction or foraging habits of native animals.

2. Conservation areas should be left unfenced but should not be developed. Some eastern regions contain important habitat for relict subpopulations of iguanas and waterfowl. Spectacular, large native trees persist in the east, and on the west end there is sandy nesting habitat for iguanas living in adjacent limestone areas. Nature trails might be considered.

3. A coastal reservation should be managed for iguanas that nest behind dune ridges or in coastal strand communities, and for sea turtles that nest in sandy seaward regions. Clearing of vegetation and allowing of foot traffic in the dunes should be avoided, and buildings (except shade shelters) should not be permitted in an effort to avoid disruption of nesting sites. The beaches should be left in pristine condition. Designation of a coastal reservation as public domain would ensure that Anegada retains its trademark vast white beaches that attract visitors to the BVI.

The advantage of setting aside these lands must be properly explained to the Anegadians. Establishment would benefit iguanas and Anegadians alike. The iguanas would benefit from management policies that promote their recovery, and the park could provide economic benefits to the Anegadian community in the form of jobs and business expansion. Within the park, positions might include interpretive naturalists and guides, wardens, maintenance personnel, park restoration staff, and construction workers. Outside the park there would be increased demand for service-oriented businesses such as grocery stores, restaurants, bars, gift shops, dive businesses, rental shops, hotel facilities, and taxis. Other benefits might include government training and development of a natural history and historical museum associated with the park (Goodyear & DeRavariere 1993).

It takes capital to develop attractive tourist destinations. Anegadians should quickly gain assistance (e.g., development grants) from the world conservation community to develop a park that they can staff and run if they choose to conserve iguanas and their habitat. This would make Anegada the only British Virgin Island on which profits could be made largely by locals as opposed to outside investors.

In 1997 the National Parks Trust began a headstart program (hatching and raising iguanas in captivity for several years before release to the wild) on Anegada and planning a feral cat removal program—valuable steps to reducing predation pressures on young iguanas. However, without controlling livestock and restoring habitat, it is unlikely that the island can sustain higher numbers of adult iguanas. It is urgent that decisions to allocate and manage land for iguanas are made rapidly. Without action, the iguanas on Anegada could be extirpated within the next decade.

Aruba Island Rattlesnake

R. Andrew Odum and David Chiszar

Common Name: Aruba Island rattlesnake
Scientific Name: *Crotalus unicolor*
Order: Squamata
Family: Viperidae
Status: Threatened under the Endangered Species Act of the United States; Critically Endangered (CR) in the 1996 IUCN Red List of Threatened Species.
Threats: Habitat loss from human development; feral ungulates; indiscriminate killing by people.
Habitat: Caribbean desert from sea level to 188 m in elevation.
Distribution: Restricted to ~45 km² area in the southeastern portion of Aruba, Dutch Caribbean (within the area circumscribed by a line drawn from Boca Prins to Daimari, southwest to Savenata, east to a golf course and back to Boca Prins).

DESCRIPTION

The Aruba Island rattlesnake is a moderate size species (<1 m) described by Van Lidth de Jeude in 1887. It is distinguishable from its presumed parent species, *Crotalus durissus*, by having fewer ventral scales and an indistinct pattern (Klauber 1972). Dorsal color varies from light brown, tan, or gray to bluish slate. Ventral color may be white, cream, bluish, or salmon.

NATURAL HISTORY

The Dutch island of Aruba is 27 km north of the Paraguana Peninsula of Venezuela. Total landmass is ~193 km². The sea between the island and the mainland is fairly shallow (<250 m), which may have allowed the formation of a land bridge during recent ice ages. Indeed, the fauna of Aruba is more reflective of the fauna of northern Venezuela than of the other Netherlands' Antilles islands of Bonaire and Curacao. The rattlesnake is an example of a species that is present on Aruba but not on the other islands.

The island is within the band of continuous northeasterly trade winds, which vary between 10 and 25 knots and create a comfortable climate with a mean temperature of ~27°C. Approximately 430 mm of rainfall occur annually, mostly between October and February. The habitat is thus limited to species that are tolerant of very dry conditions.

Exactly when the founding stock for the Aruba Island rattlesnake arrived on Aruba is unknown. There are no fossil records for *C. unicolor*. Also, no comparative genetic studies of *unicolor* and mainland *durissus* have been undertaken to determine the likely source for the Aruba population. Original founders may have traversed a land bridge from Venezuela or Colombia, or rafted to the island from elsewhere in Central or South America. The modest degree of differentiation between *C. unicolor* and *C. durissus* raises questions as to whether *C. unicolor* should be regarded as a distinct species. Campbell and Lamar (1989), for example, suggest the *C. unicolor* be considered a subspecies of *C. durissus*.

Aruba Island rattlesnakes eat lizards and rodents. Similar to other rattlesnakes, they depend on a "sit and wait" strategy for prey acquisition (O'Connell et al. 1982; Goode et al. 1990). This cost-effective strategy limits the amount of energy expended on foraging, thereby allowing infrequent feedings.

On Aruba, the rattlesnakes breed between December and February during the slightly cooler months of the rainy season. Birth occurs during early summer, and litter size varies from 4 to 15. In captivity, breeding is less seasonal and appears to be correlated with cooling in the animal's environment. Another difference between captive and free-ranging animals is the interbirth interval. In the wild, snakes breed every second, third, or fourth year depending on the availability of prey, which is largely correlated with rainfall. In captivity, where prey is provided on a continual basis, females can produce litters yearly (Fitch 1970; Carpenter 1980; Crews 1987; Chiszar et al. 1994).

CONFLICTING ISSUES

The historical decline of *C. unicolor* has not been closely monitored. The species was considered very rare by 1980 (Peterson personal communication). Interviews with local residents in 1986–1988 indicated that the rattlesnake was more widely distributed in the past. Its current range is limited to the undeveloped southeastern area known as the Mondi. Reinert et al. (1995) estimated that there were as few as 250 adults left on the island.

The main threat to this species is human encroachment on unsecured habitat. Aruba currently has 70,000 permanent residents and at least 800,000 visitors annually (Arikok 1998). Interactions between humans and wildlife are common, particularly for tourists who rent four-wheel-drive vehicles to gain access to the Mondi. The situation for *C. unicolor* is exacerbated by human attitudes toward snakes in general and venomous snakes in particular, which many people feel compelled to kill on sight. Introductions of goats, sheep, and donkeys have altered species composition of the vegetation, thereby impacting food sources for the rattlesnake's natural prey, a native mouse *(Calomys hummelincki)*.

The first formal conservation effort for *C. unicolor* commenced in 1982 as a Species Survival Plan (SSP) of the American Zoo and Aquarium Association (AZA), which has been at the core of all subsequent conservation efforts. Initially focused on preserving the species in captivity within American zoos, the program broadened its scope in 1986 to include conserving *C. unicolor* on Aruba (Odum 1996).

In general, the SSP efforts to save *C. unicolor* on Aruba have been very successful. The first challenge was to establish and maintain a working relationship with the Aruban government. In 1986 initial contacts were made to investigate the possibility of a cooperative conservation effort between the SSP and Aruba. The SSP sought to contribute in areas where it could facilitate and assist (not direct) conservation for the animal in the field. Biological expertise not available on Aruba was offered free of charge to the Aruban government. Also, the SSP assumed the role of advocate for the species. This role may have been the single most important contribution to the successes achieved so far.

Since the late 1980s the SSP has had ongoing fieldwork on the island to obtain data about the life history, range, population size, microhabitat preference, ecology, and threats to *C. unicolor*. Along with this work were numerous formal and informal education programs, an international symposium on rattlesnakes, lobbying efforts, and a public relations campaign for the preservation of Aruba's natural heritage. It soon became evident that the most important factor necessary to save the rattlesnake was preservation of habitat. By 1992 a Conservation Action Plan (CAP) was developed for the rattlesnake by the AZA SSP, IUCN Conservation Breeding Specialist Group, local Aruban conservation organizations, and the Aruban government. Based on the scientific studies performed by the SSP, the plan included the CAP recommendations for preservation of land vital for the current population of *C. unicolor*. This preserve was created in 1997 as Parke Nacional Arikok. The park encompasses ~18% of the land mass of the island and includes the majority of the snakes' habitat as described in the CAP.

Aruba is a member of the Kingdom of the Netherlands, and Dutch influence in Aruban society is substantial. The official business language is Dutch, and the government, education, banking, and legal systems are modeled after those in Holland. More important to conservation of the rattlesnake are Dutch customs and folkways. The Dutch are well known as a liberal society that teaches tolerance for ethnic, religious, and political pluralism. These attitudes of tolerance are reflected in Aruban culture and extend to biological diversity, including the rattlesnake, thereby countervailing against the primal fear of snakes (Wilson 1998).

Over the last 25 years economic development has been rapid and the standard of living for Arubans has improved considerably. The population has also increased and the ethnic demography has changed. Ten years ago it was uncommon to find a non-Aruban hotel or restaurant worker. Today, many workers are immigrants from the Philippines, Indonesia, Colombia, and elsewhere in South America who do not have the same cultural ideals of tolerance. The rapid increase in jobs for tourism and the reopened Coastal Oil refinery in St. Nichols have been significant aspects of this growth. Many of the new immigrants loathe snakes, which they kill on sight.

Land is a very limited resource. Great political pressure has been placed on developing the snakes' habitat for housing and commercialization. Aruba is a European-type democracy with many political parties, and the stewardship of the government changes every few years. With each new government have come new threats to the remaining habitat for *C. unicolor* in the form of spoils to political party supporters or to pressure groups. Yet an environmental movement on Aruba gained strength and ultimately became powerful enough to prevent widespread habitat destruction.

With environmental, commercial, and other interests at odds over the use of the rattlesnake's habitat, the issue was left to government to resolve. In an innovative decision, a commission was established in 1995 to develop a comprehensive land management plan for the entire island. This commission was composed of government officials from several departments, commercial interests, biologists, geologists, engineers, environmentalists, and private architectural contractors. The final plan was a compromise, but it encompassed all the recommendations in the CAP for establishing a national park and wildlife preserve. One reason this approach was possible is that most of the land on Aruba is owned by the government and leased to the private sector. Thus private ownership, which is often an obstacle to securing undeveloped areas for conservation, was only a minor influence on Aruba.

Feral goats, sheep, and donkeys have had a negative impact on the rattlesnake by directly killing snakes and by altering vegetation, which has decreased rodent populations—the primary food of the rattlesnake (Reinert et al. 1995). Removal of these ungulates from the snake's habitat would aid recovery of the ecosystem and the rattlesnake population. This is not easily accomplished because of cultural attitudes. Historically many Arubans were herders, and wealth was measured by the size of their herds. Most Arubans have abandoned this lifestyle for business opportunities created by the strong tourism industry. Yet some elders cling to the "old ways," keeping large herds of animals that are allowed to range freely on the island. When attempts have been made to exclude goats, several herders have destroyed fences to gain grazing access for their animals. They have also used their political influence to prevent any sanctioned goat eradication program. Ultimately this problem will resolve itself when the old herders retire and are not replaced.

Another governmental problem was internal restructuring. As conservation gained political importance, each government department attempted to stake out its own territory, leading to confusion and inefficiencies. In 1986 the jurisdiction for the rattlesnake resided in one department. Today at least three separate departments have responsibilities over the rattlesnake and its conservation, requiring delicate coordination for all SSP efforts. This problem should be resolved when the national park is developed and fully staffed.

Nongovernmental organizations play an important role in the conservation of *C. unicolor*. In 1986 there was a budding environmental movement, which has grown substantially over the last 12 years. AZA biologists and educators have supported these organizations and the environmental movement in general as a key component of the conservation effort for *C. unicolor*. The two most important conservation organizations on the island are STIMARUBA (Love Aruba) and FINAPA (Aruba Foundation for Nature and Parks).

FUTURE AND PROGNOSIS

The prognosis for *C. unicolor* has steadily improved over the last 10 years. The formation of Parke Nacional Arikok is perceived as a major step toward long-term preservation of a viable population of *C. unicolor* on Aruba. The national park is well protected by law and should be secure for the future. Still, increasing human populations in a very limited area and changing political environments may jeopardize the snake, and improper management of the park might have a detrimental impact. In 1999 a field research project is planned to determine the effects of feral ungulate grazing on the Aruban ecosystem. It is hoped that the results of this study will provide the final justification to exclude ungulates from parkland. Concurrent with this study will be a project to evaluate the effects of translocating snakes. This is a common management technique in dealing with "problem" snakes that are found near human residences. Effects of translocation on resident and translocated *C. unicolor* are not yet known (Dodd & Seigel 1991; Reinert 1991).

The influx of new workers, who are not aware of the endangered status of the rattlesnake and are not as tolerant as native Arubans, needs to be addressed. An education and public relations campaign will be necessary to prevent indiscriminate killing of animals. Education programs for children should continue, to ensure that attitudes remain favorable toward the snake and toward wildlife in general.

The captive snake population is still necessary to protect *C. unicolor* from extinction. Its role will be as an emergency backup for reintroduction if the wild population falls rapidly. Related to maintenance of the captives is the need to assess their ability to survive if they are ever reintroduced. This requires study of anatomy, physiology, behavior, and immunology and represents a new frontier within the disciplines of conservation biology and restoration ecology (Chiszar et al. 1985; Chiszar, Murphy, & Smith 1993; Chiszar, Smith, & Radcliffe 1993). Another important factor will be the maintenance of a good relationship between the AZA and Arubans. The

AZA could also continue to provide expertise in disciplines not represented on the Island; and, programs designed to transfer knowledge to Arubans should continue so that they can monitor and manage the wild population effectively.

Asian Elephant

Raman Sukumar

Common Name: Asian elephant
Scientific Name: *Elephas maximus*
Order: Proboscidea
Family: Elephantidae
Status: Threatened on the 1996 IUCN Red List of Mammals; Appendix I of CITES.
Threats: Historic threats include the capture of elephants for domestication, and the loss of habitat to expansion of agriculture and human settlement. Loss and fragmentation of habitat continue throughout most parts of the Asian elephant's range, and illegal hunting for meat, hide, and ivory is a major threat in several regions.
Habitat: Tropical and subtropical forests; moist and dry deciduous forests; dry thorn forests; grasslands; and montane forests up to about 2,500 m in elevation in South and Southeast Asia. The highest densities are found in regions with a variety of vegetation types.
Distribution: Historically, from the Tigris-Euphrates basin in western Asia eastward along a narrow coastal belt into the Indian subcontinent and continental Southeast Asia. In the northeast, elephants were distributed up to and possibly beyond the Yangtze River in China. Several islands such as Sri Lanka, Sumatra, and Borneo also had elephants. Today their distribution is fragmented among 13 Asian countries— India, Nepal, Bhutan, Bangladesh, Sri Lanka, Myanmar, Thailand, Laos, Kampuchea, Vietnam, China (only southern Yunnan), Malaysia (peninsular Malaysia and Sabah), and Indonesia (Sumatra and Kalimantan).

DESCRIPTION

Among land mammals, the Asian elephant is second only to the African elephant (*Loxodonta africana*) in size. An adult female stands about 245 cm in height and weighs 3,000 kg, whereas an adult male is about 30 cm taller and weighs 4,500 kg. Elephants in eastern populations are smaller than those in other parts of the range. The characteristic feature of the elephant is its trunk, which is an elongation of its nose and upper lip. The Asian elephant has a single "finger" at the tip of its trunk. Only the male of the species carries tusks (Shoshani & Eisenberg 1982; Sukumar 1989).

NATURAL HISTORY

Of the estimated 45,000 wild Asian elephants remaining, India has the largest numbers (25,000), followed by Myanmar (5,000). Other countries believed to have over 2,000 elephants are Sri Lanka, Sumatra (Indonesia), and Thailand (Sukumar & Santiapillai 1996).

The ecology of the elephant is strongly influenced by its very large size (Owen-Smith 1988). The Asian elephant may consume up to 2% of its body weight as dry fodder (8% fresh weight) each day (Sukumar 1989). It is a generalist feeder, eating a wide variety of plant species and parts. In tropical rain forests, elephants consume a variety of palms, lianas, fruits, and herbs; in tropical dry forests they feed mostly on grass, bamboo, and the leaves and bark of shrubs and trees. Seasonal shifts in diet are pronounced in drier tropical forests, with more browse consumed during the dry season and more grass during the wet season.

Asian elephants live in matriarchal family units consisting of one or more adult cows, their daughters, and their prepubertal sons. Family sizes average 6 to 10 individuals. Several family groups form "clans" that show coordinated seasonal movements. Males typically disperse from the family when they are 10 to 15 years old and establish their own home range independent of the natal family. The annual home range sizes of family groups vary from under 50 km^2 in southeastern Sri Lanka (Wickramanayake personal communication) to nearly 1,000 km^2 in southern India (Baskaran et al. 1995). The home range size of adult males falls in the range of 100–400 km^2 in southern India (Sukumar 1989; Baskaran et al. 1995). Elephants communicate through chemical signals (Rasmussen 1997) and a variety of sounds.

The Asian elephant is a polygynous species, that is, a dominant male mates with more than one female. Mating success is determined by dominance hierarchies among males and by females' choice of males, both of which are influenced by the phenomenon of musth in adult bulls (Eisenberg et al. 1971). Musth is a state of sexual arousal accompanied by physiological and behavioral changes. A bull is in musth only once a year; the length of time varies from a few days to 3 or 4 months, depending on the age and social status of the bull (Sukumar 1994).

Female elephants usually give birth to their first calf when they are 12 to 17 years old. Males are capable of reproducing when they are 15 years of age, though they may be excluded from mating until 20 to 25 years for social reasons. Typically, a cow produces a calf every 4.5 to 5.0 years (McKay 1973; Sukumar 1989). Mortality rates over 5 years average only 1% to 3% per annum in females and up to 6% per annum in males under natural conditions. In elephant populations where tusked males constitute a large proportion of the male segment, such as in southern India, ivory poaching may cause male mortality rates to rise to 20% per annum (Sukumar 1989). Population growth rates in Asian elephants under natural conditions do not seem to exceed about 1.5% per year. Unlike the African elephant, however, Asian elephant populations apparently do not suffer from drastic declines owing to environmental factors such as drought.

CONFLICTING ISSUES

Since the late 1970s, the IUCN/SSC Asian Elephant Specialist Group has emphasized the decline of the Asian elephant population and the need for its long-term conservation. Based on several field studies in Sri Lanka, India, Sumatra, Thailand, and Malaysia, the Specialist Group developed an action plan for Asian elephant conservation in 1990 (Santiapillai & Jackson 1990). The plan covered the status and distribution of elephants in all 13 range countries, the primary threats to survival, and a set of recommendations for conservation action. Several countries, however, have not been able to respond adequately or rapidly enough to meet the challenges of elephant conservation.

Sri Lanka was one of the first countries to set up a system of protected areas with a focus on elephants, but they have been under great pressure from development projects such as dams and agricultural expansion. Peninsular Malaysia has an ongoing elephant conservation program that involves the capture of small herds or solitary elephants and translocation to more viable habitats. Vietnam has developed an elephant conservation plan, but it was established only after wild elephant numbers had been reduced to below 150. Myanmar set up its first elephant sanctuary in 1997 in an attempt to rescue its dwindling elephant population. In 1992 India launched a major elephant conservation program—Project Elephant—with an initial budget of U.S. $8 million for 5 years.

About 30% of the Asian elephant's present range is included in protected area systems. Of the remaining area, a certain proportion is still under some form of protection, as in the case of Reserved Forests in India. The rest of the area is under private ownership, the control of village councils (e.g., Northeastern India), local tribes and chieftains (Myanmar, Indochina), military or rebel groups (Kampuchea), and so on. The stakeholders in elephant conservation thus include a broad cross-section of society, including local farmers and commercial plantation companies whose cultivated lands may be raided by elephants, or timber corporations (as in Myanmar) that depend on a supply of elephants for their operations. Habitat conservation and elephant-human conflicts face a number of nonbiological challenges associated with economic development, local community rights, motivations and attitudes of local residents and authorities, and lack of financial resources.

Cultural and religious factors in many Asian countries have contributed to a positive attitude toward elephants, even among local villagers who suffer from elephant depredations. In India, for instance, farmers in most areas have traditionally revered the animal as representing the elephant-headed god Ganesh. Such reverence also exists in other countries, including the Aceh province of Sumatra and in Sri Lanka. Values are changing, however, as Asian societies undergo rapid socioeconomic change.

The two main conflicting issues are (1) land development that destroys and fragments habitat, and (2) elephant-human conflicts (crop depredation and human mortality) resulting in local antagonism toward the elephant and its conservation.

Development of land where wildlife can reside is a common theme in conservation. In the case of the Asian elephant, the issue is complex because it occurs in different contexts in the various regions. In many parts of India, such as the south and the northwest (where most of the elephant's range is government-managed Reserved Forest), the issue today is not loss of habitat per se but the threat of fragmentation from roads, railway lines, dams, and canals. Fragmentation has combined with agricultural development to reduce elephant habitat. In other parts of India, such as the northeast, considerable loss of forested habitat continues as a result of forest clearing and shifting cultivation. Here, large areas of forest are under the control of village or district councils.

In recent years, efforts to prevent loss of forest or to preserve wildlife corridors have included outright purchase of land, rejection of local developmental plans by the federal government, or public interest litigation through the judiciary to halt harmful developmental projects. For example, during the 1980s a successful legal campaign by local conservation groups stopped the Silent Valley dam project in the southern state of Kerala, India. In the state of Tamilnadu, environmental clearance was denied to a project scheduled to widen a canal (which would have obstructed the movement of elephants associated with the Mudumalai Sanctuary). In the northeast, the government has purchased forest land from village councils in the state of Meghalaya and declared them a protected area.

In countries such as Malaysia (peninsular Malaysia and Sabah) and Indonesia (Sumatra), forest is being opened up at a very rapid rate for commercial plantations, chiefly rubber and oil palm. In Sumatra the lowland forests are also being cleared for resettling migrant people from the island of Java. This has resulted in considerable loss of habitat for elephants (Santiapillai & Ramono 1990) and escalation of crop depredation by elephants. In an effort to control this conflict, over 600 elephants have been captured during the past decade.

Elephant-human conflict is widespread (Blair et al., 1979; Sukumar 1989; Williams & Johnsingh 1997; de Silva 1998; Nath & Sukumar 1998). Such conflict is particularly acute in countries such as India and Sri Lanka that have large and high-density elephant populations. Economic loss from damage to cultivated crops reaches millions of dollars each year, particularly when commercial crops such as rubber and oil palm are involved. In India alone over 200 people are killed each year by wild elephants, with a large proportion of the incidents occurring in cultivated fields and settlements when elephants come at night to raid crops. Conflict occurs even in countries with small elephant populations, such as Bangladesh and Vietnam, or those with low elephant densities, such as in Sumatra and Malaysia.

The ecological reasons behind crop depredation by elephants are only partly understood (Sukumar 1989, 1995). Elephants seem to raid crops more frequently in regions where they have lost habitat or a part of their home range, where the habitat is fragmented, and where they have to traverse cultivation to reach water sources. In addition, the higher palatability and nutritive value of cultivated plants, compared with wild plants, may motivate elephants to forage in agricultural land. Adult male elephants also have a greater propensity to raid crops than do female-led family groups; this may have its roots in the evolutionary forces driving male and female behavior in a polygynous mammal (Sukumar & Gadgil 1988). All this means that a certain degree of crop depredation by elephants and conflict with people is inevitable, even when habitats are intact and rich in forage resources.

In most places people still use traditional methods, such as firecrackers and other noise-making devices, to drive elephants from their fields. These techniques have had very limited success. It is expensive to dig and maintain ditches along the forest-cultivation boundary, especially in areas of high rainfall. In recent decades, high-voltage electric fences have been used in many countries. The key to the success of electric fences is proper maintenance and innovation against an intelligent animal (Nath & Sukumar 1998). Upkeep is a serious problem in countries such as India because villagers are not interested in the upkeep of government fences. There is monetary compensation for crop losses in most Indian states, but administrative problems have caused dissatisfaction among farmers.

FUTURE AND PROGNOSIS

The following recommendations are suggested to help overcome the conflicts surrounding land development and to promote the long-term conservation of the Asian elephant.

1. Land use policies in elephant habitat must be made clear to prevent further fragmentation of habitat or escalation of elephant-human conflicts. Development of land for crops such as oil palm, rubber, or sugar cane should not take place in areas where such plants are likely to attract elephants. Land use policies should also be pragmatic and aim at maintaining larger, intact habitats, even if this means giving up smaller, fragmented forest patches for development.

2. Future development of any kind within elephant habitat should recognize the need to maintain corridors for the free movement of elephants. Corridor areas should be legally designated or purchased. Alternatively, local residents may be provided with incentives for maintaining their lands as corridors.

3. Local communities must be involved in programs to keep elephants away from agricultural land, such as maintaining electric fencing. This might involve giving them responsibility for maintaining barriers and making provision for fuelwood, fodder for livestock, and other human needs.

4. Where levels of elephant-human conflict are low, a system of compensating farmers for crop losses is the most economically sensible option.

5. Elephant population management—including the capture and translocation of herds from fragmented patches, the capture of notorious raiding bulls, or the occasional destruction of an animal that is a threat to human life—is essential in order to minimize conflict. This would also promote relations between local residents and wildlife managers (Sukumar 1991).

Asiatic Lion

Ravi Chellam and Vasant K. Saberwal

Common Name: Asiatic lion
Scientific Name: *Panthera leo persica*
Order: Carnivora
Family: Felidae
Status: Endangered under the IUCN; listed on Appendix I of CITES and Schedule I (the highest level of protection) of the Wild Life Protection Act (1972) of India.
Threats: Historically, the major threats to Asiatic lions have been hunting and large-scale habitat destruction owing both to industrial expansion and the presence of numerous temples within the Gir Forest. Genetic and demographic problems associated with a single, small population are a longstanding threat, as is the potential for a catastrophic event eliminating this single population.
Habitat: In the past, Asiatic lions occurred in habitats ranging from moist deciduous forests to hot deserts. The Gir Forest is a dry deciduous forest with grasslands, thorn forest, and riverine forest; the latter is a critical habitat during the dry season.
Distribution: The last surviving free-ranging population of Asiatic lions is basically restricted to the 1,412 km² of Gir National Park and Wildlife Sanctuary in the Saurashtra Peninsula of the state of Gujarat in western India. Groups of lions have dispersed and settled in at least two forested areas, the Girnar hills and the coastal plantations, each with less than 20 animals. Neither satellite population is self-sustaining over the long term.

DESCRIPTION

The body color of adult lions varies from pale yellow to white on the underside, to light tan to dark brown above and on the flanks. Adult males have a mane, which varies in color from black to golden yellow. Dark spots that cover the bodies of cubs gradually fade and disappear by the age of 3 years. The tails in adults end in a tuft of black hair. Adult males weigh 160–190 kg, whereas adult females weigh 100–120 kg—smaller, on average, than the African lion. There are other anatomical features that differentiate the two subspecies: a belly fold in the Asiatic lion, not always found in the African lion; a less luxuriant mane in the male Asiatic lion; and minor differences in skull structure.

NATURAL HISTORY

The Gir lions are a forest-dwelling population (contrary to the normal association of lions with open habitats) that prey on deer (the only lion population to do so). In the 1970s livestock (75%) and wild prey (25%) were the key components of the diet (Joslin 1973), proportions that had reversed by the late 1980s (Ravi Chellam 1993). This dramatic shift in predation patterns is owing to the increased population of wild ungulates, especially chital (*Axis axis*), a consequence of reduced livestock grazing within the protected area. Chital is the most commonly eaten prey, and sambar (*Cervus unicolor*) is the most preferred prey species. The Gir's wild ungulate population is now estimated at over 40,000 animals. Seasonal home ranges of adult males are about 140 km^2 and those of females are between 65 and 85 km^2 (Ravi Chellam 1993).

A group of related lionesses and their offspring form the core of a pride, to which is attached a group of males. These male coalitions hold territories for as long as 5 years. Unlike in Africa, it is rare in Gir to see adults of both sexes together unless they are mating or sharing a large kill, such as sambar or domestic buffalo. The presence of dense stalking cover may enable males to hunt more successfully in the Gir Forest than in the open African savannah. With the medium-sized chital (adult body weight ~50 kg) being the most common prey for females, it is a good strategy for males to be nutritionally independent of females, thereby allowing the rest of the pride adequate access to the kill.

Lions mate throughout the year. A birth peak occurs from February to April. Litter size ranges from one to five. Lionesses start breeding when they are 3 years old, whereas males need to obtain a territory, at approximately 5 years of age, before breeding successfully. Adult sex ratio is 1 male : 2.2 females (Ravi Chellam & Johnsingh 1993).

CONFLICTING ISSUES

The range of the Asiatic lion has shrunk dramatically over the past two centuries owing to the hunting and habitat destruction just discussed. The fact that the species survives today is largely due to the efforts of the Nawab (ruler) of Junagadh. In the early twentieth century he declared the Gir Forest his hunting preserve, an area within which the lion would receive full state protection. At the time, the lion population is reported to have dipped to 12 to 20 animals. It has recovered to approximately 300 animals in the Gir National Park and Sanctuary.

The recovery of the subspecies is remarkable, particularly given the high incidence of human dependence on natural resources in and around Gir. The Asiatic lion remains a highly endangered species, almost entirely confined today to the approximately 1,500 km^2 of the Gir protected area. Given the home range requirements of the species there is little scope for the population to increase within the confines of the Gir Forest, leading in turn to a spillover of lions into a human-dominated landscape.

The Gir situation typifies the larger story in Indian conservation: an endangered species is limited now to a relatively small area surrounded by cultivation and human habitation. In most parts of India, the proximity of human and wildlife populations results in considerable tension between resident peoples and wildlife managers. Around the Gir Forest, lion attacks on cattle and humans have intensely alienated local residents from government-initiated conservation efforts (Saberwal et al. 1994).

The Gir Forest itself is inhabited by a buffalo-herding pastoralist community, the Maldharis. Two thousand individuals with approximately 15,000 livestock are resident within the Gir Forest today. Calls for a relocation of these Maldharis have been justified on the grounds that their presence represents a disturbance to lion habitat. The revival of the ungulate population in the 1980s following the relocation of about half the Maldhari population resident in Gir at the time is often used as clinching evidence of the disturbance caused by the Maldharis. It is more likely that the condition of the forest improved owing to the exclusion of the 60,000 to 70,000 cattle that seasonally grazed there. One also needs to note that the Maldhari relocation of the 1970s has drawn sustained criticism owing to the sharp drop in quality of life of the relocated Maldharis. We do not see the immediate need to relocate the Maldharis currently living in Gir, especially because there is a need to implement other conservation measures of greater importance. Our focus below is therefore restricted to the villages that adjoin the Gir Forest.

Lions have been located at distances exceeding 40 km from Gir's nearest boundary (Ravi Chellam 1993). These animals move through an intensely cultivated, inhabited landscape. Most villagers own at least one cow or bullock, with more wealthy families owning small herds of cattle. Where a farmer is unable to afford a protective cow-shed, cattle may be tethered outside a house at night. Unproductive animals are simply let loose. Given the religious significance of the cow in Indian culture, there is no danger of these cattle being slaughtered.

However, as far as the lion roaming the countryside is concerned, there is a good chance of coming across unguarded cattle. For lions that live on the edge or outside the protected area (a consequence of being forced out of a larger home range by a younger, more fit male, or of subadults simply not managing to acquire a home range within the Gir Forest), areas outside the protected area are the only potential source of food. Lion movement through densely settled areas leads inevitably to a high incidence of lion-human interactions.

Lion attacks on humans have tended to increase after periods of severe drought, as recorded at the turn of the century as well as recently after the extended drought of 1985–1987. Between 1988 and 1991, 120 attacks on humans were recorded, including 20 fatalities, compared with 65 injuries and 8 deaths in the preceding decade (Saberwal et al. 1994). The spate of attacks has slowed considerably since 1991.

A potential explanation for these figures is that during times of drought there is an abundance of livestock, both dead and alive, on which lions can feed. Once a drought ends, villagers are more protective of their cattle in their attempt to rebuild their herds. There is a corresponding drop in cattle availability, an increase in the persistence of lion attempts to secure carefully guarded prey, and, consequently, greater chances of contact with humans.

Perhaps the most significant change in lion behavior following the drought of 1985–1987 were instances of lions feeding on human bodies (in 7 out of 20 human fatalities). Such behavior suggests the possibility of attacks as actual predation, rather than merely acts of defense following a chance encounter. Although the argument is speculative, the possibility that lions are losing their fear of humans—simply as a result of the growing levels of interaction between the two—is cause for concern.

Villagers are vocal in airing their grievances. Owing to the fact that electricity for running wells is often diverted to industry during the day, cultivators are forced to irrigate their fields at night. Because of a potential lion presence in the village at night, many villagers stress the enforced need to water their fields in groups of four or five, and they see this as wasteful of their time. Villagers repeatedly comment on the curtailment of human activities outside the house once the sun has set (Saberwal et al. 1994).

Villagers also highlight the difficulties of obtaining monetary compensation for livestock losses. To obtain compensation the villager must get an assessment from a forest officer of the condition and value of the animal prior to death. Many villagers talk of having to bribe forest officials to obtain the required attestation. In counterpoint, foresters accuse villagers of taking useless cattle to the edge of the forest in hopes of the animal being killed by a lion, then seeking compensation for what was originally a useless animal.

There are as yet limited reports of retaliatory lion killings by villagers. It is unclear, therefore, as to how villager hostility toward lion conservation has translated among the local population overall. Nevertheless it is important to note the high levels of animosity among villagers toward lion conservation. Villagers observe that conservationists might feel less strongly about lion conservation were their children to face the nightly threat of a lion attack.

Given that the region is drought prone and that the lion population is increasing, the problem is unlikely to go away. The small and isolated nature of the population adds to the urgency of the problem and the need to find pragmatic and workable solutions.

FUTURE AND PROGNOSIS

A number of steps need to be taken toward improving conditions in areas adjoining the Gir Forest. The means of obtaining compensation should be simplified, with greater transparency in evaluating the value of killed cattle and payment of compensation to villagers. Compensation levels should be greatly increased. On the whole, a more humane approach must be adopted while dealing with villagers, especially in cases of attacks on people.

Villagers need to derive tangible benefits from lion conservation, including improved education and health facilities, and alternative sources of fodder and fuel. Recent development measures undertaken in villages around Gir are a step in the right direction, although their impact will only be known as they unfold over the coming years.

There is clearly a need to deal with the locally abundant lion population in the Gir Forest. Captive breeding facilities in India and outside the country may be willing to accept animals owing to a paucity of pure-bred Asiatic lions in captivity. This short-term solution is contingent on the continued availability of space in these facilities.

A more urgent consideration is the introduction of a culling, and maybe even sport hunting, program. Any such program should be carefully designed to target specific age/sex categories of the population (i.e., those segments disproportionately involved in lion-human conflicts) and even more carefully monitored. Given the prevailing philosophies governing wildlife conservation in India, this option is bound to meet with resistance. Overcoming this resistance may be necessary to significantly reduce human-wildlife conflicts. Inability to reduce these conflicts could lead to the killing of lions by villagers, analogous to the anonymous poisoning of tigers reported from other parts of the country.

Merely protecting and managing the Gir population, though crucial, is insufficient as a long-term conservation strategy. The establishment of at least one more lion population in the wild is necessary to guard against chance genetic and demographic impacts on the Gir population. Such an initiative is currently under way, with a free-ranging lion population to be established in the Kuno Wildlife Sanctuary in central India. The initial focus is on relocating tribal villages from the core of the area to forested areas on the fringe. Subsequently, the prey base will need to be enhanced before a group of lions can be moved from Gir (Ravi Chellam et al. 1995).

The lion translocation project is a long-term conservation initiative. With continued administrative and political support, and a large slice of luck, the lion's roar, surely the most memorable, if heart-stopping, sound in the Indian jungle, may well reverberate in parts of its former range from where it has been absent now for more than a hundred years.

Aye-aye

Eleanor J. Sterling and Anna T.C. Feistner

Common Name: aye-aye
Scientific Name: *Daubentonia madagascariensis*
Order: Primates
Family: Daubentoniidae
Status: Listed on Appendix I of CITES; Endangered under the IUCN.
Threats: Habitat destruction; killing as a crop pest or as a harbinger of bad luck.
Habitat: Moist and dry forests of Madagascar.
Distribution: In the rain forests on the northwestern and northeastern sides of the island of Madagascar.

DESCRIPTION

The aye-aye, a nocturnal lemur, is one of the world's most unusual-looking primates. Most obvious are its flat face, large, naked ears, and long bushy tail, but it is probably best known for its extraordinary, elongated middle finger. The aye-aye also has a curious constellation of features not found in other primates, including continuously growing front teeth, abdominal nipples, claws on all digits but the thumb, and a nictitating membrane ("third eyelid"). It has one of the largest hands relative to body length of any primate. Weighing 2.5 to 3 kg, it is much larger than other nocturnal primates (Feistner & Sterling 1995). Its dental formula is also unusual: 1/1 incisors, 0/0 canines, 1/0 premolar, and 3/3 molars. Aye-ayes have dark, coarse, white-tipped outer guard hairs and a layer of soft, short underfur.

NATURAL HISTORY

The aye-aye is the only living representative of the family Daubentoniidae. As with the majority of lemurs, aye-ayes are only found on the island of Madagascar in the Indian Ocean. They inhabit moist and dry forests and can live in or near coconut plantations. Their diet includes few species but varied food types, including seeds, larvae, fungus, and nectar. They use their chisel-like teeth and probe-like middle finger to gain access to structurally defended natural foods, such as wood-boring larvae and hard-coated seeds and cultivars such as coconuts and sugar cane (Sterling 1994a; Sterling et al. 1994). Aye-ayes are largely nongregarious and may rely on scent-marking for recognition of others (Price & Feistner 1994). They can range quite far each night in search of food and mates, spending up to 80% of the night traveling and feeding (Sterling 1993). They do not have the restricted mating season of other lemurs but apparently reproduce year-round (Sterling 1994b). Females first reproduce at 3 to 4 years of age and have one baby at a time. The interbirth interval appears to be 2 to 3 years, so aye-ayes have low reproductive rates. Females "park" their infants in the nest when they are foraging.

In one study (Sterling 1993), aye-ayes were shown to prefer to feed in large-diameter, tall trees with large crowns and dense undergrowth. Tall trees with many vines were preferred for nest sites. Aye-ayes frequent plantations of coconuts and sugar cane on a regular basis and often nest in forest surrounding villages. Recent information on many aspects of aye-aye biology can be found in Feistner and Sterling (1994).

CONFLICTING ISSUES

The aye-aye is the only extant member of its family, but in the recent past a more robust form, *Daubentonia robusta*, inhabited the dry southwestern forests (Lamberton 1934; Simons 1994). The extinction of this species could be attributed to late Holocene climatic change, to human intervention, or to a combination of both (MacPhee & Raholimavo 1988).

Habitat degradation and cultural beliefs are the two most important threats to extant aye-aye populations. Deforestation rates in Madagascar are difficult to measure, but Green and Sussman (1990) estimate that at current rates, only the steepest slopes will remain in 35 years. As wildlife populations increasingly face habitat fragmentation owing to deforestation, aye-ayes will increasingly come into contact with local human populations. Unlike many other lemurs, aye-ayes are not systematically hunted by the Malagasy, either for food or for pets. However, villagers kill animals in some regions in accordance with local taboos, known as *fady*.

The "conflict" between humans and aye-ayes in Madagascar differs from other adversarial situations in that aye-ayes are not threatening people's food supply to any great extent, nor do they pose a physical threat to humans. The conflict takes place on a much more spiritual/cultural level.

The aye-aye has held a powerful place in the belief system of the Malagasy for centuries. Voyagers have collected myths and stories about the aye-aye since the late 1700s. Grandidier (1908) remarked that it was difficult to procure animals from the wild because the Malagasy he worked with believed that anyone who touched an aye-aye would die within the year. In a number of villages on the eastern and northwestern coasts, if an aye-aye is seen near the village, the residents think they must move to prevent bad luck. In most of these villages, legend has it that if the aye-aye seen is old, then an older person will die; if the aye-aye is young, a child will die. In the northeast, as recently as 1986, aye-ayes were killed to prevent bad luck and taken to a rival's yard to bring misfortune on him. In the northwest in particular, when an aye-aye was killed it was often placed on a well-traveled road or at the confluence of more than one route. By displaying the animal on the route, it was thought that travelers would carry the aye-aye's bad luck away from the village. This theme varies from region to region, but the essence is the same in many places: aye-ayes will bring grave misfortune when seen.

Contrary to popular belief, however, the aye-aye is not always thought to bring bad luck. In the southeast, they are believed to bring good fortune. Some Malagasy on the eastern coast believe that the aye-aye has a human origin. When they encounter a dead aye-aye in the forest, they bury it with all the ceremony accompanying the funeral of a grand chieftain (Lamberton 1910; Fanony personal communication). These same people believe that they will die within a year if they cause the death of an aye-aye. The roots of the belief that aye-ayes are reincarnated humans may lie in the fact that aye-ayes often nest in cemeteries, which are frequently the only remaining forested regions near a village.

There are currently no clear factors—either ethnic or geographic—that explain which people believe the aye-aye is good, evil, or both (Sterling 1993). Inevitably, as Malagasy continue to move to new regions and interact with people from other ethnic groups, the distinctions in belief systems will be further blurred.

Never thought to exist at high population densities, the aye-aye was presumed extinct in the 1930s (Petter & Peyrieras 1970). In the 1950s, researchers rediscovered a population of aye-ayes on the eastern coast of Madagascar and helped to create a reserve there. Subsequent development activities in this reserve convinced conservationists that the aye-aye was on the brink of extinction, and the government of Madagascar designated the island of Nosy Mangabe as a Special Reserve to serve as a refuge for the species (Adriamampianina 1984). Scientists released nine aye-ayes (four males and five females) onto the island in 1966–1967. Since the 1960s other aye-aye populations on the mainland have been located, and in the mid-1980s a Man and the Biosphere Reserve was established at another east coast site.

FUTURE AND PROGNOSIS

The aye-aye's attraction to cultivars such as coconuts and sugar cane is a striking feature of its ecology, as comparatively few primates live in or around villages (Richard et al. 1989). This, in combination with the marked increase in human populations over time, has led animals in previously isolated rain forest habitat to become more vulnerable to contact. At present there are no systematic measures of how many animals are killed by villagers each year. Identifying the extent and distribution of confrontation between villagers and aye-ayes may be important for a species management plan.

Another critical element for a species management plan, population density estimates, are unavailable for the aye-aye, as populations are very difficult to count owing to the aye-aye's mobile, nocturnal, and secretive habits. Indirect census methods are also inappropriate, as there are no reliable proximate signs that are useful with other primates. Given their nongregarious behavior and low reproductive rates, aye-aye population density in any given region is bound to be low. With the high rates of habitat fragmentation and destruction in Madagascar, aye-ayes living at low population densities are vulnerable to extirpation.

Several studies have uncovered good news regarding the future of aye-ayes. Once thought to have one of the most restricted distributions among extant lemur species, it now appears that the aye-aye may be one of the three to four most widely distributed genera of primates in Madagascar. It is known to occur in at least 16 protected areas, most of which are of an acceptable size for harboring a viable population of aye-ayes. Maintaining these reserves, and limiting contact between aye-ayes and human populations, should be one key recommendation for a management plan, as the strength of people's beliefs about aye-ayes should neither be underemphasized nor belittled.

The international profile of the aye-aye has also become more prominent. Recent reviews of the lemurs of Madagascar have highlighted the animal's unique taxonomic status as the only living representative of an entire primate family (Mittermeier et al. 1994). Moreover, the development of a captive breeding program has heightened awareness about this species (Feistner & Carroll 1993; Feistner et al. 1996). The captive breeding program helps to establish a safety net for aye-ayes (Winn 1989; Carroll & Haring 1994). As of July 1998 there were 34 aye-ayes living in captivity (including those in two institutions in Madagascar), of which 12 (35%) were captive-bred. Moreover, second-generation breeding has occurred.

The breeding program also provides important opportunities for research. For example, the extraordinary foraging behavior and independent digit control of aye-ayes has been investigated (Milliken et al. 1991; Feistner et al. 1994; Erickson et al. 1998). The regular breeding of aye-ayes since 1993 has allowed the study of infant development (Feistner & Ashbourne 1994; Winn 1994a), an aspect little studied in the wild. Other aspects of aye-aye behavior are gradually being better understood by studying captive animals (Dubois & Izard 1990; Curtis & Feistner 1994; Price & Feistner 1994; Sterling et al. 1994; Winn 1994b). The possibility for both Malagasy people and others around the world to see aye-ayes at close range in captivity also provides important educational opportunities, further raising awareness about this species and its habitats.

Basking Shark

Sarah L. Fowler

Common Name: basking shark
Scientific Name: *Cetorhinus maximus*
Order: Lamniformes (mackerel sharks)
Family: Cetorhinidae
Status: Listed as Vulnerable by the IUCN (1996). Protected from direct fishing, land-ing, and sale in U.S. Atlantic waters (1997); fully protected in Florida State waters (to the 3-mile limit on the East Coast and 9 miles on the Gulf Coast) and in British coastal waters. Once the Barcelona (1995) and Bern (1997) Convention listings are fully ratified and implemented, it should soon be protected in the Mediterranean. The shark is Protected from directed fisheries in New Zealand (1991), but landing and sale of bycatch (i.e., nontarget catch) is permitted.
Threats: Fisheries—target fisheries have caused population declines, and bycatch in other fisheries may also deplete populations. Collisions with boats are another threat.
Habitat: The shark has been mainly recorded from surface sightings in temperate coastal, inshore, and oceanic waters, but it is sometimes taken as bycatch in deep-water fisheries.
Distribution: Basking sharks inhabit continental and insular shelves of temperate and boreal waters. They are usually sighted close to the coast but occasionally well off-shore. Most U.K. records are correlated with water temperatures of 8° to 14°C (Marine Conservation Society database), but sightings off the New England northwestern At-lantic coast occur in water ranging from 11° to 24°C, with peak densities at 22° to 24°C (Owen 1984).

DESCRIPTION

Sharks are cartilaginous fishes closely related to skates and rays, and more distantly to chimaeras (holocephalan fishes). They are quite dissimilar to bony fishes (teleosts).

The basking shark is the second largest fish in the world (after the whale shark). Born at between 1.5 m and 1.7 m total length, it reaches lengths of over 10 m. The body is fusiform (cigar-shaped) with a conical snout, a large dorsal fin (up to 2 m high), and a crescent-shaped tail. The distance between the tips of the upper and lower caudal (tail) fin roughly equals the height of the dorsal fin. Five very long gill openings extend from the upper sides to the lower surfaces of the throat. The mouth opens very wide during filter feeding. Large gill arches support long gill rakers and gill lamellae for oxygen exchange. The distinctive "strap gills" are one of the main features identifying elasmobranchs, the sharks and rays. The numerous tiny teeth are not used for feeding. Basking sharks have a characteristic mottled dark to light grey dorsal pattern but are paler underneath. Distinctive scars (from boat collisions or parasite damage) and natural patterns may be used to recognize individuals.

This shark is named from its habit of "basking" on the surface in good weather; the dorsal fin, upper tail lobe, and tip of the snout may break the surface during feeding.

NATURAL HISTORY

The basking shark is a plankton feeder, usually sighted on the surface in areas of dense zooplankton where filter feeding is most effective. Preferred feeding areas are often along coastal fronts, inshore areas off headlands and islands, and in bays where "tide lines" are formed by strong currents. Surface sightings are most common from spring to autumn in most parts of the world, and they occur in higher latitudes as temperatures rise and plankton blooms occur. North Atlantic populations are not visible on the surface during winter. However, a few sluggish, torpid individuals have been caught from deep water in winter (e.g., in the Gulf of St. Lawrence, Newfoundland [Lien & Aldrich 1982], and in the Firth of Clyde, Scotland [Fairfax 1998]). Some European winter specimens had shed their gill rakers and were growing new ones ready for spring (Parker & Boeseman 1954). When zooplankton are scarce during winter, basking sharks may enter a dormant state in deep water, as indicated by their huge oil-rich liver. In California, however, they are seen most commonly from October to March (Phillips 1947).

Very little is known of basking shark behavior, as biologists have been limited to making observations of summer surface aggregations and feeding individuals. The sharks are passive filter feeders and move slowly with mouths agape. Their gill bars hold gill slits open to allow gill rakers to strain zooplankton and small fish from the seawater. Small groups have been seen following "nose to tail"; this may be courtship behavior (Earll 1990). Mating has not been observed but occurred in females taken from Scottish surface waters in early summer (Matthews 1950). Basking sharks occasionally leap clear out of the water, possibly to remove external parasites or to interact with other sharks.

Basking sharks, like many other large sharks, appear to migrate and segregate seasonally by sex and age. Scottish summer surface harpoon fisheries mainly took females (a ratio of 18 females : 1 male; Watkins 1958), whereas bycatch in deepwater Newfoundland gill nets took twice as many males as females (Lien & Fawcett 1986). Newborns are rarely seen, and they may also segregate from the populations taken in fisheries. Most basking shark wintering locations are unknown.

The basking shark is ovoviviparous: this means that the pups hatch inside the uterus and are probably nourished during pregnancy by the continuous production of unfertilized eggs (ovophagy) until they reach a length of 1.5–1.7 m, larger than any other known shark. The only pregnant female recorded in the literature gave birth to six young (Sund 1943). From information about growth in this species, coupled with data from better-studied lamnid sharks, it is thought that males mature at a length of 4–5 m when they are probably 12 to 16 years old. Females probably mature at 8.1–9.8 m when they are 16 to 20 years of age. Longevity may be 50 years for a maximum length of 10 m (Pauly in press). Length of gestation is unknown, but because of the very large size of the pups it is thought to be over 12 months (18 months to 3 years have been suggested; Compagno 1984). Like several other species of large sharks, females are likely to "rest" between pregnancies to recover condition before bearing the next litter.

CONFLICTING ISSUES

Sharks have only recently come to the attention of conservationists and managers, and shark conservation is a relatively new concept. Because elasmobranchs (sharks and rays) constitute only 1% of commercial fisheries yields, there has been little incentive to invest in management, research, and monitoring. Although some fisheries managers have considered sharks and rays to be similar to other commercial fishes, their ecology and K-selected life history have more in common with mammals than with the bony fishes and shellfish that support most fisheries.

Although there is not much data on shark fisheries, most have followed a "boom and bust" pattern—a relatively short period of high landings followed by a sudden steep decline and possible collapse of the fishery (Camhi et al. 1998). Several shark fisheries are regularly cited as examples of this (e.g., Roedel & Ripley 1950; Olsen 1959; Gauld 1989; Anderson 1990; Walker 1993), including basking shark fisheries on the western coast of Ireland (McNally 1976) and a basking shark eradication program in British Columbia (Clemens & Wilby 1961; Darling & Keogh 1994).

The sudden growth of shark fisheries in the mid-1980s heightened the need for shark fisheries management and species protection. This growth was at least partly triggered by the liberalization of Chinese policies, an economic boom in Pacific Rim nations, and the sudden and massive expansion of a small traditional market for shark fin soup. Shark fins quickly became one of the most expensive fisheries products in the world, with dried fins selling for up to U.S. $250 per pound and a single bowl of shark fin soup for U.S. $90 (Rose 1996). The result was a massive growth in fisheries targeting sharks and increased finning of shark bycatch. Markets also developed or expanded for shark meat, liver oil, cartilage, jaws, and even skin. FAO Fisheries Statistics for 1995 recorded landings of sharks and rays totaling 754,864 metric tons. Because of widespread underreporting of bycatch and landings, the total catch is likely twice that reported (Bonfil 1994). Estimated numbers of sharks and rays killed by fisheries range from 40 million to over 200 million annually.

The basking shark is valuable for its huge oil-rich liver, which may account for 17 to 25% of the shark's weight and is used in cosmetic and aviation industries (Kunzlik 1988). Its meat is valued in some markets, its cartilage is processed into health food supplements (Fleming & Papageorgiou 1996), its skin may be utilized, and its huge fins have high value in international trade. Lum (1996) reported that basking shark fins from Norway are the most expensive in Singapore, and they can cost U.S. $300 per kilogram (dried), or U.S. $66 per bowl in restaurants. Because of their enormous size and scarcity, single basking or whale shark fins are reportedly worth U.S. $10,000 or more.

Increased exploitation of sharks and acknowledgment, in a few countries, of their vulnerability has resulted in the introduction of some shark fisheries management. A U.S. Atlantic Shark Management Plan aims to rebuild and manage stocks (Anonymous 1993). Australia, New Zealand, and Canada have introduced some shark fisheries management, and management plans are in preparation in South Africa and Mexico (Camhi et al. 1998). However, these represent a small proportion of the 125 nations currently fishing for or trading in shark products (Rose 1996). The oceanic fisheries taking millions of sharks annually are completely unmanaged. More countries provide certain shark species with legal protection than those that manage their shark fisheries.

In October 1998 an intergovernmental consultation meeting at the Food and Agriculture Organization (FAO) of the United Nations agreed on an International Plan of Action (IPOA) for the Conservation and Management of Sharks. This document sets out a conservation and management strategy for fisheries and shark populations and encourages nations all over the world to adopt and implement it. It is too early to determine how effective this voluntary agreement will be, once it is adopted by the FAO Conference in late 1999.

One problem with managing shark fisheries is that recovery of depleted populations is an extremely slow process. Some populations of large sharks only increase at 5–10% yearly (Camhi et al. 1998). Under these conditions, commercial fishermen are usually reluctant to accept significant long-term quota reductions or other regulations on a valuable fishery. Unmanaged shark fisheries also provide a useful fallback for commercial fishermen; when other fish stocks decline or quotas are filled, they can redirect their efforts to these unregulated stocks. Although some recreational fisheries take very large numbers of other sharks (Scott et al. 1996), this is not an issue for the basking shark.

It is too early to determine how effective legal protection of shark species will be, especially for those taken as bycatch and those of high value for their fins (e.g., basking sharks), jaws (e.g., white shark), or other products. Enforcement of existing legislation is patchy in some countries. It is difficult to determine whether protection is effective for rare species, species that are difficult to see or to count accurately, species that are unmonitored in most fisheries landings (which rarely distinguish between species of sharks or rays), and species that are unmonitored in international trade.

Monitoring or controlling international trade in shark products is politically difficult. Proposals for listing threatened species of sharks on Appendices to the Convention on International Trade in Endangered Species (CITES) have been and continue to be considered by some parties to CITES (Anonymous 1999). An Appendix I listing prohibits all international trade. An Appendix II listing requires international trade to be monitored. Most opposition to CITES listings for sharks or related species arises from the reluctance of some parties to accept that CITES (a wildlife convention) is appropriate for managing marine fishes, even where there are presently no other means of monitoring or regulating the species in international trade.

FUTURE AND PROGNOSIS

The prognosis for the basking shark is better than that for many other vulnerable sharks. It is widely distributed in temperate waters worldwide and only subject to localized directed fisheries. Many other sharks are endemics that are under severe fishing pressure throughout their range. Unlike many other species of sharks, basking sharks apparently do not depend on vulnerable inshore pupping and nursery grounds. As a plankton feeder, rather than a top predator, the basking shark is not among the millions of predatory sharks caught annually in large-scale line fisheries. Finally, the basking shark is receiving legal protection in some centers of its distribution.

On the other hand, not enough is known about this very large, K-selected animal, and continued bycatch in other fisheries may be affecting populations (which are virtually impossible to monitor) even where they are protected. Additionally, target and utilized bycatch-fisheries continue to take this slow-moving, vulnerable animal as it swims and feeds in surface waters. The high value of basking shark fins, liver oil, cartilage, and meat makes even basking sharks that are legally protected a tempting target for fishermen, especially as other fish stocks decline.

In all countries that regularly record basking sharks within their territorial waters it would be appropriate to enact legal protection for this vulnerable species. Additionally, monitoring of international trade in the easily recognizable giant fins of basking sharks would provide a better indication of regional and global catches of this species than can be obtained from other records.

Black-footed Ferret

Brian Miller, Richard P. Reading, and Tim W. Clark

Common Name: black-footed ferret
Scientific Name: *Mustela nigripes*
Order: Carnivora
Family: Mustelidae
Status: Endangered under the Endangered Species Act of the United States; CITES Category I; Extinct in the Wild (EW) on the 1996 IUCN Red List (although there are presently small reintroduced populations in Montana and South Dakota, USA).
Threats: Destruction of its prey, prairie dogs (*Cynomys* spp.), owing to poisoning, conversion to agriculture, and sylvatic plague; susceptibility of ferrets to canine distemper and plague.
Habitat: Prairie dog colonies of North America.
Distribution: Reintroduction sites in Montana, South Dakota, and Arizona of the United States, but formerly extended across the western Great Plains of North America from southern Canada to northern Mexico.

DESCRIPTION

The black-footed ferret, described by Anderson et al. (1986) and Miller et al. (1996), has a typical mustelid (weasel family) body plan, with a long, thin torso and short legs. The vertebrae are very flexible, enabling ferrets to maneuver in a prairie dog tunnel. Small, mouse-like ears are rounded, and the skull is short and broad. The dental formula is 3/3 incisors, 1/1 canines, 3/3 premolars, and 1/2 molars. Black-footed ferrets are approximately the same size as the domesticated form of European ferrets (*M. putorius*). Adult males average 1040 g, whereas adult females average about 710 g. Black-footed ferrets are buckskin in color, with a darker saddle in the middle of the back. Their undersides are lighter than their sides or back, as is the fur on the muzzle and throat. The most distinct markings are the black mask across the eyes, and dark legs.

NATURAL HISTORY

Nearly 30,000 years ago, modern black-footed ferrets could be found on prairie dog colonies throughout the Great Plains of North America. Historically, the black-footed ferret range was contiguous with three of the five species of prairie dogs: black-tailed (*Cynomys ludovicianus*), white-tailed (*C. leucurus*), and Gunnison (*C. gunnisonni*) (Biggins et al. 1997). Modern museum specimens verify that ferrets inhabited the grasslands of southern Alberta and Saskatchewan, Canada, through 13 western U.S. states (Anderson et al. 1986). No modern museum specimens have been located in Mexico, but ferrets likely lived in northern Mexico as well (Miller et al. 1996). There are now reintroductions taking place on prairie dog complexes in Montana, South Dakota, and Arizona in the United States (Reading et al. 1997).

The black-footed ferret is highly specialized to exploit the prairie dog ecosystem. Without prairie dogs there are no ferrets. About 90% of the ferret's diet is prairie dog, and ferrets live in the prairie dog burrows (Clark 1989). Because of their extreme specialization on prairie dogs, ferrets are classified as a single species with no recognized subspecies (Anderson et al. 1986).

Black-footed ferrets probably breed during March and April in the wild, and females do not ovulate until copulation. Gestation lasts 6 weeks, with the young being born in a prairie dog burrow. Completely helpless at birth, the young first appear above ground at 8 weeks of age. The average litter size in the wild is 3.3 (Forrest et al. 1988). By 4 months of age, young ferrets disperse and begin to live as solitary carnivores.

Because black-footed ferrets occupy and mark their home ranges, young that cannot find a vacant area must leave in search of new territory. Those dispersers often navigate across large, unfamiliar spaces without prairie dogs, which leaves them vulnerable. Between 60 and 80% of the young of the year starve or are killed by larger predators (Forrest et al. 1988). Securing a home range is therefore critical to survival and reproduction. Black-footed ferrets that secure a home range may live 2 or 3 years in the wild.

Females adjust their home range size according to changes in prey density, but males' home range size appears to be more strongly based on behavioral spacing mechanisms and access to females (Miller et al. 1996). Although sex ratio at birth is equal, the adult ratio is 1 male : 2 females (Forrest et al. 1988). The average density on a white-tailed prairie dog complex in Wyoming was 1 black-footed ferret per 50 hectares of prairie dogs (Forrest et al. 1988).

CONFLICTING ISSUES

Clark (1989), Miller et al. (1996), Biggins et al. (1997) and described the decline of black-footed ferrets. Ferrets once roamed across the 40–100 million hectares of prairie dog colonies that existed throughout the North American west. However, conversion to agriculture and federally sponsored poisoning campaigns reduced the area of prairie dog colonies to 600,000 hectares by 1960—a decline of over 98%. As prairie dogs disappeared, so did ferrets. As it now stands, the government has the conflicting goals of restoring an endangered species while subsidizing the habitat destruction that puts the species in peril.

Poisoning programs began around the start of the twentieth century to provide more grass for livestock and continue today (Miller et al. 1996). The destruction of prairie dogs eliminated many black-footed ferrets directly, and it reduced and isolated the remaining populations of ferrets. Remaining colonies were then more susceptible to extinction by stochastic events such as disease, genetic problems, demographic variability, and natural catastrophes (Wilcox & Murphy 1985). Exotic diseases are also implicated in ferret decline. Black-footed ferret mortality to canine distemper is 100%, and both prairie dogs and ferrets are susceptible to sylvatic plague (Biggins et al. 1997).

Although conservation activities began in 1964, by the end of the 1970s many people believed black-footed ferrets were extinct. But soon after a ranch dog killed a ferret west of Meeteetse, Wyoming in 1981, a small population was discovered. That population never comprised more than 43 adults (Forrest et al. 1988), and in 1985 it declined rapidly. The cause of decline was identified as canine distemper, although sylvatic plague was also present and probably played a role (Biggins et al. 1997). The few remaining animals were captured to start a captive breeding project. In the winter of 1985–1986, less than a dozen remaining animals represented the entire species (Miller et al. 1996).

Captive breeding has been successful, and the first reintroductions were attempted in Wyoming in 1991–1994. Additional reintroductions are under way in Montana (1994–present), South Dakota (1994–present), and Arizona (1996–present), with small wild populations of ferrets surviving in Montana and South Dakota (Biggins et al. 1997).

Several groups and individuals are involved with, interested in, or oppose black-footed ferret recovery efforts. The most important stakeholders include ranchers, wildlife managers from state and federal agencies, researchers from a variety of institutions, and conservation organizations. Many of the problems facing ferret conservation efforts are based on conflicting values and attitudes among key stakeholders (Miller et al. 1996; Clark 1997).

Most ranchers oppose ferret recovery efforts. Their opposition stems from negative attitudes toward prairie dogs, concern over issues of states' rights versus federalism, private property rights, and management of public grazing lands (Reading 1993). Ranchers generally perceive prairie dogs as pests that compete with livestock for grass (Reading & Kellert 1993). This attitude persists despite research demonstrating that the level of competition is only 4–7%, that prairie dog poisoning is not cost effective, and that prairie dogs are important to the maintenance of grassland ecosystems (reviewed in Miller et al. 1996). Ranchers' attitudes are reinforced by government policies that have institutionalized poisoning.

Ranchers in the United States also fear loss of control over public grazing lands and an attack on their rural western lifestyles, resulting in negative attitudes toward the U.S. Endangered Species Act (ESA) and the endangered black-footed ferret (Reading & Kellert 1993). These fears are based on real trends. Ranchers have witnessed a gradual erosion of control over public lands owing to the development of other interests, including conservation, that now compete for influence in public lands management. Conflicts developed because ranchers and local residents are largely antagonistic toward ferret recovery efforts, whereas conservationists, wildlife managers, and urban residents largely support those efforts (Reading & Kellert 1993).

Education efforts focused primarily on providing information to the general public but did little to address the values and attitudes of antagonistic stakeholders. These programs had little success in increasing support, probably because knowledge is only one of several factors influencing values and attitudes (Reading & Kellert 1993). More effective efforts in Montana focused on shared values, such as appreciation for wildlife, and involvement in positive experiences associated with the reintroduction program. A proposal has been made to convert poisoning costs into a positive incentive for ranchers who manage for wildlife and livestock, but so far it has not been implemented (Miller et al. 1996).

Black-footed ferret recovery also suffers from unproductive conflict between governmental agencies, conservation organizations, and researchers over management of the program. Key stakeholders—including federal agencies, state agencies, nongovernmental conservation groups, and researchers—all possess different underlying values, which are expressed by differences in problem definitions, goals, methods of operation, philosophies, cultures, and ideologies (see Reading 1993; Clark 1997).

In general, western state agencies of the United States assert they should manage their wildlife. Their expertise has traditionally focused on game management and enforcement. Hunters and ranchers (the latter can grant hunting access) are important constituents, and state management styles are generally conservative. The state of Wyoming preferred a hands-off policy for the small wild black-footed ferret colony before that population crashed in 1985. Federal agencies often place a greater emphasis on nongame species, and they have a wider range of constituents. The U.S. Fish and Wildlife Service was legally mandated to manage black-footed ferrets by the ESA, but that agency has a nonconfrontational culture. When the wild black-footed ferret population was discovered in Wyoming in 1981, the USFWS passed program control to the state even though the state lacked resources, expertise, and the ability to manage outside its boundaries. Conservation organizations and researchers differ from agencies in their greater sense of urgency and their concern for nonconsumptive uses of wildlife, endangered species recovery, and a more active, "hands-on" approach to endangered species management. The constituency of researchers is largely made up of their peers. These differences inevitably evolved into conflicts, even though all the stakeholders supported ferret recovery (Miller et al. 1996; Clark 1997). Without effective methods of resolving conflicts, stakeholders struggled for control of the program to advance their particular cause (Clark 1989). In some cases, this conflict escalated to goal inversion in which the goal of ferret recovery was superceded by the goal of attaining control (Miller et al. 1996; Clark 1997).

The state of Wyoming organized ferret recovery as a classic, and rigid, government bureaucracy (Clark 1997). Such bureaucracies are efficient for routine tasks but not well matched to the complex, urgent, uncertain, and often conflict-laden environment characteristic of endangered species recovery (Clark 1997). The typical bureaucratic response when conflict arises is not to democratically incorporate various views and ideas but to centralize control around fewer people of similar philosophy. The resulting "group think" stifles innovation, reduces program flexibility, and weakens the capacity to respond rapidly to new problems.

With Wyoming at the helm, control became a dominant theme in the ferret program, and resulting delays hindered recovery (Clark 1997). In fact, delays nearly resulted in extinction when the program did not react quickly to the population crash in 1985, despite reliable data indicating a crisis. Additional delays inhibited implementation of more efficient reintroduction techniques until black-footed ferrets left Wyoming. Other problems characteristic of the program's early years included poor information flow, high costs, reduced use of outside expertise, inefficient use of scientific research, and frequent conflicts that sapped time and energy (Miller et al. 1996).

FUTURE AND PROGNOSIS

More recent changes in organization have followed pressure for reform and a major program review. Today, program organization is functioning more smoothly, wild black-footed ferrets are breeding on prairie dog colonies in Montana and South Dakota, and scientific research is being used to improve reintroduction techniques. Endangered species management is a complex, dynamic, and uncertain endeavor requiring an open, task-oriented environment with free access to information and interdisciplinary approaches. Recent trends toward changing the organization of the ferret recovery program are promising.

On the other hand, prairie dog poisoning and attitudes toward prairie dogs remain problematic. Poisoning continues to dominate the prairie dog management paradigm of most state and federal government agencies in the United States (Miller et al. 1996). As a result, few prairie dog complexes remain that are large enough to support a viable population of ferrets. Prairie dogs have the capacity to recover quickly, but there must be reduced poisoning pressure. In addition, sylvatic plague can destroy prairie dog complexes, and canine distemper is 100% fatal to ferrets. The effects of inbreeding on long-term viability of ferrets remain unknown, and in captivity some individuals reproduce better than others, resulting in an over-representation of certain animals in the population (Biggins et al. 1997). We believe these social and biological challenges facing ferret recovery can be met, but it will not be easy.

Boreal Toad

Erin Muths and Paul Stephen Corn

Common name: boreal toad
Scientific name: *Bufo boreas boreas*
Order: Anura
Family: Bufonidae
Status: The Southern Rocky Mountain Population (SRMP) of the boreal toad is Endangered in Colorado and New Mexico and has been ruled "warranted but precluded" by the U.S. Fish and Wildlife Service (USFWS 1995).
Threats: Causes of the decline have not been identified with certainty, but stress and decreased resistance to disease such as "redleg" (*Aeromonas* bacterial infections) may play a role (Carey 1993).
Habitat: In the southern Rocky Mountains, boreal toads are found in alpine and subalpine zones between 2,300 and 3,700 m in elevation.
Distribution: The boreal toad occurs along the western coast of North America from Alaska to central California and in the Rocky Mountains from Alberta to New Mexico (Stebbins 1985). Another subspecies, the California toad (*B. b. halophilus*) occurs in southern California; three sibling species, the Yosemite toad (*B. canorus*), black toad (*B. exsul*) and Amargosa toad (*B. nelsoni*), occur in California and Nevada with restricted or relict distributions.

DESCRIPTION

Adult male boreal toads are identified by the presence of nuptial pads (dark, cornified patches that help hold on to the female while mating, or in amplexus) on the dorsal surfaces of the first two digits of the front feet or by vocalization when handled. Boreal toads are dimorphic (differ by gender) in size. In Rocky Mountain National Park, males averaged 71.5 mm snout-vent length (SVL; $N = 1255$, range 52–86 mm) and 39 g mass ($N = 980$, range 21–74 g). Adult females, captured during the breeding season, averaged 82.9 mm SVL ($N = 183$, range 62–101.2 mm) and 68 g mass ($N = 120$, range 26–142 g). Like most toads, boreal toads have glandular skin with prominent parotoid glands behind the eyes. Cranial crests are absent. Boreal toads are light brown or gray to black on the dorsal surface and lighter with black spots on the ventral surface. Coloration is variable, especially on the ventral surface. A light-colored dorsal vertebral stripe is usually present, but it may be broken or very faint.

Analysis of mitochondrial DNA from toads throughout the range of *B. boreas* revealed significant variation (Goebel 1996), such that the Southern

Figure 2. Boreal Toad Observations

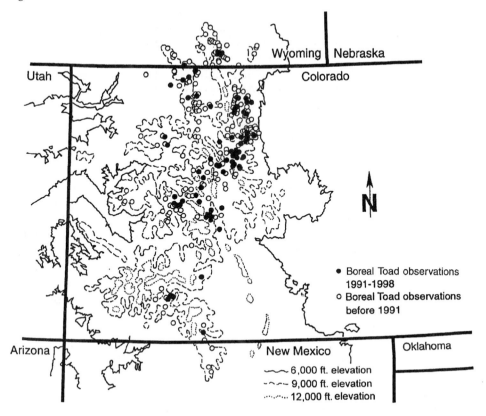

Rocky Mountain Population may warrant recognition as a distinct species. At present, the Southern Rocky Mountain Population is considered an evolutionarily significant unit (Goebel 1998).

NATURAL HISTORY

Boreal toads were formerly thought to be common in the mountainous areas of southern Wyoming, Colorado, and northern New Mexico (Figure 2), but significant declines have been documented in the last 10 to 15 years (Carey 1993; Corn 1994; Stuart & Painter 1994; Corn et al. 1997). Despite the recent declines, boreal toads still occur in all ranges of the southern Rocky Mountains, except the Grand Mesa in western Colorado. They occur in the Medicine Bow Mountains in southeastern Wyoming and possibly in one county in New Mexico. Surveys in 1996 and 1997 identified 33 breeding sites in Colorado and one in Wyoming (Loeffler 1998; Figure 2), but no breeding has been observed in New Mexico since 1986 (Stuart & Painter

1994). Boreal toads in Utah and southeastern Idaho were not included in the original petition for listing but are genetically related closely to toads in Colorado and should be considered part of the Southern Rocky Mountain population (A. Goebel et al. personal communication).

Toads breed in beaver ponds, glacial kettle ponds, large drainage lakes, and other ephemeral water sources (Stebbins 1954; Hammerson 1992; Corn et al. 1997). Adults use willow and upland forests (e.g., Engelmann spruce) after the breeding season (Jones 1997; Corn & Muths unpublished data). Campbell (1970a) observed boreal toads overwintering in rock-lined chambers in a stream bank. Toads also use ground squirrel burrows as hibernacula, or places to hibernate.

Boreal toads are long-lived amphibians (probably longer than 15 years). Males and females generally do not breed until they are 3 and 4 years old, respectively. The timing of spring breeding relates to snowmelt and can be as early as April or as late as July depending on elevation (Corn et al. 1997). Occasionally toads produce egg masses in late July or August, well after the normal breeding activity peak for their populations (Livo & Fetkavich 1998). Males aggregate at pond margins to wait for females where they emit a soft chirping call, sometimes referred to as a "release" call if they are clasped mistakenly by another male. Males do not produce a chorusing vocalization as do other male toads found in the Rocky Mountains, such as Woodhouse's toad (*Bufo woodhouseii*).

Eggs are laid in the water in long, single strands, often wound around pond vegetation. Development and growth of eggs, tadpoles, and juvenile toads are temperature dependent. Typically in the Colorado mountains, eggs are laid in late May and early June and hatch in 1 to 2 weeks. Tadpoles feed on algae and metamorphose in 3 to 7 or more weeks. Juvenile and adult toads eat a variety of insects and other invertebrates (Campbell 1970b).

Vulnerable lifestages of boreal toads include tadpoles, metamorphs, and recently metamorphosed toadlets. Reproductive failure occurs frequently in high elevation populations when tadpoles fail to metamorphose before the onset of winter. Tadpoles are somewhat unpalatable but face numerous predators including garter snakes, aquatic insects, tiger salamanders, and birds (Beiswenger 1981; Livo 1998; Livo & Jones personal communication). Metamorphosis brings reduced swimming and locomotive ability and increased risk of predation (Arnold & Wassersug 1978). Recently metamophosed toadlets are vulnerable to desiccation and must find a place to overwinter where they are protected from desiccation and freezing. Adult toads are somewhat protected from predation by poisonous secretions from their skin glands, but ravens can be important predators (Olson 1989; Corn 1993). Raccoons, domestic dogs, and foxes have been identified as occasional predators. The extent and impact on toad populations from predators is unknown, but Olson (1989) observed 20% of breeding adults at one aggregation killed by ravens in Oregon. A high rate of predation could be a serious problem in populations that have already declined (Corn 1993).

CONFLICTING ISSUES

Several potential causes for the decline of boreal toads have been investigated. Blaustein et al. (1994) found that ambient ultraviolet-B radiation decreased the hatching success of boreal toad eggs in Oregon, but Corn (1998) did not observe any UV-B related mortality in a similar study conducted in Rocky Mountain National Park, Colorado, and further showed greater than 100% variation in ambient UV-B from year to year (Corn 1998). Kiesecker and Blaustein (1995) observed a synergism between the water fungus *Saprolegnia ferax* and UV-B radiation, but *Saprolegnia* has not been studied in the southern Rocky Mountains.

Acid precipitation is not thought to have contributed to declines of boreal toads (Corn & Vertucci 1992; Vertucci & Corn 1996). Habitat destruction has not been identified as a cause of significant declines in the southern Rocky Mountains but is of concern as development expands, encroaching on potential boreal toad habitat.

Because the boreal toad is not a federally listed endangered species, most of the money available for research and management is controlled by the Colorado Division of Wildlife (CDOW) and comes primarily from the Great Outdoors Colorado monies, generated from state lottery dollars. This allows CDOW some flexibility in administration, but it also potentially constrains other work proposed by other state or federal entities.

An interagency recovery team was formed in 1994. Declines had been documented and very few breeding sites were known (Carey 1993). Initial recovery plan documents were written under circumstances that dictated aggressive action to save the toad. Over the last 5 years, more effort has gone into surveys and research. Surveys have yielded more than 30 breeding sites in 12 counties (Loeffler 1998). This increase in knowledge has influenced the goals and direction of the Recovery Team but has also provided a basis for disagreement.

There is general agreement on recovery goals but continued discussion on the means of obtaining these goals. Is intensive management, perhaps including aggressive reintroduction and translocation, required? Or should management be more cautious, with continued emphasis on research? Because it is still not known why toads have declined, we tend to favor the latter approach.

To date, the Boreal Toad Recovery Team has worked through differences of opinion and focused on compromise with an agenda that benefits toad recovery in the state. Although various disagreements continue to recur and delay action, the toads are not suffering a rapid decline. This suggests that crisis management and disagreements can sometimes lead to alternative solutions to problems.

One issue that has been debated within the Recovery Team is whether or not, and when, to use translocation as a management tool. Translocation has been debated in the literature for a number of years (e.g., Griffith et al. 1989; Burke 1991; Dodd & Seigel 1991; Reinert 1991; Thomas & Whitaker 1994). The State of Colorado is responsible for the boreal toad and wants to avoid federal listing. The State is interested in proactive management actions such as translocation. The scientific advisors on the recovery team, taking into account genetic considerations, the potential for introduction of disease, and the fact that more breeding populations are found each year, have recommended a more conservative approach: no additional translocations unless known conditions deteriorate significantly. Several small-scale translocations have taken place in the past and have provided useful information about effective methods.

After some debate and a re-evaluation of the current situation of the Southern Rocky Mountain Population of the boreal toad, the Recovery Team reached a compromise. In general, translocations will not be used unless existing toad populations decline precipitously. However, a well-monitored experimental translocation project on the Grand Mesa has been initiated with a 3-year commitment to pre-translocation surveys of the habitat and surrounding area. Surveys in the last decade have yielded no toads on the relatively isolated Grand Mesa, although it is a historic locality for *B. boreas*. If no toads are found, a well-designed, experimental translocation of multiple age classes (eggs, tadpoles, and metamorphs) will ensue. This project is designed to (1) identify which life stage survives most successfully following translocation, and (2) determine the length of time before the translocated population is self-sustaining.

FUTURE AND PROGNOSIS

Additional problems and conflicts are possible because there is little specific information about habitats required by toads away from breeding sites. This lack of information promotes management decisions that may or may not be beneficial for maintaining toad populations. The problem is exacerbated because the lack of federal listing allows states with different philosophies to manage toads differently, and federal agencies can choose to ignore Recovery Team recommendations. Although federal listing would result in more uniform application of management decisions, such listing may not be necessary and could potentially derail the current recovery efforts owing to delays in producing a new recovery plan as directed by the ESA. A conservation agreement between federal land management agencies and CDOW, signed in 1998, may alleviate some of the problems of multiple managers. Improved communication and cooperation among states that provide habitat to the Southern Rocky Mountain Population of boreal toads are also necessary to promote recovery.

ACKNOWLEDGMENTS

Thanks to Therese Johnson and Lauren Livo for commenting on earlier drafts of this entry.

Brown Bear

David J. Mattson

Common Names: brown bear, grizzly bear
Scientific Name: *Ursus arctos*
Order: Carnivora
Family: Ursidae
Status: The brown bear is a game animal where it is relatively numerous and lives outside of protected areas (e.g., most of Russia, Romania, Bulgaria, Croatia, Slovenia, Slovakia, Serbia, Sweden, parts of Finland, parts of Turkey, Hokkaido [Japan], Canada, Alaska [U.S.]). Where it is isolated and occurs in small numbers, the brown bear is usually protected (e.g., Spain, Austria, France, parks in Italy, Norway, Poland, contiguous United States).
Threats: Death caused by humans responding to depredations on livestock or agricultural crops, perceived or actual threats to human safety, and traditional medicine markets for body parts of bears; anything, such as roads, that promotes increased exposure of brown bears to humans and increases the chance that humans will kill bears if given the opportunity; and declining supplies of high-quality foods such as spawning salmonid fish and stone pine (*Pinus* spp.) seeds.
Habitat: Potentially all vegetation types of temperate, boreal, and arctic regions where humans are scarce, except most hot deserts and grasslands that lack substantial numbers of native ungulates.
Distribution: Much of northern Russia, Alaska (U.S.), and western and far northwestern Canada; sizable populations and extensive ranges in Romania, Bulgaria, Finland, Sweden, Croatia, Slovenia, Hokkaido (Japan), and Transcaucasus republics of the former U.S.S.R.; extensive ranges but uncertain numbers in Turkey, China, and central Asian states of the former U.S.S.R.; limited and typically isolated ranges in Spain, France, Italy, Poland, Norway, Slovakia, Belarus, Latvia, Estonia, Greece, Albania, Iraq, Iran, Afghanistan, Pakistan, India, Mongolia, and the contiguous United States.

DESCRIPTION

Brown bears have several distinguishing physical features (Craighead & Mitchell 1982; Mattson 1998). Unlike most species of obligate carnivores, brown bears do not have premolars that can efficiently shear meat. Instead, they have teeth well suited for grinding and manipulating food in the mouth. Brown bears also have long (5–11 cm) unretractable claws and a large muscle mass over the shoulders that constitutes their characteristic "hump." This combination of muscles and claws is thought to increase the efficiency with which brown bears dig, especially compared to other ursids. Brown bears walk flat-footed and are slow-moving, especially compared to large canids or cats that typically walk and run on their toes. Brown bears also are among the largest of the carnivores (90–550 kg as adults), although average body size varies several-fold among populations. Males are larger than females. The color of brown bear pelts varies from black to blond and commonly exhibits lighter markings on the chest and torso. Guard hairs usually have a lighter tip ("grizzled") that causes the pelage, or coat of hair, of brown bears to glisten when backlit.

There has been considerable confusion over the taxonomy, or classification, of brown bears owing to the substantial variation in their pelage and the size and shape of their skull. Because of this confusion, most biologists have deferred to geneticists to define "evolutionarily significant units," or clades. Although no such comprehensive analysis has been done for Asia, two clades of brown bears (eastern and western) have been defined for Europe (Kohn et al. 1995) and four clades (Alaskan, northwestern Canadian, Alaskan Islands, and Rocky Mountain) have been identified for North America (Waits et al. 1998). Although several classifications of subspecies have been proposed for Asian brown bears, all recognize the distinctness of bears that live in the Gobi Desert, the Pamir Mountains, and the Tibetan Plateau (Chestin et al. 1992). Of these evolutionary lineages, brown bears of the Rocky Mountain and western European clades and of the Gobi Desert, Pamir Mountains, and Tibetan Plateau subspecies are in greatest peril (Servheen 1990).

NATURAL HISTORY

Until 100 to 300 years ago brown bears occupied most temperate, boreal, and arctic regions of the Northern Hemisphere. They were absent only from the hottest and driest deserts (except the extraordinary bears of the Gobi) and from areas in North America where their dispersal in the wake of melting Pleistocene icecaps may have been impeded by competition from resident black bears (*Ursus americanus*) and by mortality and competition from native Americans.

Brown bears are highly intelligent omnivores. They characteristically make sophisticated decisions about what, when, and where to eat, based on considerations of nutrient and energy content and risks of conflict with humans and other bears. Because brown bears have a simple digestive tract, highly digestible, high-energy foods are typically the most important part of their diets (Mattson 1998). These foods include tissue from ungulates such as moose (Eurasian elk, *Alces alces*) and elk (Eurasian red deer, *Cervus elaphus*), spawning salmonids, fleshy fruits of numerous shrub species, and the largish seeds of stone pines. Rodents and starchy roots are important foods for some bear populations or during years when higher quality foods are less abundant.

Brown bears undergo unique physiological changes each year associated with hibernation (Nelson et al. 1983). Most of hibernation is spent inactive in dens, although the physiological state of hibernation is not synonymous with the physical act of denning. Large reserves of body fat are critical to surviving hibernation, which may last as long as 6 months. During this time brown bears do not eat nor do they urinate, defecate, or lose any bone mass.

Together with other ursids, brown bears have the lowest reproductive rate of carnivores. Brown bears typically first reproduce at 4 to 7 years, normally have two cubs per litter, and usually produce litters only once every 3 years (Craighead & Mitchell 1982). The breeding season occurs early in the year, usually May through July. Development of the embryo is delayed to permit emergence of young during spring. At birth, cubs are the smallest of any nonmarsupial mammal when compared to the size of their mother. Mortality rates for brown bear are highest during the early years, but they potentially live to be 20 to 30 years old.

The densities and range sizes of brown bears vary by orders of magnitude (e.g., 50–2,500 km^2 ranges), primarily depending on the abundance of high-quality foods. In general, brown bears are densest and occupy the smallest ranges in coastal areas and are most dispersed and wide-ranging in drier, interior regions (Miller et al. 1997). Males use ranges several times larger than those of females. Neither sex is territorial. Females typically occupy ranges in or near the ranges of their mother, whereas young males may disperse up to 100 km. Despite this, brown bears are much less successful than carnivores such as wolves (*Canis lupus*) or mountain lions (*Felis concolor*) in colonizing new ranges.

CONFLICTING ISSUES

The primary threat to brown bears continues to be mortality caused by humans (Servheen 1990). In all but the most remote regions, bears past the age of weaning die because humans kill them, primarily with guns. In most former brown bear ranges, bears were given little protection before the 1950s. In many areas it was de facto or even formal policy to eradicate them (cf. Couturier 1954; Brown 1985). Prior to the 1900s most people thought that brown bear numbers were inexhaustible, that brown bears were a threat to the persistence and expansion of human settlement and agriculture, that they had no value, or even that they were evil.

Measures to protect diminished brown bear populations began in the 1930s (Czechoslovakia) and 1940s (Bulgaria and Poland) and became common in Europe (e.g., Spain, Greece, and Norway) and the contiguous United States by the 1970s (Servheen 1990). Even so, there is still no meaningful protection for severely threatened populations in Iran, Iraq, Pakistan, and China. Populations, such as those of eastern Europe, that remained sizable even at their lowest ebb responded dramatically to this protection.

On the other hand, small populations of brown bears in western Europe are still declining (e.g., the eastern Pyrenees and southern Norway) (Elgmork 1996; Parde 1997).

Factors that govern the rate at which humans encounter bears and the odds that the bear will die in an encounter (lethality) are critical (Mattson et al. 1996). Several factors have an important effect on the rate of encounter between a bear and humans. These include (1) the number of people living in or visiting an area, (2) the amount of road and trail access into bear habitat, (3) the degree to which human habitations and facilities are attractive to bears because of the availability of human-related foods, (4) the degree to which bears are attracted to livestock ranges, (5) the location of human facilities in areas inherently attractive to bears, (6) the lack of high-quality natural foods, which may induce bears to wander nearer to human facilities, and (7) the extent to which a bear has become tolerant of humans. Less important are factors affecting the lethality of encounters, which include (8) whether or not humans are armed, (9) the economic value of bear body parts (e.g., gall bladders for traditional Asian medicine), (10) human hostility because of perceived losses owing to bear management, (11) knowledge about bears, (12) the cultural value placed on bears, and (13) the aggressiveness of the bear toward people (Highley & Highley 1994; Kellert et al. 1996; Mattson et al. 1996).

Brown bears can co-exist within a range of human densities (Mattson 1990). In North America, humans have been extremely lethal to brown bears. Thus brown bears here survive only in areas where there are a few people and little chance of encountering them. By contrast, brown bears in Europe co-exist with much higher densities of humans, presumably owing to a combination of greater wariness on the part of bears and lower lethality on the part of humans.

Changes in the abundance of natural foods also have affected brown bear populations (Mattson 1990; Servheen 1990). For example, drainage of wetlands (loss of sedges) in the European part of Russia and harvest of oak forests (loss of acorns) in Spain and France probably had a negative effect on vulnerable bear populations in these regions. Additional examples include diminished salmon runs in Hokkaido (Japan) and the northwestern United States and the loss of whitebark pine (*Pinus albicaulis*) (loss of pine nuts) in some parts of western North America owing to the exotic white pine blister rust (*Cronartium ribicola*).

Ultimately most of the threats to brown bears arise from traditional management and land use practices and from related human values (Primm 1992; Kellert et al. 1996). Opponents of brown bear conservation are typically a mix of wealthy individuals with a stake in continued extraction of resources from brown bear habitat, ideologues who espouse the dominion of humans over nature, and those who see their livelihoods threatened. Opposition to the protection of brown bears and their habitat consequently arises from perceived threats to the continued creation of wealth, perceived threats to traditional values and practices, and outright hostility. In contrast, proponents of brown bear conservation typically consist of those who believe in the innate value of nature and the dependence of human well-being on the health of natural ecosystems. Bear hunters span these two constituencies, because they have a stake in healthy populations and value bears primarily because they can kill them.

Even with policies that favor conservation, problems can arise because of divergent values held by people who live in brown bear habitat (Kellert et al. 1996; Primm 1996). People often feel threatened by environmental policies that seemingly discount not only the high value they place on things such as sheep or timber, but also their entire world view. People who promote brown bear conservation, as well as the agencies responsible for implementing policies, often come to be viewed as "the enemy." Hostility to bears can escalate beyond a reasoned response to situations such as bears killing sheep because the bears come to symbolize a larger assault on an entire way of life (Primm 1996).

Policies favoring brown bear conservation are also often thwarted by the combined influences of local politicians and the cultures of agencies responsible for management (Primm 1992; Mattson & Craighead 1994). Owing to pressure from their publics, local politicians typically provide little support for brown bear conservation and may even be overtly hostile. This is currently the case in areas that support brown bears in the contiguous United States. Politicians from these regions often work hard to prevent implementation of ostensibly strong policies such as the U.S. Endangered Species Act by threatening agency budgets or intimidating agency decision-makers (Primm 1992). Many agencies responsible for bear management have cultures that value hunting or the extraction of resources such as timber. These values are, to some degree, incompatible with providing absolute protection to an animal such as the brown bear. Consequently, implementation of policies to protect bears or bear habitat has often met considerable resistance from within the agencies themselves (Mattson & Craighead 1994).

FUTURE AND PROGNOSIS

Long-term prospects for currently vulnerable brown bear populations are uncertain. Populations in Spain, France, Italy, Norway, northern Idaho (U.S.), northwestern Montana (U.S.), north-central Washington (U.S.), and possibly Greece are highly vulnerable to extirpation owing to their small size and growing isolation (Servheen 1990; Mattson et al. 1995). Without radical management intervention and changes in the way humans interact with bears, they stand a good chance of disappearing within the 21st century. The fates of other, larger but vulnerable populations depend more on habitat trends.

Virtually everywhere except Siberia, human numbers are projected to increase in and near brown bear habitat. This increase is predicted to be most dramatic in (1) affluent countries, such as the United States, where the aesthetic qualities of brown bear habitat are inducing exponential human immigration, or (2) third world countries, such as in central Asia, where human birth rates are high. Simultaneously there will likely be increases in access to brown bear habitat. Road construction will probably be dramatic over the next few decades in southern Canada and parts of Russia in conjunction with aggressive timber harvest programs. This combination of increased numbers of people and local access to bear habitat does not bode well for peripheral and isolated bear populations. On the other hand, there has been some decreased access and plans for additional removal of roads from public lands in the contiguous United States.

Several critically important brown bear foods are also likely to diminish in the 21st century owing to pathogens and global climate warming (Mattson 1990). High-elevation or high-latitude foods, such as stone pines, are likely to move northward and experience major, local declines as favorable conditions vanish from the tops of mountains. Army cutworm moths (*Euxoa auxiliaris*), an inhabitant of alpine areas and an important food for interior North American brown bears, will be similarly affected. Salmon spawning grounds around the northern Pacific are expected to shift northward and be substantially reduced owing to warming of oceanic waters. Whitebark pine will continue to decline in North America as a result of the effects of white pine blister rust. There is no instance where high-quality foods will likely increase.

The 20th century has seen a substantial decline in per capita lethality to brown bears and a substantially increased willingness to protect bears and their habitat (Clark & Casey 1992; Kellert et al. 1996). There is no reason to expect that this trend will be reversed. Thus the fate of vulnerable brown bear populations will probably depend on whether people can reduce their lethality to bears soon enough to counter the combined effects of increasing numbers of people and loss of high-quality foods.

Chinese Alligator

John Thorbjarnarson, John L. Behler, and Xiaoming Wang

Common Name: Chinese alligator
Scientific Name: *Alligator sinensis*
Order: Crocodylia
Family: Alligatoridae
Status: Class I protected species in China; Appendix I of CITES; Critically Endangered under the IUCN. Approximately 400 survive in the wild.
Threats: In the past, widespread killing was a problem; currently, the most critical problems are an almost complete disappearance of natural habitat and environmental contamination.
Habitat: Wetlands of the alluvial plains in the lower Yangtze River valley.
Distribution: Formerly widely distributed throughout the lower Yangtze River valley and parts of the Shaoxing River; current distribution within the southeastern corner of Anhui Province.

DESCRIPTION

The Chinese alligator is one of the world's smallest crocodilians, with adult males rarely exceeding 2 m in total length (TL); females are about 75% of this size. Like the American alligator (*Alligator mississippiensis*) of the southeastern United States, Chinese alligators have characteristically broad snouts. Young animals are darkly colored with lighter bands on the body and a head speckled with yellow/white. Adults are almost uniformly dark gray/black (Brazaitis 1973).

NATURAL HISTORY

The Chinese alligator inhabits the area of climatic transition between subtropical and temperate regions of eastern China. From late October through mid-April, alligators apparently are dormant in subterranean dens dug into the edges of ponds, marshes, or rice paddies. Alligators emerge from their burrows in early May. Courtship, bellowing, and mating appear to peak in mid-June. During this time males may move from pond to pond searching for females (Chen & Li 1979). Females nest in late June and early July, and eggs hatch in September.

Chinese alligators are mound nesters; nests are 40–70 cm high, mounded from leaf litter and vegetation scraped together by the female (Chen 1985). Nests are constructed near permanent water, and owing to the highly disturbed habitat they are easily found by people. Where available, alligators appear to prefer small islands for nest sites, as these may be the least disturbed habitats available. The sex of hatchling Chinese alligators is determined by the temperature of incubation (Lang & Andrews 1994), but the exact relationship is not yet fully understood.

Chen (1985) estimates that females become mature in 6 to 7 years, and Chen (1991) reported an average female size of 135.0 cm (117.5–144.0 cm, $N=19$). Mean clutch size in captivity (ARCCAR 1983–1991) was 26.7 (Webb & Vernon 1992). Mean egg mass is 40–45 g. Incubation is reported to last 60 to 70 days (Chen 1985). Hatchlings measure an average of 20–21 cm TL and weigh 30 g (Huang 1982; Chen 1991).

Hsiao (1935) reported that Chinese alligator stomach contents included the remains of rats, beetles, water bugs, and fishes. Chen (1985) found the diet of alligators comprised river snails (Pomacea; 41% by weight), spiral-shelled snails (22%), rabbits (16%), freshwater mollusks (8.3%), freshwater shrimp (4.1%), frogs (2.6%), fish and Odonate larvae (2.3% each), and insect remains (1.9%).

Chinese alligators are prolific burrowers, digging into the margins of reservoirs, rice paddies, ponds, and marshes with their front legs and head. Burrows are frequently 10–25 m in length and used as hibernacula (hibernation sites) and retreat sites. Chambers are large enough to allow the alligator to turn around in the den, so it can enter and leave head-first (Chen 1991).

CONFLICTING ISSUES

Alligators were declared a Class 1 protected species by the Chinese government in 1972. However, despite official protection, many animals were still killed at that time (Watanabe 1982). The Chinese conservation program for alligators has focused principally on captive breeding. The largest breeding center, the Anhui Research Center for Chinese Alligator Reproduction (ARCCAR), established in 1979, is located 5 km south of Xuancheng city in southern Anhui Province. The original breeding stock comprised 212 animals collected from the wild from 1979 to 1982. ARCCAR initially had significant problems with breeding and husbandry protocols (Watanabe 1982) but presently runs a successful breeding program (Figure 3). The first captive-born hatchlings were produced in 1983 (Chen 1990; Webb & Vernon 1992), and annual production as of 1997 is approximately 2,500 eggs with a 90% hatching rate (Wan et al. 1999). In 1992 the ARCCAR was registered under CITES, allowing its captive-bred alligators to be traded internationally under the regulations of Appendix II. International trade was seen as a means to raise funds for supporting the increasingly expensive operation of the breeding center. However, following registration, little or

Figure 3. Current Distribution of the Chinese Alligator (shaded area)

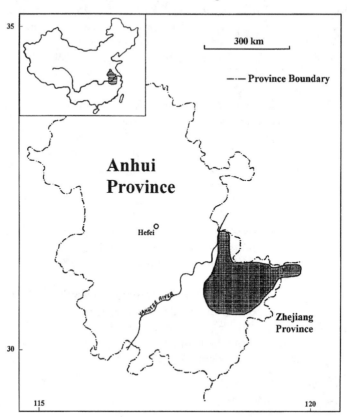

no trade has been reported. A smaller breeding center, the Yingjiabian Alligator Conservation Area, was also established in 1979 in Changxing County, Zhejaing Province.

In 1982, Anhui Province created an alligator reserve in a region of rice cultivation and tree and tea farming settlements in five counties of southern Anhui Province. In 1986, this area became a national reserve for Chinese alligators. The National Chinese Alligator Reserve covers a total of 433 km² and contains the largest surviving groups of alligators. The total population size of wild alligators is thought to be approximately 400 (Wan et al. 1999), most of which are distributed among 13 sites in the reserve. Although the reserve is fairly large, the amount of habitat actually preserved for and used by alligators is extremely small. The human population of the five-county region is 2.2 million (mean density = 243 people per km²) and has grown considerably over recent decades. Most alligators live in small ponds in densely populated areas that are heavily cultivated, principally for rice.

Conflicts between rural farmers and the few surviving groups of alligators are severe. The protected areas consist only of the ponds themselves, with little or no land surrounding them. Burrowing even a small distance into the shores of ponds will take alligators into agricultural fields. Moving between ponds means walking through rice fields. Ducks and other domestic animals use the edges of these ponds and are potential prey for alligators. One of the key factors that may determine the ability of alligators to survive in these areas is the presence of small islands that provide terrestrial retreats that buffer alligators to some degree from human impacts.

The principal factor contributing to decline of the species has been habitat loss. The middle and lower Yangtze at one time had large expanses of wetland marsh and lake habitat. From the late 16th century to the early 20th century (Quing Dynasty), there was a large migration of people from the north into the lowland Yangtze region (Huang 1982; Chen 1990) who converted the river's floodplain into agricultural fields. Virtually the entire area is now under cultivation (Watanabe 1982). The last remaining alligator populations in Anhui Province are in areas where agriculture was not initiated intensively until the 1950s and 1960s.

Beginning in 1958, sodium pentachlorophenate was applied to agricultural fields to kill snails that served as vectors of schistosomiasis (human liver flukes), and this inadvertently poisoned alligators (Chen 1990). Also, after 1949 large amounts of chemical fertilizers and insecticides were used in cultivated fields, reducing or in some cases virtually eliminating the prey base on which alligators depend (Chen 1990).

The loss of natural vegetation throughout eastern China has also exacerbated periods of drought and flooding. It is thought that large numbers of alligators drowned in their burrows along the Yangtze River during floods in the winter of 1957 (Watanabe 1982). Flood and drought also may force alligators to move overland, where they are captured or killed (Chen 1990).

Where they exist, remnant populations of alligators are not tolerated by local people because they prey on small domestic animals (particularly ducks) and their burrowing interferes with the complex water control structures that are necessary for rice cultivation. Historical records from the Ming Dynasty (1368–1644) suggest that large numbers of alligators were trapped and killed in the past because of their burrowing, and their meat was used for wedding feasts (Chen 1990). More recently, Hsiao (1935) states that when it is "convenient," alligators are killed by farmers for food. Other accounts suggest that during most of the 19th and 20th centuries alligators were not eaten by people, but if killed were chopped up and fed to ducks (Anonymous 1992) or to pigs (Webb & Vernon 1992). However, B. Chen (personal communication) reports that after 1986 people in Anhui began eating alligator meat on the basis of claims that it was dragon meat.

A recent threat may be related to the rapid economic growth in China.

This is especially notable in the area around Shanghai but also to a significant degree in southern Anhui Province. Increased development of agricultural areas may threaten the already tiny pockets of habitat left for the alligator. The growing buying power of many Chinese people may also exacerbate the threat of capturing alligators for sale—dead or alive.

FUTURE AND PROGNOSIS

Chinese conservation efforts for alligators have focused on captive breeding and have been very successful in that respect. Breeding centers in Anhui and Zhejiang Provinces now support a population in excess of 6,000 alligators, and surplus animals have been sent to many zoological parks and breeding programs in other parts of the country. A captive breeding program in North America (Behler 1977, 1993) currently involves 20 zoos. With such a large captive population, Chinese alligators are not threatened with extinction. However, survival of alligators in the wild is tenuous, and the clear conservation priority is to protect the remaining wild population and evaluate the feasibility of establishing new ones through releases of captive-bred animals.

The Chinese have protected wild alligators by making them a Class 1 protected species and establishing an alligator reserve in Anhui Province. However, the reserve as currently managed will probably not support viable alligator populations. Most subgroups contain fewer than 20 individuals, with very little habitat. The size of the officially protected sites should be enlarged and buffer zones should be created by including terrestrial as well as aquatic environments. Habitat modifications of existing protected sites, such as the creation of small islands in ponds, should also be considered. Because enlarging the protected sites will require the loss of arable land, this will create conflicts with agricultural communes. In many areas local villagers take great pride in having remnant alligator populations in their midst while unwittingly contributing to their demise. Public education programs in these villages, based on the alligator as a unique cultural treasure, will be essential to long-term conservation. Although the challenge is difficult, conservation programs must strive to develop mechanisms for improving the quality of the alligator's habitat while at the same time benefiting local communities.

Developing economic benefits for the surrounding communities can be problematic. One method used with other crocodilians is the commercial sale of skins and/or meat. There is interest in developing such an approach in China; however, currently such a program has very limited potential because no concrete mechanisms for commercial sale exist, and generating false expectations of commercial use would be counterproductive. Chinese alligator skins are not commercially valuable, and the sale of juveniles as pets has a very limited market. The sale of meat within China may eventually have the greatest potential. However, the sale of meat should be based on both detailed economic and biological evaluations, and it should be adopted only when the survival of wild populations is secure. Ecotourism is another alternative to consider. An important conservation priority will be to evaluate the feasibility of releasing captive-bred alligators to establish new populations.

Dalmatian Pelican

Henry M. Mix and Axel Bräunlich

Common Name: Dalmatian pelican
Scientific Name: *Pelecanus crispus* (Bruch 1832)
Order: Pelecaniformes (tropicbirds, gannets, cormorants, darters, pelicans, frigatebirds)
Family: Pelecanidae (pelicans)
Status: Classified by the IUCN (1996) as Vulnerable; considered Vulnerable in Europe (Crivelli 1994a); included in Appendix II of the Convention on the Conservation of European Wildlife and Natural Habitats (Bern Convention), in Annex I of the European Union's Wild Birds Directive, in Appendix I of CITES, in Appendix II of the Convention on the Conservation of Migratory Species of Wild Animals (Bonn Convention), and in the Agreement for the Conservation of African-Eurasian Migratory Waterbirds (AEWA) under the Bonn Convention (Crivelli 1996).
Threats: Drainage and degradation of wetlands throughout the range of the species; compounded by shooting, persecution by fishermen who regard the species as a competitor, collision with power lines, contamination by heavy metals and pesticides, and disturbance at colonies (Collar et al. 1994; Crivelli 1996).
Habitat: Inhabits fish-rich, open inland wetlands and shallow brackish coastal lagoons from sea level to 850 m above sea level. Nests in colonies among reeds (on floating islands) and on bare earth islands. Sometimes fishes inshore on sheltered coasts.
Distribution: Breeds within a highly fragmented range from eastern Europe to East Asia and Central Asia.

DESCRIPTION

The Dalmatian pelican is the largest of the eight pelican species. It is 160–180 cm in length, has a wingspan of 310–345 cm, a bill length of 37–45 cm, and weighs 10–13 kg (Cramp 1977). Sexes look alike. Bare skin around the eyes is yellow to purple, the gular pouch is yellow to orange, and the legs are lead gray. It has a pale yellow iris. Plumage is mainly silvery white (breeding) or grayish-white (nonbreeding). On the underwing the primaries appear dark gray and the secondaries dusky white, whereas on the upperwing the primaries are black and the secondaries gradate to white from the outside to the inside. The bird's scientific name, *crispus*, is derived from its short, untidy, curly crest. Juveniles can be distinguished by their grayish-brown upperparts, gray bill, and gray gular pouch.

NATURAL HISTORY

The Dalmatian pelican is gregarious throughout the year. Pelicans often fish cooperatively, forming a semicircle to drive fish forward. The diet of the Dalmatian pelican consists entirely of fish. Dementiev and Gladkov (1951) estimated that two adults and two juveniles ate 1,080 kg of fish in 8 months, an average of 1,123 g daily per bird. They form monogamous pair bonds of at least seasonal duration. Both parents tend the young, with the family group continuing to associate after fledging. Breeding and feeding sites are often several kilometers apart.

The pelicans breed on the ground, on floating islands in reedbeds, or on earthen islands covered with halophytic vegetation (plants that grow in salty soil) that are completely isolated from mammalian predators (Crivelli & Michev 1997). Nests at Ayrag Nuur, Mongolia, were located in shallow open water several hundred meters from the shoreline (personal observation). The nest, a heap of grass, reeds, twigs, and small branches, is 1–1.5 m in diameter and 1–1.5 m above the water. The clutch consists of 2 to 3 (rarely 4) eggs and averages 1.8 eggs (Crivelli 1996). One brood is raised per year. Incubation lasts 30–32 days, the fledging period is 11–12 weeks, and young birds become independent at roughly 100–105 days. They probably mature at around 3 years of age.

Dalmatian pelicans are migratory or partially migratory. European breeders winter in the eastern Mediterranean from the Balkans to lower Egypt (Cramp 1977). A large southwestern and south Asian group winters in the fish-rich wetlands of the southern Caspian region and from Mesopotamia through southern Iran to Pakistan and India (Perennou et al. 1994). Birds from Mongolia seem to winter in Hong Kong's Deep Bay (Carey personal communication), and at several sites along the east coast of China (Simba Chan personal communication). Small numbers are occasionally reported from other sites in China. Breeding and migration seasons vary among populations, with central Asian populations arriving and breeding later in the year than European breeders. This is owing to the more continental climate in the Central Asian breeding grounds.

CONFLICTING ISSUES

The current global population consists of 12,000–16,000 individuals (midwinter counts; Rose & Scott 1997) and a breeding population of approximately 3,500–4,550 pairs that are distributed in Montenegro (former Yugoslavia, 10–20 pairs), Albania (40–70), Greece (500–550), Romania (70–150), Bulgaria (70–90), Turkey (100–150), Iran (5–10), the former USSR (2,700–3,500 in the Russian Federation, Turkmenistan, Ukraine, Uzbekistan, and Kazakstan), and Mongolia (<10) (Collar et al. 1994; Crivelli 1994b, 1996; personal communication). The species formerly bred in Hungary and China as well.

Since the 19th century, several breeding colonies in Europe have vanished from Hungary, the former Yugoslavia, Greece, Romania, Ukraine, and the Russian Federation (Crivelli & Michev 1997). Today about one-quarter of the *P. crispus* world population breeds in Europe. Only 40% of European breeding sites are currently protected (Crivelli 1994a). The European population is well monitored and can be considered stable. However, colonies may be expanding (Greece, Russian Federation) or shrinking (Albania) locally. The extent of remaining suitable habitat limits recovery, and the species remains vulnerable to further decline and extinction in Europe (Crivelli & Michev 1997). An action plan for conservation of the species in Europe was recently published (Crivelli 1996).

The stronghold of Dalmatian pelican lies within its central range, with Kazakstan holding more than 40% of the world population (Crivelli 1994b). An overview of the status and recommendations for conservation measures for pelicans in the former USSR was published by Krivenko et al. (1994).

Populations breeding in eastern Kazakstan and in Mongolia (and formerly in China) are considered discrete groups with no contact between them (Zhatkanbaev & Gavrilov 1994). The eastern subpopulation (east of the Altai Mountains) remains little studied. It is likely that Dalmatian pelicans in Mongolia were always restricted to the semi-arid Basin of the Great Lakes in the western part of the country. The only proven breeding records come from this region, where five large, shallow lakes account for approximately 70% of the surface area of the approximately 3,000 lakes in Mongolia (MNE 1996). Until the 1960s, several hundred pairs were recorded from this area. Until the 1970s or early 1980s, the largest colony was found at Har Us Nuur (Tseveenmyadag personal communication). It is not known when this breeding site was abandoned. Tseveenmyadag and Bold (personal communication) assume that the species has declined continuously since the early 1980s. Today the last confirmed breeding site is a small colony with up to eight nests at Ayrag Nuur, where breeding was confirmed in 1995, 1998, and 1999; however, very little is known about breeding success (Bräunlich 1995; Liegl personal communication; Köppen personal communication).

Estimates of 30 to 50 pairs (Bold & Tseveenmyadag in Crivelli 1994b) and 200 individuals (Shiirevdamba 1997) in Mongolia and a maximum of 500 individuals in eastern Asia (Rose & Scott 1997) seem optimistic. Based on estimates in 1998 and 1999, the known population of Dalmatian pelican in China (or eastern Asia) is probably not more than 130 (Carey personal communication; Simba Chan personal communication). The current known population of *P. crispus* in eastern Asia thus consists of 31 birds (24 adults and 7 young) at the only known breeding site in Mongolia in 1999 (Köppen personal communication) and 100–200 wintering birds in Hong Kong and China. There is reason to believe that this population is close to extinction and should be considered critically endangered.

The main reason for the rapid decline in Mongolia is increased hunting by Mongolian herdsmen, although the species has been protected under Mongolian law since 1953 (Shiirevdamba 1997). The nomads of western Mongolia traditionally use the upper mandible of pelican beaks to groom their horses. The roots of this tradition are unclear, but Mongolians believe that using the beak makes their animals stronger and faster. It is not known why pelican beaks are attributed miraculous and stimulating effects. The use of pelican sweat blades is unusual in Mongolian horse culture, and this tradition is apparently restricted to western Mongolia. Further east, the blades are often replaced by beak-like carvings made from softwood. Pelican sweat blades are family treasures and are only displayed on special occasions and feast days. Traditionally the beaks are handed down within a family, normally from father to son, for many generations.

Pelican beaks formerly had a nonmaterial value and were not available for sale. Their cultural value was greater when a family possessed them over generations. However, recently there has been a general and rapid change in values in Mongolia, and pelican beaks are rapidly becoming an expensive prestige item. The most intensive hunting pressure seems to be caused by nonresident Mongolians, mostly well-equipped urban people who try to make "fast money" by selling pelican beaks. For example, in 1995 one beak was offered for the equivalent of two camels, one horse, and two bottles of vodka. Hence the price of one beak (U.S. $200–300, exceptionally up to $1,000) is many times the annual income of a Mongolian herdsman. As a result the pelicans are hunted intensively by Mongolians, who otherwise rarely take birds. During the national Naadam festival in 1995, an estimated 100 pelican sweat blades were displayed by owners of racing horses in Dzavhan Sum, Uvs Aimag (Bräunlich 1995).

The colony at Ayrag Nuur might be influenced by changes in the water regime too, since its location changed within a few years corresponding to water level fluctuations. This could negatively influence breeding success.

FUTURE AND PROGNOSIS

Wherever they occur, Dalmatian pelicans rank among the specially protected species. Despite these favorable legal conditions, implementing practical conservation programs in geographically remote breeding areas is usually difficult. Unstable political situations in many regions of the pelicans' range could undo the results of long-term conservation measures within a short period. For example, a colony in the Albanian Karavasta Lagoon declined by 75% after only 2 years of political unrest (Xharo personal communication). In Greece, where breeding numbers increased significantly during the last decade due to effective conservation efforts (Crivelli personal communication), 200 nests were destroyed in a colony in a National Park during an intentional disturbance by a fisherman in 1998 (Anonymous 1998).

Most of the world population of *P. crispus* lives in continental, eutrophic or mesotrophic (i.e., having a high or medium amount of dissolved nutrients, respectively) shallow lakes. These rank among the most threatened lakes because their comparatively small water bodies are very sensitive to changes in the water regime (Dokulil 1994). For example, ongoing water extraction in the middle reaches of the Amu Darya River, and the associated ecological disaster of the disappearing Aral Sea, threaten two very important breeding sites. Moreover, overexploitation of fish resources and poaching are common in many breeding, staging, and wintering areas.

The only chance to safeguard the last known colony of the eastern subpopulation, which is found in Mongolia, is to guard the area continuously during the breeding season (May to August). Herdsmen from the vicinity of the lakes could be enlisted for such seasonal work. Additionally, running a small field station close to the breeding area could provide an opportunity to compile valuable data on the biology and ecology of the pelicans. Students from the University of Khovd, roughly 200 km from the colony, would also gain practical training. Conservationists should try to stimulate the establishment of new colonies and the extension of the existing one by providing artificial nests. However, initial results from such recovery efforts cannot be expected for 8 to 10 years, because the demography of this long-lived species would prevent more rapid increases (Crivelli personal communication).

Initiatives by single herdsmen to use alternative materials (wood and horn) for sweat blades could be extended and encouraged by means of a public awareness campaign. However, this model would only work if the beak imitations were to develop the same cultural value as the originals. Martin and Vigne (1995) give an example of an alternative material that was introduced to replace an endangered species product.

Another possible strategy would be saturation of the regional market for beaks, using beaks from more common pelican species found elsewhere. This unconventional, pragmatic approach is promising, given the low human population density in the Great Lake Basin. It would require the distribution of several hundred pelican beaks free of charge, thereby reducing hunting pressure on the species in the region. First, a careful investigation must be carried out to determine whether such an initiative would strengthen commercial attitudes toward pelican beaks instead of acting as a conservation instrument. Furthermore, a feasibility study must appraise whether exploitation of a beak surplus (owing to natural mortality in large colonies of common pelican species) would be possible.

Captive breeding of *P. crispus* is presumably not an appropriate conservation approach, although pelicans breed relatively easy in zoos. Currently 244 captive individuals are registered worldwide in the International Species Information System (ISIS). The actual number of captive Dalmatian pelicans may be twice that in Europe alone, since many facilities do not participate in ISIS. As such, zoos could play a proactive role in pelican conservation by raising public awareness and collecting funds to develop conservation initiatives for the survival of this charismatic species in the wild.

ACKNOWLEDGMENTS

Helen Ferguson kindly helped with the English translation of this entry.

Douc Langur

Dean Gibson

Common Name: northern douc langur, douc or red-shanked langur
Scientific Name: *Pygathrix nemaeus nemaeus*
Order: Primates (primates)
Family: Cercopithecidae (cercopithecines)
Status: Listed as Endangered by the IUCN (The World Conservation Union) in 1972 (Wolfheim 1983); listed in Appendix I of CITES and as Endangered under the U.S. Endangered Species Act since 1975; given the highest rating for conservation efforts by the IUCN Action Plan for Asian Primate Conservation (Eudey 1987). Vietnam produced a Red Data Book in 1992 listing *Pygathrix nemaeus nemaeus* as Endangered and *Pygathrix nemaeus nigripes* as Vulnerable (Lippold 1995).
Threats: Hunting; deforestation; human encroachment.
Habitat: Douc langurs are found in a variety of forest types including primary and secondary moist, semi-deciduous, and evergreen forests from sea level to 2,000 m in elevation (Lippold 1977, 1998; Wolfheim 1983). Although doucs are primarily a forest species, they have also occasionally been seen in banana plantations (Lippold 1977).
Distribution: Forests of Laos, Cambodia, and Vietnam (Wirth et al. 1991). An early report suggests that the species may have occurred on the island of Hainan in the South China Sea (Napier & Napier 1967); however, this report has never been substantiated (Groves 1970).

DESCRIPTION

The red-shanked douc langur, with its color complexities, is stunningly beautiful. It is a relatively large langur, with males weighing approximately 10–12 kg and females approximately 8.2 kg (Rowe 1996; Kirkpatrick 1998). Doucs have a soft, flesh-toned face (which bronzes when exposed to sunlight) surrounded by wispy, white whiskers and beard, and a grizzled gray/ black and white coat pattern that is accented at the neck with bands of russet red and white. Their almond-shaped eyes are deep brown and their mouths are fleshy, rounded, and white. The hind limbs are black from the hip to the knee and coordinated with burgundy knee socks, while the forearms are white from wrists to elbows. The long, thin tail is white, and the hands and feet are black. Males can be distinguished from females by the presence of two white spots, one on either side of the white rump patch. Douc infants are born with a reddish and grizzled black natal coat. The back is grizzled black similar to the adult coat. However, the head coloration is a mixture of dark reddish brown and black. In contrast to the adult coat, the majority of the infant's face is dark gray to black. The legs and forearms are entirely red, and the upper arms are reddish gray. The tail, ventral, and lateral areas are white to gray. The hands and feet are dark gray and lack much hair. The douc's natal coat color changes quickly to that of an adult coloration between 8 and 18 months of age.

Two subspecies of douc langurs are currently recognized; the Northern or red-shanked douc (*Pygathrix nemaeus nemaeus*), and the Southern or black-shanked douc (*Pygathrix nemaeus nigripes*). The black-shanked douc langur is similar to the red-shanked with the exception of having a darker bluish face, entirely black legs, and gray arms.

NATURAL HISTORY

Douc langurs are among the most endangered and unstudied primates of the Old World (Bennett & Davies 1994). Small populations survive in the fragmented forests of Laos, Cambodia, and Vietnam (Wirth et al. 1991); but owing to continuous hunting pressures and habitat destruction, the original distribution of douc langurs will probably never be known. For example, the species may have occurred on Hainan Island (Napier & Napier 1967) but today is absent, either having been extirpated from or never actually occurring on the island (Wang & Quan 1986). The status of the species and its habitat in Cambodia and Laos is not known (Wirth et al. 1991; Wolfheim 1983), although forests along the border between Vietnam and Laos are reportedly in pristine condition and likely to support douc populations (Freed personal communication). The status of the douc langur in Vietnam has been better documented. Recent survey work indicates a patchy but more widespread distribution than was previously thought (Kirkpatrick 1998; Lippold 1998). Current douc langur population estimates are between 2,000 and 20,000 throughout their range in Southeast Asia (Traitel 1996). This represents a drastic decline from a 1987 estimate of 72,000 (Kirkpatrick 1998). Field surveys to better document population size and distribution are critically needed.

Detailed behavioral and ecological information on douc langurs is sparse. Douc langurs are leaf-eating monkeys with a specialized gut morphology consisting of a large, multi-chambered stomach with foregut bacterial microflora for the digestion of cellulose (Oates & Davies 1994a). Given their diet, doucs spend long periods of time virtually inactive while they digest their leafy meals. Although they are fond of new leaf growth, these arboreal primates also feed on unripe or green fruits, buds, and flowers (Lippold 1977). Unlike many other primate species, douc langurs occasionally share food (Gochfeld 1974).

Doucs are usually found in single or multi-male groups of 3 to 15 individuals with a sex ratio of 1 male : 2 females (Lippold 1977; Napier & Napier 1985); however, groups of up to 50 individuals have been reported (Lippold 1977, 1995). Variations in group size may be dependent on quality of habitat and disturbance levels, with larger groups being found in forests that have had little to no disturbance (Lippold 1998). Solitary animals from both sexes have also been recorded (Lippold 1977), which suggests that both sexes migrate from natal groups.

In the wild, sexual maturity is estimated at approximately 4 years of age for females and 5 years for males (Napier & Napier 1985). Estrus cycles occur every 28 to 30 days, and infants are born after a gestation of approximately 210 days (Lippold 1981). Females in the same group often give birth at the same time, which is likely associated with peak fruiting season (Lippold 1977).

As in other langur species, douc females demonstrate "aunting" behavior by sharing in the care of infants. Douc females not only carry and groom another female's infant (Napier & Napier 1985), but in captivity they have also been observed nursing another's infant (Lippold 1979). Adult douc males are tolerant of infants and juveniles and have been observed providing paternal care (Lippold 1979).

CONFLICTING ISSUES

Vietnam, once completely forested, has lost 90% of its forest cover as a result of decades of war and deforestation. During the Vietnam War alone, forests were affected by over 13 million tons of bombs and 72 million liters of herbicides (Eames & Robson 1993). In 1968 an estimated 2,600,000, bomb craters, some measuring 10 m deep, existed throughout the forest and landscape (Orions & Pfeiffer 1970; Lippold 1977). Many of these areas are recovering from chemical defoliation, and only 1% of remaining Vietnamese forests are considered pristine (Eames & Robson 1993).

In 1986 the Vietnamese government initiated efforts to conserve its remaining biodiversity by establishing a protected area system that includes over 10,945 km² (3% of the land area) of parks and reserves (Eames & Robson 1993; Lippold & Thanh 1998). However, the park system offers little safety because of poor law enforcement, inadequate funding, and insufficient staff, training, and equipment (Lippold 1995; Traitel 1996; Lippold & Thanh 1998). Although doucs are known to occur both within and outside of protected areas in Vietnam, including secondary forests, they do not occur in regenerating forests that have been sprayed with the defoliant Agent Orange (Lippold 1995). As a result of war and the use of defoliants, douc populations (which may have already been low) (Wolfheim 1983) have been reduced to small, isolated forest patches (Nadler 1995). These remnant and fragmented populations are threatened with continued hunting and habitat disturbance, inbreeding depression (i.e., decreased fitness due to breeding of close relatives), and stochastic events. Overall, the chances for the survival of these small and fragmented populations are slim.

Loss of habitat to subsistence farming and commercial agricultural development is thought to be the greatest threat to langur populations (Mittermeier & Cheney 1987). Deforestation through logging, clearing for agriculture, human settlement, gold mining, and forest exploitation for fuelwood, timber, and palm extraction continues in Vietnam at a rate of 100,000 hectares per year (Eames & Robson 1993; Lippold & Thanh 1998).

Logging may reduce the food supply for arboreal leaf-eating monkeys; however, Davies (1994) found that a 50% reduction in tree numbers owing to logging did not result in a 50% loss of food supply (leaves). Indeed, an almost doubling in the productivity of remaining trees provided an almost constant supply of young leaves (Davies 1994). High juvenile and infant langur banded-leaf monkey (*Presbytis melalophos*) mortality occurred during the logging operations; however, there was an overall population increase 5 to 6 years after the logging (Davies 1994). Langur species are not equally sensitive to logging, and although selective logging may cause a temporary population decline, by itself it rarely causes extinction (Oates & Davies 1994b). These observations suggest that a well-structured forest management plan that adequately addresses logging, exploitative practices, and natural disasters could allow langurs and local subsistence farmers to co-exist. As douc langurs are known to occur in a variety of forest habitats in Vietnam, Lippold (1998) suggests that the species is ecologically and socially flexible. This would indicate that controlled logging disturbances may not be detrimental to the overall population in Vietnam. However, further research into forest disturbance, forest recovery dynamics, and the ecology of douc langurs is necessary to explore this option.

Historically, douc langurs have also been subjected to hunting pressures. Subsistence and military hunting—including target practice (Lippold 1977; Wolfheim 1983)—as well as hunting for medicinal purposes (Wirth et al. 1991), have contributed to the douc langur population decline. In 1990 the human population of Vietnam was estimated at 68.5 million with a growth rate of 2.5% (Eudey 1991), of which 80% were subsistence farmers (Eames & Robson 1993). Today, with such a high human population growth rate, hunting pressures have become even more severe. Protected wildlife is eaten in fashionable restaurants and is commonly seen for sale in street markets, and an estimated 12 times the official amount crosses the borders into China and Laos illegally for food, pets, and medicinal purposes (Lippold & Thanh 1998). Moreover, poaching for the live pet and tourist trade continues to rise (Wirth et al. 1991; Freed personal communication).

To help curtail poaching pressure, a rescue and rehabilitation center for confiscated animals was created in 1995. The Endangered Primate Rescue

Center at Chu Phong National Park assists conservation efforts for endangered Vietnamese primates (Traitel 1996). It provides care for primates that have been confiscated from hunters, private parties, or foreigners attempting to export them. As the reserve guards and Forest Protection Departments become better educated regarding Vietnam's wildlife laws and begin to enforce them, increasing numbers of primates are brought to the Center (Nadler 1995). The Rescue Center is effective in the short term by curtailing the loss of douc langurs and providing a foundation for in-country captive breeding efforts. The long-term goal of the Center is to develop captive breeding programs to produce animals for eventual release into protected areas once wildlife laws are enforced (Killmar personal communication). Given the difficulty of maintaining doucs in captivity, it is unclear if maintaining captive populations in Vietnam will succeed in the long term. Caring for these delicate primates, which have complex nutritional requirements and deteriorate quickly when stressed, can be labor intensive and costly. On average, the Rescue Center acquires one primate a month (Nadler 1995), whereas many more confiscated injured animals continue to be released into the forest by forestry and police personnel (Lippold & Thanh 1998). Additional rescue support and conservation education for police and forestry personnel are needed.

FUTURE AND PROGNOSIS

The people of Vietnam stand to lose a national treasure, if wildlife protection laws are not enforced for the douc langur. The government of Vietnam has demonstrated its recognition of the importance of Vietnam's biodiversity. The establishment of protected areas was the first step in providing douc langurs with a chance of survival. However, several existing protected areas may be too small to maintain viable populations. Even though the Vietnamese government is slowly improving protection, hunting and encroachment remain problematic (Wirth et al. 1991; Eames & Robson 1993). The government must balance (1) species and habitat conservation along with a growing human population, most of whom are subsistence farmers, and (2) insufficient resources and funding. The nation's struggling economy creates added pressure to exploit natural resources (Traitel 1996). Without conservation education along with alternative economic options for poverty-stricken people, the problems of deforestation, forest exploitation, and hunting will likely become worse as the human population continues to increase.

Overall, the long-term conservation of douc langurs will depend on effective protected area management, the elimination of hunting, managed deforestation and exploitation, and additional ecological research and survey work in Vietnam, Laos, and Cambodia.

Eastern Barred Bandicoot

Peter Myroniuk and John Seebeck

Common Name: eastern barred bandicoot
Scientific Name: *Perameles gunnii*
Order: Marsupialia (marsupials)
Family: Peramelidae (bandicoots)
Status: Endangered in the state of Victoria, Australia; Threatened under the Victorian Flora and Fauna Guarantee Act (1988); Vulnerable under the Australian Government's Endangered Species Protection Act (1992); Critically Endangered in the 1996 Action Plan for Australian Marsupials and Monotremes (Maxwell et al. 1996); Vulnerable under the IUCN (IUCN 1996).
Threats: Clearing of land for agriculture; introduction of domestic stock and rabbits; altered fire regimes; introduction of feral predators (e.g., domestic cats and dogs, and the red fox, *Vulpes vulpes*); motor vehicle kills; agricultural chemicals; diseases spread by cats; urbanization.
Habitat: Grasslands and grassy woodlands of western Victoria and northern and eastern Tasmania. Critical habitat for the species includes remnant grassy woodlands and grasslands as well as highly modified grassy woodlands and grasslands that provide cover, nesting opportunities, and a food supply.
Distribution: On mainland Australia the species' former distribution extended across the vast grassy woodland plains of southeastern Victoria, but today it is relegated to a small area around the town of Hamilton and a few, small reintroduced populations in the state. It formerly occurred in southeastern South Australia but is now extinct in that state. In northern and eastern Tasmania it occurs in grassy woodlands (Figure 4), but its range has contracted in recent times (Dreissen & Hocking 1991).

DESCRIPTION

There are 21 recognized species of bandicoots, 10 in the family Peramelidae (which includes eastern barred bandicoots) and 11 in the family Peroryctidae. Adult eastern barred bandicoots may weigh over 1 kg (average about 800 g), the average head-body measurement is about 300 mm (range 270–350 mm), and the tail is about 110 mm long. The fur is yellowish-brown, with four pale bars on the hindquarters.

NATURAL HISTORY

Eastern barred bandicoots are opportunistic feeders of the grassy woodland/grassland. They use their long foreclaws and pointed snout to dig small, distinctive, conical holes in the litter and topsoil layers to gain access

Figure 4. Historic and Current (including some reintroduced populations) Distribution of Eastern Barred Bandicoots in Victoria, Australia

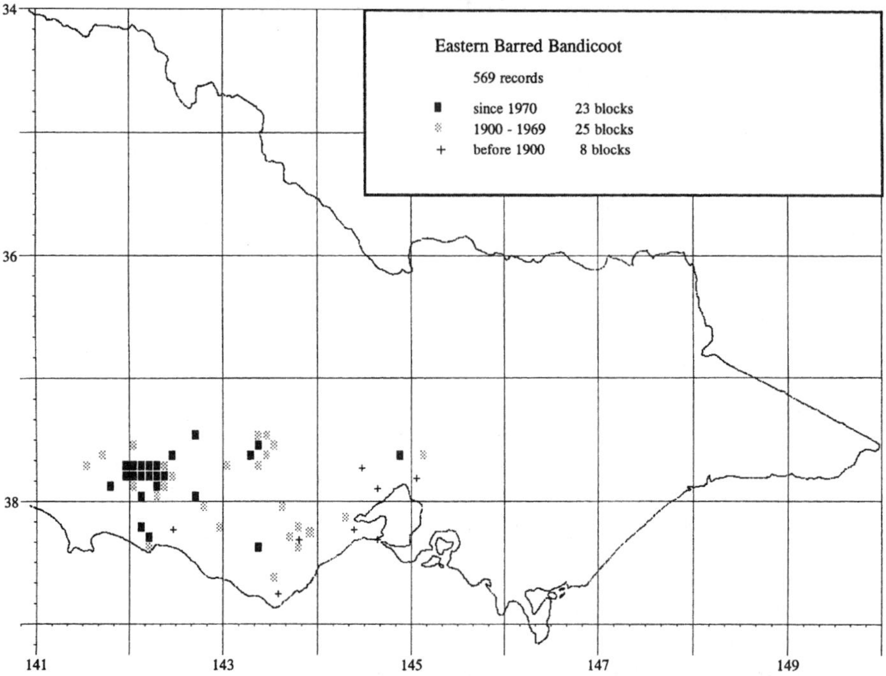

Atlas of Victorian Wildlife Database, 1998. Arthur Rylah Institute for Environmental Research, Heidelberg, Victoria, Australia.

to a variety of invertebrates (e.g., earthworms, crickets, beetles, and caterpillars) and plant material (e.g., bulbs, tubers, and fungi); berries and other fruits are eaten in season as well (Dufty 1991, 1994b). Structure of the habitat (rather than botanical composition), amount of vegetative cover, available nesting opportunities, and food availability are critical features of bandicoot habitat. Tasmanian bandicoots are genetically distinct at the subspecies level from Victorian bandicoots (Robinson et al. 1993).

Eastern barred bandicoots are crepuscular (active at dawn and dusk) and nocturnal, resting during the day in shallow nests lined with grass. They are generally solitary except for females with young-at-foot and do not appear to be territorial, although they may defend a core area of their home range. Males' home range is much larger than females' and may include the home range of several females (Dufty 1991, 1994a, 1994b).

The eastern barred bandicoot is a marsupial. Females give birth to young about the size of a large kidney bean (approximately 10–12 mm) after a gestation period of approximately 12.5 days (close to the shortest known for any mammal). At birth, the young makes its way to the mother's pouch, where it will spend up to 60 days attached to one of eight teats. Females can give birth to litters of up to five, but the average litter size is two to three. Young emerge from the pouch between 60 and 70 days after birth (Dufty 1995; Seebeck 1995).

Eastern barred bandicoots have one of the highest reproductive potentials of any mammal. Females may breed from about 3 months of age and males at about 4 months of age. Reproduction can occur throughout the year but is usually depressed during drought conditions. In good years, each female could potentially produce three or more litters per year. In the wild, eastern barred bandicoots rarely live beyond 2 years of age; however, in captivity they can live up to 5 years.

CONFLICTING ISSUES

In Victoria, the wild population of the eastern barred bandicoot has been reduced to a tiny relict colony in and around the city of Hamilton. That habitat is highly modified, including suburban gardens, agricultural pastures, modified woodlands, and remnant native grasslands and woodlands. Since the early 1990s eastern barred bandicoots have been reintroduced into several sites within their former range as part of a long-term species recovery effort. There are currently seven reintroduced populations and a captive population maintained in several facilities. However, none of the reintroduced populations can be considered viable in the long term unless predator control is effective, population sizes increase, and land management practices change (particularly on private land).

Concern about the decline in bandicoot numbers was noted as early as 1937 (Harper 1945), although it was not until the 1970s, after several populations had declined significantly, that the threatened status of the bandicoot was appreciated (Seebeck 1979). By that time, more than 99% of the known range in Victoria had been fragmented (Seebeck 1979).

When Europeans settled Victoria in the 19th century, eastern barred bandicoots covered some 23,000 km^2 and numbered in the tens of thousands. By 1982–1983, however, there were only an estimated 1,750 bandicoots surviving (Lacy & Clark 1990). By 1985 there were 633 (Brown 1989), and in 1988 less than 300 animals remained (Dufty 1988). This indicates a decline of about 25% per year. Population viability analysis (i.e., computer modeling of population dynamics) in 1989 predicted that the eastern barred bandicoot would be extinct on the Australian mainland within 10 years if direct action was not taken (Lacy & Clark 1990). It is now believed that the wild population of Victoria is very close to extinction.

During the early 1980s, recovery efforts focused on the remnant wild population at Hamilton. Several organizations and individuals tried to halt the decline and restore habitat for the eastern barred bandicoot. Principal among these was the state government's Department of Natural Resources and Environment (and its predecessors). They focused on maintaining and expanding habitat in an effort to increase the overall population size as well as address attitudes toward bandicoots and domestic pets through community education programs. The majority of bandicoots live on private land, so community extension programs encourage private landholders and the local municipal council to manage their properties with an eye toward conservation of eastern barred bandicoots. The Department had the knowledge, structure, resources, and instruments (e.g., legislation and management plans) to undertake bandicoot recovery, but it lacked general community support.

The remnant wild population of eastern barred bandicoots occurred on the rural/town fringe, in a community consisting predominantly of agriculturalists and various support industries. Local knowledge of bandicoots was high, but many considered their conservation a low priority (Reading et al. 1995). As the market for sheep and wool dropped, local communities such as Hamilton suffered economically. Thus wildlife conservation issues are a very low priority for many people in the area. Nevertheless local lobbying by the Hamilton Institute of Rural Learning, a local community group, was successful in persuading the government to build a protective enclosure for eastern barred bandicoots. Interested members of the local community subsequently formed Friends of the Eastern Barred Bandicoot, a conservation group. They focus on habitat maintenance and restoration, public education, monitoring of bandicoot numbers, fundraising, and predator control.

A small, captive breeding program commenced in 1988 at Gellibrand Hill Park (now known as Woodlands Historic Park) some 20 km northwest of Melbourne in an effort to increase bandicoot numbers and establish a second population for insurance. The Department of Natural Resources and Environment, which initially coordinated the breeding program, shifted the responsibility to the Zoological Parks and Gardens Board (ZPGB), and the program's primary emphasis shifted from managing the wild population to captive breeding and reintroduction.

Despite almost 20 years (1972–1991) of research and wild population management, including the establishment of a captive colony, bandicoot numbers still declined. The original Management Plan for bandicoot recovery (Brown 1989) called for Recovery Teams. At the time, the teams were highly optimistic of success. However, after 2 years of plan implementation the bandicoots were still declining in Victoria. Possible reasons were identified (Reading et al. 1991; Backhouse et al. 1994). One was a lack of authority at the Recovery Team level. Because decisions had to be endorsed by other senior managers within government, delays in funding and decisionmaking occurred.

The Recovery Team, acting on genetic advice, considered taking a large number (50–100) of bandicoots from Hamilton for a captive breeding program. These high numbers were not acceptable to the local community, which saw the plan as an abandonment of the wild population that would add to its further reduction. At this time captive husbandry techniques were not fully developed, and the strategy was seen as high risk. The strategy that was subsequently adopted was a compromise between community concerns and the resources available to the recovery program.

By 1990 wild bandicoot numbers were critically low. Estimates put the number at between 100 and 200. The negative effect of stochastic events—such as individual road kills (particularly of females), disease (e.g., toxoplasmosis), inbreeding, aging, and population sex structure—became more pronounced. In November 1991 there was a critical review of bandicoot recovery (Reading et al. 1991). This analysis produced a revised recovery plan that placed greater emphasis on captive breeding (Backhouse 1992).

The 1991 review proposed a radical organizational restructuring of the recovery program. The result was the establishment of a Core Decision Group possessing appropriate leadership, authority for funding and other resources, and decisionmaking and evaluative processes with clear lines of authority. Its charter was to develop and implement a Strategic Plan. Senior managers from within the Department and the Zoological Parks and Gardens Board formed the Core Decision Group. A strategic planner was appointed and charged with preparing a revised recovery plan. The long-term aim of the revised plan was to ensure the survival of the eastern barred bandicoot by achieving a stable wild population of 1,000 individuals in three separate locations within their former range and establishing a stable metapopulation of 200 animals in captivity. The short-term aim was to remove from the wild as many animals as possible and to increase the captive population to 175 individuals from at least 31 founders. To assist in achieving these targets, a viable, intensively managed, captive breeding program was proposed.

As a consequence of these changes, additional animals were captured from the wild and some were transferred from the existing captive colony at Woodlands Historic Park to establish a more intensive captive breeding program based within the Zoological Parks and Gardens Board. The emphasis was on breeding sufficient numbers to provide for their subsequent reintroduction to selected sites within the species' former range. A number of sites had been identified, and site preparation commenced.

Active communication and involvement in the recovery program, including the decisionmaking process, generally ameliorated community concerns. Community representatives were invited to hold positions on the recovery team and associated working groups. Community input was sought in revising and developing recovery plans and actions. A separate working group addressed community issues as well as improved communication. A comprehensive newsletter was distributed widely throughout Hamilton, Victoria, and other states.

Of the seven current reintroduction sites, five are on conservation reserves and the others are on private land. Community support is apparent at all sites, and it includes volunteers from many walks of life (who help with population monitoring), "Friends" groups, sporting shooters who assist with vermin control, and the landholders and their neighbors.

FUTURE AND PROGNOSIS

Overall, organizational and community conflicts have been ameliorated by reviewing and improving organizational structures, decisionmaking, and communication processes on an annual basis. As a result, a greater sense of commitment and priority toward bandicoot conservation has emerged. However, one cannot underestimate the intricate and complex interactions among the individuals participating in this effort. Dynamic, forward, and lateral-thinking leadership mixed with individuals that are orientated toward and committed to recovery goals is essential. In addition, program personnel should maintain a sense of urgency, while avoiding restrictive organizational norms or cultures.

In any recovery program that has many competing issues and conflicts, an adaptive management style is required. Decisive leadership, timely decision-making, full commitment from organizations and individuals, and adequate funding and resources are essential for any degree of success in species recovery efforts.

During the important first decade of eastern barred bandicoot recovery in Victoria many of the elements described here have been put in place, and the status of eastern barred bandicoot conservation has improved. There is now a new set of conflicting issues related to land management—in particular, predator control. Community support continues to grow, as does the bandicoot population, but continuing vigilance is essential to ensure that the recovery program succeeds in the long term. Recommendations for addressing the future conservation challenges for eastern barred bandicoot recovery are as follows:

1. ensure that a minimum of five reintroduced populations are demographically and genetically viable for the long term by managing them as one metapopulation, with exchange of individuals among the subpopulations,

2. develop and implement aggressive feral animal control,

3. encourage and establish additional populations of bandicoots on private land, and

4. encourage greater community knowledge and involvement in bandicoot conservation.

Ethiopian Wolf

Claudio Sillero-Zubiri

Common Name: Ethiopian wolf, Abyssinian wolf, Simien fox, Simien jackal
Scientific Name: *Canis simensis* (Rüppell 1835)
Order: Carnivora
Family: Canidae
Status: Officially protected in Ethiopia under the Wildlife Conservation Regulations of 1974, Schedule VI; Critically Endangered under the 1996 IUCN Red List of Threatened Species.
Threats: Habitat loss and fragmentation; human interference. Recent population decline is owning to a combination of factors that result from the previously mentioned causes; these include disease epizootics (epidemics in wildlife populations), hybridization with domestic dogs, road kills, and shooting.
Habitat: A very localized endemic species confined to isolated pockets of afroalpine grasslands and heathlands above 3,000 m in elevation, where they prey on afroalpine rodents.
Distribution: Endemic to Ethiopia, where it is restricted to a few mountain ranges at altitudes of 3,000 m to 4,600 m. Two subspecies, remnant populations of *C. s. simensis*, occur north of the Rift Valley in Gondar (Simien), Wollo, and North Shoa (Menz). The more numerous *C. s. citernii* is found in the Bale Mountains National Park (largest population) and Arsi Mountains south of the Rift Valley.

DESCRIPTION

The Ethiopian wolf is a medium-sized canid with long legs and a long muzzle, roughly resembling a coyote (*Canis latrans*) (Sillero-Zubiri & Gottelli 1994). The skull is elongated, with a slender, protracted nose and small and widely spaced teeth, presumably an adaptation to grab small mammals. The dental formula is 3/3 incisors, 1/1 canines, 4/4 premolars, and 2/3 molars—total 42. Male Ethiopian wolves are 20% larger than females. Adult males average 16.2 kg (14.2–19.3 kg) and females 12.8 kg (11.2–14.15 kg). Adult wolves have short fur of a distinctive bright tawny-rufous color with a dense whitish underfur. The throat, chest, a band around the ventral part of the neck, the underparts, and inside of limbs are white. The ears are pointed and broad. The tail is a thick black brush with the proximal third white underneath.

NATURAL HISTORY

Canis simensis is one of several species of mammals restricted to the Ethiopian afroalpine grasslands and heathlands (Yalden & Largen 1992). Ethiopian wolves are highly specialized to prey on the small mammals of the afroalpine grassland community, such as the endemic giant molerat (*Tachyoryctes macrocephalus*), grass rats, and Starck's hare (*Lepus starcki*) (Sillero-Zubiri & Gottelli 1995a). In Bale, rodents accounted for nearly 96% of all prey found in feces. Wolves are most active during the day, synchronizing their activity with that of above-ground rodents (Sillero-Zubiri et al. 1995). Small wolf packs occasionally kill mountain nyala (*Tragelaphus buxtoni*) and reedbuck calves (*Redunca redunca*).

Ethiopian wolves live in discrete and closely knit social packs that communally share and defend an exclusive territory. In optimal habitat, packs consist of 3 to 13 adults (mean = 6), containing 3 to 8 related adult males, 1 to 3 adult females, 1 to 6 yearlings, and 1 to 6 pups (Sillero-Zubiri & Gottelli 1995b). Pack adult sex ratio is biased toward males (2.6:1). Annual pack home ranges are small in optimal habitat (average 6.4 km²) but larger in areas of lower prey biomass (average 13.4 km²). Home ranges are discrete, occupying all available habitat, and are stable in time, drifting only during major pack readjustment following disappearance of a pack or significant demographic changes (Sillero-Zubiri & Gottelli 1995b). Dispersal is tightly constrained by the scarcity of suitable, unoccupied habitat. Males do not disperse and are recruited into multi-male philopatric packs; two-thirds of the females disperse at 2 years of age and become "floaters," occupying narrow ranges between pack territories until a breeding vacancy becomes available (Sillero-Zubiri, Gottelli, & Macdonald 1996).

Wolf packs patrol and scent-mark their territory boundaries (via raised-leg urinations, defecations, and scratches) regularly. Aggressive interactions with neighboring packs are common, highly vocal, and always end with the smaller group fleeing from the larger (Sillero-Zubiri & Macdonald 1998).

The dominant female of each pack may give birth once a year between October and December, with only about 60% of dominant females breeding successfully each year (Sillero-Zubiri, Gottelli, & Macdonald 1996). All pack members guard the den, chase potential predators, and regurgitate or carry rodent prey to feed the pups. Subordinate females may assist the dominant female in suckling the pups. After weaning, and until 6 months of age, pups subsist almost entirely on solid foods supplied by helpers. Breeding females typically are replaced after death by a resident daughter, resulting in a high potential for inbreeding. Extra-pack copulations and resulting multiple paternity may be the mechanism by which inbreeding depression is circumvented. Up to 70% of matings (*N* = 30) involve males from outside the pack (Sillero-Zubiri, Gottelli, & Macdonald 1996).

CONFLICTING ISSUES

Because most rare canids occur at low density in fragmented populations, large areas and many populations are required to conserve them. In contrast, Ethiopian wolves have a narrow ecological niche, living at high density as a strict rodent predator in a few populations scattered in the Ethiopian highlands. This poses a conservation paradox: although there are very few wolves surviving, their protection is made easier by having to concentrate on only a handful of populations living in relatively small areas.

Gottelli and Sillero-Zubiri (1992) and Malcolm and Sillero-Zubiri (1997) have assessed population numbers and described the decline of Ethiopian wolves throughout their range. With no more than 500 adult wolves remaining in small, isolated populations, this is one of the rarest and most endangered of the large mammalian species. The main threats facing Ethiopian wolves are essentially those encountered by most wildlife today: habitat loss and human interference, both resulting from increasing high-altitude subsistence agriculture and grazing. At least two wolf populations in Gojjam and Shoa have been recently extirpated, and the ranges of other wolf populations have been reduced. Remnants of afroalpine ecosystems increasingly resemble islands hemmed in by degraded or lowland areas that act as ecological boundaries, and endemic species are increasingly at risk of local extinction as a result of this insularization (Kingdon 1990). The recent decline in the numbers of wolves seems to be owing to a combination of factors, including persecution by humans, road kills, disease, epizootics, hybridization with domestic dogs, and possible loss of genetic variability (Gottelli et al. 1994; Sillero-Zubiri & Gottelli 1994; Sillero-Zubiri & Macdonald 1997; Sillero-Zubiri et al. 1995; Sillero-Zubiri, King, & Macdonald 1996).

Wolves share their range with people and livestock. People's attitudes toward wolves vary from indifferent to negative, always in direct relation to the need for farming and grazing land. In some areas wolves are persecuted for taking small domestic stock. This is especially obvious in northern Ethiopia, where the human population is one of the largest in Africa. Persecution has been remarkable in Simien, where in the past even the national park staff perceived wolves as vermin.

Unlike the northern highlands, the Arsi and Bale Mountains have never been heavily populated. Within the Bale Mountains, Oromo pastoralists have co-existed peacefully with wolves and other wildlife (Sillero-Zubiri 1994). The pastoralists did not regard the wolf as a threat to their stock and even occasionally left their sheep unattended during the day. Similarly, wolves did not avoid human habitation and were oblivious of people walking or riding across their range.

Ethiopia has suffered the ravages of war between government forces and guerrilla movements for several decades, and only recently has the fighting stopped. War affects wildlife conservation directly through destruction of the environment and shooting of wildlife, and indirectly through policy changes that block conservation activities, deter outside funding, and divert funds from social to military activity. As a consequence of war, most afroalpine ranges in northern Ethiopia have been "out of bounds" for wildlife conservation and surveys. Simien Mountains National Park was closed for at least 6 years, park staff became targets of guerrilla activity, and local settlers encroached the park when it was not under government control.

Although they were outside the direct influence of war, wildlife conservation and park management in Bale deteriorated quickly following the end of the war in May 1991. After the overthrow of the government, peaceful co-existence between people and wildlife turned into persecution. Automatic weapons sold by runaway soldiers became widely available, and Ethiopian wolves and other wildlife became targets. The killing seemed to be part of the "retribution" carried out by local people against all elements of government, probably fueled by grudges against the park administration. At least six adult wolves were known to have been shot in late 1991. The slaughter subsided after lengthy discussions with local elders, but the threat is likely to persist until gun control is imposed.

With an ever-expanding human population, increased development, and habitat loss, Ethiopian wolves and other afroalpine wildlife are increasingly exposed to death on the roads. The wolves in Bale are particularly vulnerable to traffic kills because they regularly cross busy roads; at least six were hit by cars in Sanetti between 1984 and 1991. Two other animals were shot from the road, and two were left permanently lame from collisions with vehicles. A local superstitious belief associates wild canids crossing someone's path with bad luck and leads some drivers to shoot or run over wolves and other wild canids. Increasing use of roads owing to economic growth and increased transport may exacerbate this trend.

Even when people tolerate wolves, human presence may severely affect wolf survival through the presence of domestic dogs (Gottelli & Sillero-Zubiri 1992; Gottelli et al. 1994). Wolves are affected by domestic dogs in three ways: direct competition and aggression, dogs acting as vectors of disease, and genetic introgression (i.e., the incorporation of genetic material from another species). Dogs travel regularly with their owners into the Bale Mountains and may provide the vehicle for pathogens such as rabies or distemper to reach their wild relatives (Laurenson et al. 1997, 1998). Rabies poses a serious risk to the Ethiopian wolf, killing whole wolf packs in 1990 and 1991 and accounting for most of the recorded population decline (Sillero-Zubiri, King, & Macdonald 1996). Canine distemper may also be responsible for wolf deaths (Laurenson et al. 1998).

The Ethiopian wolf can hybridize with domestic dogs. Following hybridization, a population may be affected by outbreeding depression, or reduction in fitness. A genetic study indicated that hybridization of wolves and dogs is occurring, and some animals had an unusual appearance: shorter muzzles, heavier bodies, and different coat patterns (Gottelli et al. 1994).

FUTURE AND PROGNOSIS

Many of the problems faced by the remaining Ethiopian wolf populations seem insurmountable.

The Ethiopian wolf's restricted distribution and high density permit the implementation of concrete conservation measures. In 1983 the Wildlife Conservation Society established the Bale Mountains Research Project, which publicized the wolf's plight and started a regular monitoring program for the species. A detailed 4-year field study followed (Sillero-Zubiri 1994). Based on its findings, the IUCN Canid Specialist Group produced an action plan for the Ethiopian wolf (Sillero-Zubiri & Macdonald 1997), providing a detailed strategy for the conservation and management of remaining wolf populations. In view of the persisting human impact on the overall distribution of species and its vulnerability to extinction, this plan advocated immediate action on three fronts to conserve the afroalpine ecosystem and its top predator. Protective measures require consolidating the management of protected areas and undertaking active efforts to monitor and protect remaining populations, backed by the establishment of a population management program.

The Ethiopian Wolf Conservation Programme (EWCP), started in 1995, is seeking help to protect the afroalpine ecosystem and many of its rare highland endemic plants and animals through better management in Bale and the Simien Mountains and the establishment of other conservation areas in Menz and Wollo. The EWCP is currently monitoring Bale's and Menz's wolf populations, and it is supporting park patrols within wolf range, domestic dog control, and removal of dog-wolf hybrids. Additionally, the EWCP carries out a community conservation education campaign that targets people living inside wolf range and is aimed at improving dog husbandry and combating disease in the park and surrounding area. A large-scale dog vaccination program (targeting up to 10,000 dogs) seeks to reduce the occurrence of rabies and distemper within wolf range and is backed up by further epidemiological and demographic studies. Outside Bale, Zelealem Tefera of the Guassa Biodiversity Project is looking at the relationships between pastoralists and wildlife in the highlands of Menz, probably the second largest wolf population after Bale. The EWCP also visits wolf ranges elsewhere in Ethiopia and coordinates efforts to educate people in those areas about the plight of the Ethiopian wolf.

In order to ensure the long-term survival of the Ethiopian wolf, the action plan advocates a mixed strategy. Although captive breeding per se will not suffice to conserve the wolves, it will serve as another safeguard to avoid extinction. A small captive breeding nucleus will contribute to the conservation of genetic variability and purity. This operation may take place in a breeding facility in Ethiopia, for which funds are currently being sought. This facility will also serve as a conservation education center and provide wolves for eventual re-introduction into areas where the species has become extinct. Each wolf population, including the captive one, must be considered part of a global metapopulation, with some genetic flow occurring among them. Thus a limited number of captive-bred or wild-bred wolves may be exchanged between populations, reintroduced to areas where the wolves have been extirpated, or used to restock depleted populations. The onset of such program is dependent on funding and government approval.

If Ethiopian authorities step up park management using support from the international community, the Bale Mountains will remain the best refuge for the survival of these unique canids. Ethiopia's current progress in securing long-lasting peace and stability may help secure more international support for afroalpine conservation. It is hoped that highlighting the plight of the Ethiopian wolf (an Ethiopian flagship species) will trigger renewed efforts to conserve the afroalpine ecosystem, and thus conserve many other lesser known, endemic fauna and flora.

European Mink

Tiit Maran

Common Name: European mink
Scientific Name: *Mustela lutreola*
Order: Carnivora
Family: Mustelidae
Status: Endangered on the IUCN (1996) Red List; listed as a priority species for Europe under the IUCN Action Plan for the Conservation of Mustelids and Viverrids (Schreiber et al. 1989); under the European Union Directive on the Conservation of Natural and Semi-Natural Habitats and of Wild Flora and Fauna, listed in Annex II as a species whose conservation requires the designation of special areas and in Annex IV as a species of community interest in need of strict protection; under the Convention on the Conservation of European Wildlife and Natural Habitats (Bern), listed in Annex II as a strictly protected fauna species.
Threats: Historically, habitat loss and extensive trapping. Currently, the introduced American mink (*Mustela vison*) that out-competes the European mink (Maran et al. 1998).
Habitat: European mink inhabit small, fast-flowing rivers and creeks with lush riparian vegetation, usually in forested areas. During snow-free seasons it sometimes also dwells along the coasts of lakes and in marshes (Sidorovich et al. 1995).
Distribution: In the beginning of the 19th century, the European mink's range encompassed almost all of the European continent: from the Ural Mountains in the east to eastern Spain in the west; from central Finland in the north to the Black Sea in the south. There is no historical evidence for European mink in Sweden, Norway, Denmark, Belgium, the former Yugoslavia, Portugal, Italy, or the United Kingdom. The current range appears restricted to remnant populations in central Russia (Tver and Vologda regions), Belarus, western France, north-central Spain, and possibly Romania (Maran & Henttonen 1995; Maran et al. 1998).

DESCRIPTION

The European mink is a medium-sized, semi-aquatic member of the Mustelidae family with a long body, short legs, and a relatively short tail (Youngman 1982, 1990). The skull is narrow and dorso-ventrally flattened. The dental formula is 3/3 incisors, 1/1 canines, 3/3 premolars, and 1/2 molars (Youngman 1990). European mink have almost the same body-plan as the European polecat (*M. putorius*), though they are smaller in body size (average = males 371 cm, females 329 cm; without tail) and weight (average = males 815 g, females 540 g) (Danilov & Tumanov 1976). The body is usually chocolate brown to black, often with white markings on the neck and/or chest. The most distinctive marking is a wide, white area around the mouth. This trait is the most rapid and reliable means of distinguishing European mink from American mink; the latter have a white area only on the lower lip.

NATURAL HISTORY

Composition of the diet depends on availability of prey species and may vary among sites. However, the main prey categories remain the same: amphibians, fish, small mammals, birds, and insects (Danilov & Tumanov 1976; Sidorovich et al. 1998).

European mink breed in late March, April, and early May with a peak during the last 10 days of April. Females are induced ovulators and do not delay implantation. Average litter size at birth in captivity is 3.5 (Maran & Robinson 1996). Gestation lasts 6 weeks, and young are completely helpless during the first month. At the beginning of the second month, kits' ears and eyes open and they first emerge from the nest. Young European mink disperse and begin to live as solitary, semi-aquatic carnivores by 3 to 4 months of age. Reproductive life span is 3 to 4 years. The longest life span recorded in captivity is over 11 years.

The European mink is specialized to use riparian and river habitats, with a few reported cases in other water bodies such as lakes or marshes (Sidorovich et al. 1995). Animals rarely move more than 100–150 m from the banks of streams (Danilov & Tumanov 1976). The latter indicates relatively poor dispersal capability of the species (Danilov & Tumanov 1976; Maran & Henttonen 1995). Home range size depends on availability of food and suitability of habitat. In optimum habitats mink range 1–2.5 km along water courses, whereas in less suitable conditions the home ranges increase considerably (5–20 km) (Sidorovich & Kozhulin 1994). Optimum habitats for mink in the eastern part of their range are rivers 10–100 km long in forested landscapes (Sidorovich et al. 1995).

CONFLICTING ISSUES

Several hypotheses have been proposed to explain the decline of European mink (Maran & Henttonen 1995; Maran et al. 1998). As recently as the beginning of the 19th century, mink were widespread in Europe. Since then its range began decreasing at an accelerating rate. By the end of the 19th century, and especially during the 20th century, Europe has experienced wide-scale alteration of landscape and habitats. These changes had a major impact on most watersheds and purportedly caused local extinction or severe decline of European mink populations in several regions (Maran & Henttonen 1995).

At least in the eastern part of the range, trapping has been an important cause of decline of European mink populations (Maran & Henttonen 1995). When American mink fur became fashionable during the first half of the 20th century, trapping pressure increased markedly (Maran 1991).

American mink were introduced into Europe in the early 1920s and appear to be the cause of the most drastic impact on European mink, whose populations were already seriously depleted in several regions (Maran et al. 1998). American mink have more or less the same ecological requirements as European mink do, but they seem more flexible in their use of resources. Also, the larger body-size and higher reproductive potential of American mink favor them in competitive interactions. Aggression between mink species, where the American mink usually "wins," provides the most satisfactory explanation for the rapid disappearance of European mink from its range (Maran et al. 1998).

The European mink disappeared from most of western and central Europe before World War II or slightly thereafter, so it is not surprising that conservation actions for the species first took place in the eastern part of its range—within the political borders of the former Soviet Union. The extensive decline of mink was noticed and actively discussed in Russian zoological and conservation literature in the early 1970s. Foreseeing the fatal role of the spread of American mink (although not understanding the underlying mechanism of European mink extinction), Dr. Dimitri Ternovskij from the Institute of Biology in Novosibirsk established a breeding facility at the Institute in the early 1970s, and from 1972 to 1992 he bred 1,170 European mink from 19 founders (Maran 1996a). This breeding program had two aims: to use intra-generic hybridization (i.e., interbreeding species within a single genus) for the creation of new and valuable commercial fur animals, and to maintain a captive stock of European mink as a guarantee for the survival of the species.

From 1981 to 1989, Ternovskij released European mink onto two Kuril Islands: Iturup and Kunashir (Maran 1994). In 1988 another introduction of European mink was conducted in Tadjikistan. These introductions were controversial actions because (1) the selected sites were outside the historical range of the species, and (2) research into suitability of the sites and possible impacts to local biodiversity was never conducted. A follow-up survey indicated that at least on Kunashir Island the introduction failed owing to habitat unsuitability (Shvarts & Vaisfeld 1993). Although both the start of the breeding facility and introductions were promising, the long-term results were modest. The poor results were caused by unfavorable political and management conditions and inadequate professional performance.

Because of the similarity of American and European mink, displacement by American mink occurred in most places without notice by authorities. Further, both species were trapped for pelts and recorded under the same name—"wild mink"—by trapping organizations. Therefore, inclusion of the European mink as a protected species would have caused considerable complications to fur trappers. Thus commercial fur-trading organizations inhibited conservation activities. As a result the European mink is still not listed in the Russian Red Data Book, the legal list of protected species in Russia.

Political rivalry among research institutions (Moscow, Leningrad, and Novosibirsk) in Russia and among key researchers to obtain and maintain the lead role in research and conservation activities for European mink made countrywide cooperation impossible. All major efforts become concentrated in Novosibirsk, even though other institutions in Moscow and Leningrad made minor attempts to captively breed and release animals (Maran 1994).

It might be argued that conservation depended on strong leadership rather than effective organizations, and that rejecting attempts at wider collaboration was the only way to achieve results in the former USSR. If not, the actual activities undertaken in Novosibirsk might have been inhibited, with the leading role shifted to Moscow for political gain, but without much conservation. In addition, the importance of proper management to conservation success was not acknowledged until recently. In this light, the conservation initiative undertaken in Novosibirsk reflects well upon it, especially for an initial phase. The release of minks was basically a spontaneous action without much planning, proper study of release sites, assessment of the impact of the introduction, or follow-up monitoring. The islands selected as release sites were apparently unsuitable, and the introduction failed (Shvarts & Vaisfeld 1993).

Conservation efforts in Estonia since the early 1980s primarily followed the same approach: conservation breeding and establishment of a reserve on an island. In 1984 a conservation breeding program was initiated for the species in the Tallinn Zoo. The objective was to build a captive population, with the final aim of releasing mink into a reserve created for the species on Hiiumaa Island, Estonia. In 1986 the first success was achieved in breeding. Thereafter, more or less regular breeding took place until 1991, when several unfavorable factors brought the breeding initiative close to disaster (Maran 1996b). Also, a survey revealed the existence of a relatively abundant American mink population on Hiiumaa Island, leading to abandonment of the initial goal of establishing a reserve there.

Estonia's restored independence in 1991 allowed European mink conservationists to become acquainted with the new ideas of conservation biology. The need for international cooperation to preserve the European mink became increasingly evident. In 1992 the European Mink Conservation and Breeding Committee (EMCC) was founded as an international body to coordinate and promote conservation efforts for the species. Within 5 years the EMCC was able to re-establish the conservation breeding program and link it to the EEP (European Endangered Species Program) process, get new founders from the wild, monitor the rapidly worsening situation facing European mink in Russia and neighboring regions, and promote public awareness of European mink conservation efforts (Maran 1996b).

The main aims of the EMCC today are to increase the captive stock of mink to 400 individuals in Europe (200 in the Tallinn Zoo) and to reinitiate attempts to establish a reserve on Hiiumaa Island. The latter requires eradication of American mink from the island. The main problems are inadequate and unpredictable funding, organizational weaknesses, and lack of expertise on some issues (e.g., eradication techniques).

Since 1992, the conservation project run by EMCC has expanded to take on a European context. As such, several conflicting issues have arisen. As the European mink is a European endemic, it should be the responsibility of the entire continent. The European mink's historical range encompassed more than 16 current European countries. This means that responsibility for recovery of the species is divided by numerous political borders. In countries such as Germany the European mink disappeared long before the development of a conservation movement, and today the species is considered extinct and thus not a national priority. In countries such as Belgium, the Netherlands, and Italy the European mink was never included on lists of national fauna, and therefore legislation and conservation authorities do not regard it as important. Even worse, owing to the replacement of the European mink by its American relatives, the existence of the species has almost vanished from public memory, and current negative attitudes toward American minks as pests or overabundant fur animals are easily turned to the European mink.

These factors lead to conflicts between local and European conservation interests, and they complicate fundraising efforts for European mink conservation. The European Union has several funding agencies dedicated to addressing European-scale conservation issues. However, most of these were created to act only within the European Union, making it complicated to secure funds for projects elsewhere—such as the European mink project in Estonia.

Until recently, European mink conservation has been addressed as a one-man act in Estonia. The EMCC, though important, has acted mostly at the level of support and fundraising. The scope of actions required for successful recovery demand far more labor and expertise than is within the capacity of one person. The inability to develop a flexible and effective organizational structure for recovery is caused mainly by two factors: inadequate funding and insufficient training.

FUTURE AND PROGNOSIS

The results obtained by the EMCC, the Tallinn Zoo, and the few other zoos in Europe that captively breed European mink have been promising and form a hopeful basis for reaching the target population of 200 individuals in Tallinn and 400 in European zoos by the year 2000. Also, within a few years the American mink may well be eradicated from Hiiumaa Island, given proper management. This would permit the reintroduction of European mink there. This project, coupled with efforts to protect European mink habitat from invasion by American mink in Spain, should guarantee the survival of this seriously endangered yet characteristic species of the European continent.

ACKNOWLEDGMENT

I gratefully acknowledge financial support from the British government's Darwin Initiative.

Florida Manatee

Galen B. Rathbun and Richard L. Wallace

Common Name: Florida manatee
Scientific Name: *Trichechus manatus latirostris*
Order: Sirenia
Family: Trichechidae
Status: Vulnerable on the 1996 IUCN Red List; listed in Appendix I in CITES; Endangered under the U.S. Endangered Species Act (ESA) of 1973; Protected under the U.S. Marine Mammal Protection Act (MMPA) of 1972, the Florida Endangered and Threatened Species Act of 1977, and the Florida Manatee Sanctuary Act of 1978.
Threats: Collisions with watercraft contribute to a high rate of mortality. Creation of artificial warm-water refuges may indirectly cause mortality owing to disruption of traditional behavior patterns and hypothermia.
Habitat: Near-shore coastal areas; estuaries; inland waterways.
Distribution: Florida and southern Georgia. During summer some individuals disperse northward along the Atlantic coast through the Carolinas and along the Gulf of Mexico coast as far west as Texas.

DESCRIPTION

The Florida manatee is a subspecies of the West Indian manatee, which occurs in coastal areas of northern South America, Central America, the Caribbean, and the southeastern United States (Lefebvre et al. 1989). Manatees are streamlined and adapted to an aquatic existence, and they are unable to completely leave water. Their front limbs are modified into flippers, they have entirely lost their rear limbs, and their tail is modified into a flat, rounded paddle. Their gray skin is tough, thick, and essentially furless. External ears are absent, and their eyes are small. Adults reach a total length of 3–4 m and weigh about 550 kg (Reynolds & Odell 1991).

NATURAL HISTORY

Eating aquatic plants requires unique behaviors not found in most marine mammals, which must be nimble or fast swimmers in order to catch their prey. Manatees must remain stationary in the water to graze on rooted and floating plants, and thus they are not accomplished divers or swimmers (Hartman 1979). Aquatic vegetation is relatively low in calories and nutrients, so manatees spend 6 to 8 hours a day grazing. To conserve energy, they maintain a relatively low basal metabolic rate. They also are relatively poor producers and conservators of body heat. Physiology restricts manatees to the warm waters of the tropics and subtropics; in the southeastern United States they are at the northern edge of their distribution (Irvine 1983).

In response to seasonal changes in water temperatures, many manatees in the United States undertake a north/south migration (Rathbun et al. 1990; Reid et al. 1991). In addition, many aggregate at sources of warm water to escape the cold. Some of these winter refuges are created by natural springs, but most have been created by the effluents of power plants and pulp mills (O'Shea 1988). However, when the water temperature at these refuges drops below about 20°C for several days, manatees become hypothermic, or cold stressed, and may die (Ackerman et al. 1995). During the summer manatees are widely dispersed, making them difficult to find and observe (Rathbun et al. 1990).

Florida manatees mate mostly during the summer months and gestate their single calves (rarely twins) for about 12 months. The cow/calf bond is strong, usually lasting 1 to 2 years, and males take no part in raising the calf. Sexual maturity is usually reached in 3 to 5 years (Rathbun et al. 1995).

Historically, manatees were largely restricted to the southern third of the Florida peninsula during the winter, but there are no reliable estimates of their abundance (O'Shea 1988). Winter aggregations in Florida, however, are often in clear water, which makes counting them easier. During a 1996 winter aerial count of the entire southeastern U.S. population, 2,639 animals were spotted (Florida Department of Environmental Protection unpublished data). However, it is unclear whether manatee numbers are currently increasing or decreasing (O'Shea & Ackerman 1995).

CONFLICTING ISSUES

Much of the early concern surrounding the manatee, including its federal listing as Endangered, was based on the unsubstantiated belief that it had almost been extirpated (O'Shea 1988). With the passage of the MMPA and ESA, the U.S. Fish and Wildlife Service (FWS) became the lead federal agency in developing research and conservation measures for the West Indian manatee. In 1974 the FWS initiated a research program in Florida that became the Sirenia Project (now part of the Biological Resources Division of the U.S. Geological Survey). The Marine Mammal Commission (MMC) is a federal agency that evaluates and provides advice on federal actions under the MMPA. The first manatee recovery team produced a recovery plan in 1980, and reorganized teams have produced two revised plans, the latest in 1996 (U.S. Fish and Wildlife Service 1996). In 1984 the state bolstered its research and conservation efforts, which are now part of the Florida Department of Environmental Protection (FDEP).

Nongovernmental organizations are also involved in manatee recovery, including the Save the Manatee Club, a grass roots conservation organization. Boating interests are represented by the Marine Industries Association of Florida. The utilities industry, especially the Florida Power & Light Company, is also an active player.

Early assumptions of excessive manatee mortality in Florida resulted in an effort to document the causes of all deaths. Not only was an increasing trend found in mortality over the last 20 years, but nearly one-third of all deaths were related to human activities—mostly collisions with boats. For example, from 1986 to 1992 1,080 manatees were recovered dead in Florida, with 28.1% owing to accidents with watercraft (Ackerman et al. 1995).

Unlike many conflicts between humans and wildlife, manatees are killed unintentionally. They do not compete with people for food or directly threaten danger—they just get in the way of boats, of which there are more than 750,000 registered in Florida (plus another 250,000 that visit the state annually). The unintentional slaughter of this benign creature has created sympathy for its plight, as demonstrated by the phenomenal growth in membership of Save the Manatee Club over the past 15 years (Buffett 1996) and the proliferation of protective measures (Reynolds & Odell 1991).

As regulations have been imposed in an attempt to reduce human-related injuries and deaths, controversy has escalated. For example, is the increase in manatee mortality related to an expanding manatee population or an expanding boat and human population? The lack of reliable estimates of historical and current manatee numbers in Florida makes it difficult to answer this question (O'Shea 1995; Wright et al. 1995). The research community in Florida has also been slow to initiate studies to demonstrate the effectiveness of the numerous regulations instituted to reduce boat-related injuries and deaths. Without data, the controversy is bound to continue. In the meantime, boat-related manatee injuries and deaths continue.

The historical distribution of manatees in Florida and their north/south migration began to change when manatees learned to take advantage of artificial refuges in the 1950s and 1960s (Reid et al. 1991). However, these outfalls often are unreliable. During exceptionally cold weather some discharges do not produce enough warm water to meet manatees' needs, and sometimes sources are shut down for economic reasons or repair, resulting in manatees' deaths (Ackerman et al. 1995). Packard et al. (1989) have shown that manatees, faced with an unreliable source of warm water, may not be able to change their behaviors quickly enough to survive. The fact that manatees are adaptable is obvious—after all, they learned to use the artificial effluents. Sufficient time, however, may be needed to develop and learn new behaviors and movement patterns.

Technology can be a fickle friend. There is a growing movement to deregulate the electric power industries in many states, including Florida. In the future, economics may determine which power plants are operated and when. Although this may well serve ratepayers, it also may cause lethal problems for manatees when effluents are suddenly shut down or become unpredictable.

Uncertainties over historical and current manatee distribution and population numbers, the role of artificial refuges, and suggested solutions to deregulation are starting to be expressed (e.g., Rose 1997). The FWS, FDEP, and MMC are becoming embroiled in controversy (e.g., Frohlich 1998). The need for discussions, coordination, and leadership among all the interested and affected parties is obvious, but it has been slow in happening.

FUTURE AND PROGNOSIS

Successful manatee conservation requires teamwork. In 1997 a joint effort among the lead federal and state agencies resulted in an interagency coordinating committee to facilitate and improve coordination in the recovery program. Members include the FWS, FDEP, the Sirenia Project, and MMC.

Teamwork is difficult, particularly when team members are expected to give up some independence in the process of carrying out recovery tasks (Westrum 1994). One goal of the committee is to manage the lead agencies' agendas in a cooperative fashion by fine-tuning the recovery plan's research and management priority scheme and collaborating on priority tasks. However, the committee has not been effective in accomplishing this goal owing to organizational problems. For example, committee members are hesitant to relinquish individual agency control of data and funding decisions. There is some disagreement over which research and management tasks are highest priority, as individuals tend to feel most strongly about the tasks for which they are responsible. Interest in the committee process varies among members, and some agencies or individuals appear more devoted to succeeding than others. Because the manatee recovery program is based on shared responsibility, as opposed to being directed hierarchically by FWS, FWS is hesitant to direct the work of the other agencies. Perhaps most importantly, there is not a clear, shared goal among the committee members as to how to proceed.

Having a common goal or vision is critical to team success (Westrum 1994), as long as the vision is one that promotes the goals of manatee recovery. Although the committee members share the recovery plan as a common guide, their respective goals might differ based on their individual needs and interests. A common goal developed under these circumstances might resemble a "lowest common denominator" that meets everyone's individual needs but fails to adequately promote the overall goal of manatee recovery. Choosing a leader to direct the actions of the committee also must be a responsibility shared by all the committee members, who must all be comfortable with the expertise and abilities of the chosen individual. That person must be someone who can foster communication and cooperation among the represented agencies (Clark & Cragun 1994).

The committee is in a unique position to influence the future of manatee recovery. In terms of regulatory, research, and management authority, the agencies represented on the committee hold all the cards. Within statutory limits, they can make any decision to further any agenda they decide upon. That they all get along well enough to create this opportunity to coordinate is a rarity among domestic endangered species programs. The opportunity to make it work shouldn't be missed.

Several steps should be taken. A leader should be jointly selected to unite the members in prioritizing and coordinating recovery tasks. Their efforts must complement the recovery plan but not be limited by the plan's implementation schedule. Thereafter the committee should determine each agency's available funds, draft and adopt an agreement to share the costs of implementing priority tasks, assign tasks to the appropriate agencies or organizations, and develop a strict timetable for implementation. The committee should then meet regularly to review progress. The results of the committee's work should become the basis for revising the recovery plan. If the committee members cannot agree on goals and strategy, they should hire and share the costs of a facilitator to expedite the process.

The current recovery plan is a satisfactory base document, but it falls short of providing guidance to achieve interagency coordination, continuity, and stability in research and management actions. Given the difficulties to date in achieving consensus among the principal agencies in manatee recovery, a fresh start based on a new and substantially more aggressive approach to manatee conservation is needed.

Giant Panda

Richard P. Reading

Common Name: giant panda
Scientific Name: *Ailuropoda melanoleuca*
Order: Carnivora (carnivores)
Family: Ursidae (bears)
Status: Class I protected species under the Wildlife Protection Law of China; Endangered under the IUCN (1996); included on Appendix I of CITES.
Threats: Habitat destruction and fragmentation due to commercial logging and encroachment by cultivation, exacerbated by poaching
Habitat: Temperate forests supporting dense stands of bamboo, particularly arrow bamboo (*Bashania* spp.) and umbrella bamboo (*Fargesia* spp.).
Distribution: Wild pandas survive in six isolated regions of Sichuan, Gansu, and Shaanxi Provinces in west-central China, which are further fragmented into approximately 20 to 25 populations covering roughly 10,000 to 14,000 km^2.

DESCRIPTION

Giant pandas, with their distinctive black and white coloration and often human-like behaviors, are one of the world's most charismatic species. Although there is some debate concerning the taxonomy of giant pandas, most authorities consider it to be a member of the bear family (Ursidae). Other biologists classify giant pandas as members of the raccoon family (Procyonidae) or place giant pandas and red pandas (*Ailurus fulgens*) in an entirely separate family (Ailuridae) (see Nowak 1991).

Pandas grow to 1.2–1.7 m long with about 125 mm tails, and they can weigh 60 to over 200 kg. Males are about 10% larger than females. They have the general shape of a bear, with a relatively massive head and thick, wooly fur that is white, except for the legs, ears, patches around the eyes, and a band around the shoulders, which are black. Giant pandas possess a unique sixth digit on their forepaws that aids in grasping objects, particularly bamboo.

NATURAL HISTORY

The latest survey in the 1980s estimated that only about 1,000 giant pandas survive in the wild. Currently a joint national survey is being conducted by the World Wide Fund for Nature (WWF) and the State Forestry

Administration (SFA, formerly the Ministry of Forestry) to determine the status of panda population and habitat. These animals are distributed in six mountainous areas that are isolated from each other, and the bears are further fragmented into over 20 separate populations (MacKinnon et al. 1989). Several of the populations contain fewer than 30 individuals. The area occupied by pandas has decreased dramatically, and today they are restricted to relatively high altitudes (800–3,500 m, but usually over 1,500 m). The original distribution of giant pandas included lower elevations as well (Schaller 1993; Lu et al. in press).

Pandas feed primarily on bamboo, a group of species in the grass family, and they depend on large, contiguous stands of bamboo for survival (Schaller et al. 1985; Pan et al. 1988). The species of bamboo consumed varies by region but usually includes at least two in any one area, with the relative proportions of each consumed varying by season (Schaller et al. 1985, 1989; Pan et al. 1988; Yang et al. 1998). Bamboo is a low-quality food that fluctuates little in nutrient content throughout the year and is high in fiber (Pan et al. 1988; Schaller et al. 1989). Pandas forage for over 14 hours a day and consume a large amount of food (10–18 kg per day) (Schaller et al. 1989); however, historically this was not a problem, as bamboo originally existed as a plentiful, widespread food source and pandas had little trouble finding large patches. Smaller amounts of other food plants and some meat are also consumed (Schaller et al. 1989). In addition to bamboo, good-quality panda habitat is characterized by a canopy cover of at least 66%, slopes of less than 20%, consistent water sources (the bears must drink every day), and tree hollows or caves for den sites (Schaller et al. 1985; Schaller 1993).

The home range sizes of giant pandas vary greatly depending on the habitat (Lu 1991; Pan 1998; Schaller 1998). Pandas lead an essentially solitary life, except for females with young and during courtship and mating (Schaller 1998). The mating system is polygamous; that is, one female breeds with several males and vice versa. The mating season extends from mid-March to mid-May (Lu 1991; Pan 1998). Birthing of one to two tiny young (cubs weigh only 90–130 g) occurs in August or September (Schaller 1998). Young animals remain with their mothers for 1.5 to 2.5 years, although females may leave their cubs in the den while foraging (Lu 1991; Pan 1998).

CONFLICTING ISSUES

Giant panda numbers and range have probably been declining for centuries, although the rate of decline has risen in recent decades. Today the species is critically endangered, and numbers continue to decline despite significant effort and resources expended on their behalf (Schaller et al. 1985; Schaller 1993). The decline of giant pandas has followed the destruction and fragmentation of forests with large stands of bamboo as a result of logging and expansion of agricultural lands (Schaller 1993, 1998). Poaching is also a problem. Although these processes have been occurring for centuries, the rate of habitat loss and degree of fragmentation began reaching a critical stage in recent decades. Many remaining panda populations are too small and isolated to persist without substantial human intervention (Hu 1998; Schaller 1998). Fragmentation may be particularly problematic in small habitat patches with only one or two bamboo species because bamboo undergoes species-specific, mass flowering every 40 to 100 years followed by a mass die-off (Yang et al. 1998; Lu et al. in press). Stands require 15 to 20 years to regenerate to the size required by pandas, thus creating a local food shortage (Johnson et al. 1988).

Panda conservation involves both field conservation and captive breeding. Field efforts include the creation, expansion, and improved management of panda reserves, anti-poaching activities, public relations campaigns, and research into panda ecology (Mainka 1998). There are currently 33 panda reserves in China covering over 9,000 km², or almost 50% of the remaining panda habitat (Lu et al. in press). In addition, Chinese conservationists are striving to link existing reserves by protecting important habitat corridors and have initiated education programs to develop greater local support (Lu 1999; Li & Zhou 1998). Strict laws, including the death penalty for killing or trading pandas, have been imposed, and enforcement efforts have been increased to stem poaching (Schaller 1993; Li & Zhou 1998). Studies of panda biology and ecology have provided a base of knowledge of the species and its needs (Schaller et al. 1985, 1989; Johnson et al. 1988, 1996; Pan et al. 1988; Pan 1998; Lu 1991). Although additional ecological knowledge is necessary (Mainka in press), many conservationists suggest that there is already enough information to know that successful panda conservation requires large blocks of habitat with bamboo, protection from poaching, and cessation of capture for captive breeding and display in zoos (Johnson et al. 1996).

A Panda Rescue Program was developed by the Chinese SFA in response to massive bamboo die-offs in panda reserves during the 1980s. This program mobilized local people to assist in efforts to locate starving or abandoned animals, which were brought into captivity and added to captive breeding programs (Li & Zhou 1998). The program was largely discontinued in 1994 (Lu et al. in press).

Captive breeding programs have been undertaken by both the SFA through the breeding center in Wolong and the Chinese Ministry of Construction (MOC) through the Chinese Association of Zoological Gardens (CAZG). The captive population is considered important (1) to conserve the species in the event that pandas go extinct in the wild, and (2) as source of animals for re-introduction programs to supplement and expand wild populations (Mainka 1998). The National Conservation Plan for the Giant

Panda and Its Habitat includes re-introduction as a potential conservation strategy. However, Johnson et al. (1996) and others (see Mainka 1998) argue against re-introduction because the captive panda population is not currently self-sustaining (Lu et al. in press) and because re-introductions are usually complex, risky, and expensive ventures, especially for carnivores.

All these conservation activities occur in a socioeconomic setting that Johnson et al. (1996) suggest dominates panda conservation. They list political and policy considerations, human population growth, natural resources exploitation, economic sustainability, human attitudes toward pandas, and human-panda interactions as among the most important variables of the social context within which panda conservation efforts occur. With almost 1.3 billion people, China is the most populous nation on earth. And while China is aggressively working to address this problem, the human population continues to expand. The growing human population creates demands on the nation's natural resources, which is quickly approaching its carrying capacity. Thus China is faced with the problem of conserving its environment while simultaneously maintaining decent living standards for its people. To accommodate the latter, China has been moving from a centrally planned to a free market economy. So far the results from an economic perspective have been encouraging, and China has been experiencing a prolonged and rapid economic expansion. However, this has come at the cost of additional natural resources exploitation, habitat fragmentation, and an emphasis on values and attitudes that promote economic development over conservation. Finally, Johnson et al. (1996) suggest that administrative systems are ineffective due to resource constraints, the newness of several recent laws, and "frequent changes in government policies, such as the Great Leap Forward and the Cultural Revolution, [that] have left the general public with negative impressions" (p. 341). However, the management situation may be improving. Recent attention to capacity building, law enforcement, local support and development, and reserve management appears to have encouraging results (Lu 1999; Lu et al. in press).

Despite several problems, efforts to conserve giant pandas receive strong support and the species remains hugely popular throughout China and the world. Pandas elicit strong attitudes and values. They represent important symbols of conservation, wildlife, and the natural world. However, at a local level they are still poached or viewed antagonistically by some people who have been displaced by panda conservation activities (Schaller 1993). In addition, panda conservation—like that of many large, charismatic, and endangered animals—is complicated by the large number of agencies, nongovernmental organizations, and private interests active in recovery efforts in China and throughout the world (Schaller 1993). Important stakeholders include various institutions within the SFA, the MOC, the CAZG, and the American Zoo and Aquarium Association (AZA), the WWF, the Wildlife Conservation Society (WCS), the IUCN's Conservation Breeding Specialist

Group (CBSG), and local people living in or adjacent to panda habitat. These key actors operate under a wide variety of Chinese laws and regulations as well as international treaties, agreements, and regulations pertinent to panda recovery efforts (Johnson et al. 1996).

Several conflicts have challenged panda conservation efforts. Some of the most important conflicts relate to (1) human-panda interactions (e.g., poaching and habitat destruction), (2) captive breeding efforts and the associated Panda Rescue Program, and (3) panda exhibitions in foreign zoos, especially U.S. zoos. These conflicts are exacerbated by the extraordinarily high public appeal of pandas and, thus, the high visibility of conservation efforts.

Perhaps the most important challenge facing long-term conservation of giant pandas is to reduce conflicts with local people. Panda numbers and distributions are influenced by village densities, timber removal quotas, road impacts, and size of habitat fragments (Johnson et al. 1996). People have also poached pandas directly or indirectly killed them in traps that target other animals (Schaller 1993). As the human population has increased, panda numbers have dropped and their distribution has become more fragmented. Addressing this challenge has proven problematic, often necessitating the moving of entire villages. In addition, managers have strictly enforced panda conservation laws with harsh sentences (even the death penalty) for violators. These actions often result in local resentment of pandas and panda conservation efforts. As is often the case in endangered species conservation efforts, long-term success may only be possible through integrated programs that simultaneously target the socioeconomic problems facing local residents and the biological problems facing pandas (Johnson et al. 1996).

Captive breeding efforts have also faced conflicts. Because the work is divided between two agencies (SFA and MOC) that traditionally have not gotten along very well, cooperation has been minimal (e.g., see Wong 1991). Instead, the agencies often compete for resources, are reluctant to share information, and rarely exchange animals. In addition, Johnson et al. (1996) argue that the emphasis on captive breeding, genetics, medicine, and other high technology approaches (e.g., artificial insemination and cloning) to panda conservation, although well intended, consume money, personnel, expertise, and other resources that would be better spent on protecting habitat and conserving wild populations (see also Lu et al. in press). This is especially true because models suggest that the captive population of pandas is not yet self-sustaining.

Captive propagation of giant pandas has improved markedly in the past several years. Nevertheless, increases in the captive population have been due primarily to the capture of wild pandas rather than to captive breeding success (Lu et al. in press). Most wild pandas added to captivity came from the Panda Rescue Program. This program became increasingly controversial after researchers discovered that female pandas often leave their cubs unattended in dens for days while foraging, and unattended cubs were easily mistaken for abandoned animals (Lu et al. in press). In response, the SFA largely stopped the program in 1994 and continues to evaluate the problem but argues that the program might remain an important component of panda conservation (Li & Zhou 1998).

Giant pandas are enormously popular, and many zoos throughout the world are anxious to obtain animals for display. As such, Western zoos are often willing to pay large sums of money for the right to exhibit pandas. Conflict has erupted over concerns that (1) money paid for exhibition rights is not going to panda conservation, (2) the huge sums involved create additional pressure to capture more wild animals, (3) animals on loan to zoos are less likely to contribute to captive propagation efforts, and (4) corruption, competition between SFA and MOC to administer the program, and issues of control over the animals often seriously hinder conservation. For example, Johnson et al. (1996:365) state, "Disputes relating to conservation money and loans of giant pandas to Western zoos, which were reduced to litigation in 1988, have raised fundamental questions concerning the appropriate role of conservation groups in the management of endangered species." Supporters of panda loans point to the fundraising potential of such loans (as long as well-worded and monitored agreements are developed) and the potential to capitalize on the expertise in reproductive physiology available at many Western zoos. Recent negotiations between the Chinese authorities, the AZA, the U.S. Fish and Wildlife Service, and several individual Chinese and U.S. breeding centers and zoos have permitted panda loans to re-commence after a several-year hiatus.

FUTURE AND PROGNOSIS

Substantial progress toward panda conservation has been realized over the past several years. Important achievements include the development of a National Conservation Plan for the Giant Panda and Its Habitat, increased knowledge of panda biology and ecology, significant progress in captive propagation (although the captive panda population continues to reproduce slowly and does not appear to be self-sustaining), and improvements in several social, political, and organizational aspects of panda conservation. Alternatively, most of the major challenges facing panda conservation remain, and trends have yet to be reversed:

- the human population of China continues to increase, especially in rural regions;
- per capita resource use is growing rapidly throughout the nation, putting increased pressure on already stressed ecological systems;
- habitat loss and fragmentation continue;
- poaching continues;

- the current status and trends of pandas and their habitat remain unclear;
- many of the important, more specific objectives have yet to be determined; and
- conflicts among major organizational actors continue to hamper progress toward recovery.

Panda conservation may still be achieved, and both the people and government of China seem committed to success. Efforts on behalf of giant pandas benefit from relatively large amounts of money and other resources, compared with those available for other endangered species. However, along with generous resources comes high visibility and substantial public pressure from within and outside of China. Fear of failure in the face of high public visibility should not prevent program participants from taking justifiable risks, and participants should be buffered from external pressures and distractions.

Successful panda conservation requires continuous, long-term commitment and interdisciplinary approaches. Current studies should continue, and additional problem-oriented research should be initiated. A survey of panda numbers and remaining habitat is urgently needed, especially in regions where pandas might be locally extinct. Of primary importance is the need to develop panda conservation plans that engender local support by addressing the concerns of local people. This will probably require sustainable socioeconomic policies in addition to education programs (Johnson et al. 1996). Continued broader support at the national and international levels will help maintain sources of funding and political support for conservation efforts. Finally, the panda conservation program should maintain its current focus on reducing the causes of panda mortality and increasing habitat protection, at least in the near future (Mainka 1998). As such, the panda reserve system should be strengthened, expanded, and interconnected with effective habitat corridors.

Chinese conservationists are eagerly working toward panda recovery, and they should be commended for their efforts to dates. With additional hard work, panda numbers may well recover to levels that will ensure their survival for many years to come.

ACKNOWLEDGMENTS

This entry benefited from the helpful comments of Drs. Lu Zhi and Mead Love Penn.

Golden Lion Tamarin

James M. Dietz

Common Name: golden lion tamarin
Scientific Name: *Leontopithecus rosalia*
Order: Primates
Family: Callitrichidae
Status: Classified by the World Conservation Union (IUCN) as Critically Endangered; Endangered in Brazil and under the U.S. Endangered Species Act; listed in CITES Appendix I.
Threats: Proximate factors include long-term and widespread habitat destruction, degradation and fragmentation, and—to a lesser degree—hunting and capture for the illegal animal trade. Ultimate factors include social and economic pressures on landowners and lack of a conservation ethic in the region.
Habitat: Primary and secondary lowland forests from sea level to about 300 m in elevation. Typical habitat includes forest about 20 m in height and heavily laden with epiphytes and lianas (woody vines).
Distribution: This primate was once found throughout coastal forests from southern Espirito Santo state to southern Rio de Janeiro state, Brazil. Current distribution includes isolated forest patches in five municipalities in central Rio de Janeiro state.

DESCRIPTION

Golden lion tamarins are about 28 cm in length, including a 36 cm tail. On average in the wild, adult males weigh 620 g and females 598 g (Dietz et al. 1994). Pelage is long, silky, and golden to orange-brown. Adults have a bare face and a distinctive, lion-like mane. Their hands are long and slender with claw-like nails, all likely adaptations for manipulative foraging for small, cryptic animal prey. The dental formula is 1/1 incisors, 1/1 canines, 3/3 premolars, and 2/2 molars. Canines are long (6–7 mm)—again, a likely adaptation for predation.

NATURAL HISTORY

There are four taxa of lion tamarins, all threatened with extinction and endemic to regions of Brazil's Atlantic coastal forest. In the late 1960s, Brazilian zoologist Adelmar Coimbra-Filho called attention to declining numbers and habitat of golden lion tamarins (Coimbra-Filho 1969). Now approximately 800 individuals exist in the wild and 488 in 143 zoos (Ballou et al. 1998).

It is thought that these small-bodied primates are phylogenetic dwarfs; that is, they became smaller over evolutionary time. Although there are no fossil specimens of ancestral lion tamarins, Hershkovitz (1977) speculates that modern forms evolved from a more primitive primate that once occupied suitable habitat throughout east-central Brazil.

Wild golden lion tamarins live in relatively stable groups that average 5.4 individuals. These groups usually contain one breeding female, one or two adult males, and natal offspring from zero to three litters (Dietz & Baker 1993). Behavioral evidence suggests that golden lion tamarins are genetically monogamous (one male mates with one female). A social dominance hierarchy allows the group's dominant male to monopolize copulations during the period when the reproductive female is likely to conceive (Baker et al. 1993). Gestation lasts about 130 days (Kleiman 1977). Reproductive females usually give birth to twins in October, at the onset of the rainy season, and occasionally produce two litters per year. Infants are weaned by June, the beginning of the 3-month dry season when food resources are least abundant (Dietz et al. 1994).

Most tamarins disperse from their natal group at 3 to 4 years of age. However, unless a breeding individual is missing, established groups are generally closed to immigration, and about 31% of male and 76% of female emigrants die before finding a breeding opportunity (Dietz & Baker 1993; unpublished data). Golden lion tamarins are cooperative breeders. That is, all adults in the group help carry and provide food for all the infants in the group, regardless of genetic relatedness.

Golden lion tamarins defend territories that average 45 hectares, large areas relative to the small biomass of a tamarin group. Tamarins often use their long fingers to search for small animal prey hiding in substrates such as dead wood and palm fronds. Tamarins also eat floral nectar during seasonal periods of low fruit availability. Group sleeping sites are usually tree holes and occasionally tangles of vines or bamboo thickets. Predators on golden lion tamarins include boa constrictors, hawks and owls, small cats and humans (Coimbra-Filho 1977).

CONFLICTING ISSUES

The Atlantic coast was the first area in Brazil to be colonized by European settlers. The establishment of outposts and villages in the 15th century marked the beginning of a series of exploitative cycles in which resources were extracted from Brazil and exported to Europe. Some of these resources occurred naturally, for example, "Pau-Brasil" (*Caesalpinia echinata*), the tree species that provided the red dye used to color the robes of European royalty and from which came the country's name. Harvesting of this and economically valuable hardwoods resulted in the clear-cutting of large areas in the Atlantic forest, thereby opening the way for more colonization.

Portuguese colonists introduced sugar cane into Brazil in 1531, which resulted in the conversion of much of the remaining lowland primary forest into plantations. By 1900, clear-cutting to supply fuel for railroad steam engines and domestic use, and to clear land for pasture and crops, had reduced standing lowland forest to less than 5% of the original area (Mori et al. 1981). In 1969, Coimbra-Filho estimated that only 900 km^2 of forest and 600 golden lion tamarins remained in the lower São João valley of Rio de Janeiro.

During this period of rapid deforestation, annual export of 200 to 300 lion tamarins from Brazil was not uncommon. These animals were sent to zoos, biomedical research institutions, and the pet trade (Mallinson 1996). Despite considerable interest in golden lion tamarins, few zoos were able to breed them successfully and attrition was high (Mallinson 1996). In 1973 the captive population consisted of only 70 individuals in fewer than 20 institutions (Kleiman & Mallinson 1997).

Conservation efforts on behalf of golden lion tamarins include action to improve breeding success and management in zoos, and conservation of tamarins and their habitat in Brazil. Devra Kleiman and colleagues at Washington's National Zoo initiated research on the reproduction, nutritional requirements, social behavior, and veterinary care of golden lion tamarins in captivity. The results of this research and appropriate genetic management reversed the decline of the captive population, and by 1995 the numbers had risen to 485 tamarins in 143 zoos worldwide (Ballou & Sherr 1996).

Management of the captive population was unique and innovative in three respects. First, in contrast with policy regarding most other primates, participating zoos agreed not to sell lion tamarins, thus reducing the incentive for their capture from the wild. Second, zoos holding tamarins returned title of the tamarins in their collections to the Brazilian government. Third, participating zoos signed a collaborative agreement placing captive management decisions in the hands of an elected International Management Committee. Responsibilities of this committee were later expanded to include making recommendations to the Brazilian government on management of lion tamarins in the wild (Kleiman & Mallinson 1997).

In the late 1970s prospects for the survival of golden lion tamarins in nature were grim. The little available forest habitat was fragmented, degraded, and rapidly being cut to fuel furnaces to fire bricks used in construction in Rio de Janeiro and São Paulo. Laws prohibiting deforestation were seen as unreasonable demands on the part of the government and were largely ignored by landowners. Poço das Antas Biological Reserve, then the only protected area containing the species, was not viewed positively by the local community and lacked financial and political resources. Little was known about the tamarin's numbers and ecological requirements in nature.

Field research and conservation activities on behalf of golden lion tamarins began in Brazil in 1983 as an initiative of the Smithsonian Institution and were coordinated by a multi-institutional team of researchers, administrators, and educators loosely termed the Golden Lion Tamarin Conservation Project (GLTCP). Conflict among the eight founders of the GLTCP was minimized by frequent communication and a sense of mutual respect. Each had an area of individual expertise that was unique in the group. All agreed that the principal mission of the project was to maximize the probability of survival of a naturally evolving population of golden lion tamarins; this goal was recognized as more important than individual interests. Over time the GLTCP evolved from a loosely organized group of Americans into the Associação Mico Leão Dourado (AMLD), a Brazilian nongovernmental organization dedicated to achieving science-based goals for conserving lion tamarins in their native habitat.

Initially, conservation efforts were focused in and near Poço das Antas Reserve, at the time the only protected area for the species. Actions consisted of demographic, behavioral, and ecological studies of the wild population, re-introduction of captive-born tamarins, translocation of wild-born tamarins, and community conservation education activities in the municipalities surrounding the reserve.

Fourteen years of continuous demographic, behavioral, and ecological monitoring of about 20 breeding groups in the Poço das Antas population generated the biological information that formed the basis for population and habitat viability assessments for all four species of lion tamarins (Seal et al. 1990; Ballou et al. 1998). This research also attracted resources and prestige to the Poço das Antas Reserve, served as a platform for training Brazilian and foreign students, and provided necessary information to other GLTCP initiatives (e.g., community education).

The re-introduction of zoo-born golden lion tamarins into privately owned forests began in 1984. Although initial mortality was high, continuous monitoring allowed researchers to adjust and improve reintroduction techniques. By 1997, 147 tamarins had been reintroduced in Brazil and many of these had begun to breed. Survival of wild-born offspring was high relative to that of their zoo-born parents, and the proportion of the re-introduced population (228 tamarins) comprising of wild-born tamarins grew to 91% (Beck et al. in press). Re-introduction efforts caused the number of tamarins in the wild to increase by 27%, resulted in the protection of 3,000 hectares of additional forest through agreements with landowners, and contributed significantly to the science of re-introduction (Beck et al. 1991; Beck & Martins 1997).

In 1993, Cecilia Kierulff conducted an exhaustive census of golden lion tamarins in all available habitat islands within the species' geographic distribution. She estimated that 559 golden lion tamarins were found in 16 populations. The largest population was in Poço das Antas Reserve (347 individuals) and adjacent forests (71 individuals). Kierulff's data formed the basis for planning and executing the translocation of wild-born tamarins from small habitat patches to larger, unoccupied forests (Kierulff & Oliveira 1994).

A survey of local knowledge and attitudes concerning wildlife and protected areas was conducted in 1984 in the municipalities surrounding Poço das Antas Reserve. Forty-one percent of surveyed adults did not recognize the golden lion tamarin in a photo, and most were ignorant of the existence of the Reserve. Thus project personnel began working with local groups and institutions to motivate people to conserve forest and reduce capture of tamarins for pets. Activities included classes for schools, farm workers, guards, and local teachers; parades and festivals; and press events. Project educational materials were generally simple and low-cost, and they included press releases, public service messages for local radio and television, posters, school notebooks, and T-shirts. Field trips to the Reserve were organized for farmers, local politicians, school groups, and families. Visits were made to farms to encourage owners to conserve forest and thus qualify to receive re-introduced tamarins. After 2 years of project activities a post-treatment survey indicated an increased understanding of tamarins and the need for forest conservation. Responses suggested that television was as important in reaching the local audience as all other project methods combined (Dietz et al. 1994; Dietz & Nagagata 1995). The golden lion tamarin had been used to facilitate the creation of a conservation ethic in the region.

Under the leadership of Denise Rambaldi, AMLD personnel took advantage of this change in attitude to facilitate conservation action. Sixteen privately owned ranches received re-introduced tamarins, and seven landowners agreed to enroll their forested areas as permanent private reserves. Three ranchers planted agricultural forest corridors between their forests and those of neighbors. Two municipalities created forest reserves to protect wild groups of tamarins. In 1998 a campaign organized by the AMLD resulted in the creation of a second federal biological reserve, Reserva União, thereby doubling the amount of protected habitat suitable for golden lion tamarins.

FUTURE AND PROGNOSIS

Will golden lion tamarins survive in the wild? If populations were not fragmented and there were no further loss of habitat, golden lion tamarins would marginally meet conservation goals (98% probability of survival and 98% conservation of genetic diversity for 100 years). Therefore, to maximize the viability of both lion tamarins and forest habitat, conservation efforts must proceed within a metapopulation management context. That is, all isolated populations, including the captive population, must be included in a global management strategy so that each contributes to the viability of the whole and none are lost. Second, all remaining suitable forests must be populated with tamarins and safeguarded. Finally, the needs of tamarin populations must be reconciled with the changing social and economic needs of the people sharing the Atlantic forests of Brazil and the reality of limited resources available for conservation.

Golden-rumped Elephant-shrew

Galen B. Rathbun and Solomon N. Kyalo

Common Name: golden-rumped elephant-shrew
Scientific Name: *Rhynchocyon chrysopygus*
Order: Macroscelidea
Family: Macroscelididae
Status: Endangered on 1996 IUCN Red List.
Threats: Isolated distribution where (1) subsistence trapping reduces numbers, and (2) agricultural and urban development and tree harvesting modify and eliminate habitat.
Habitat: Dry, semi-deciduous forest and coral rag scrub (scrub vegetation growing mainly on soils made of decomposed coral).
Distribution: Coastal forests north of Mombasa, Kenya, including small and isolated sacred forests (Kayas), the Arabuko-Sokoke Forest, and Boni Forest.

DESCRIPTION

The four genera and 15 species of elephant-shrews form a well-defined Order endemic to Africa (Nicoll & Rathbun 1990). They are believed to be distantly related to aardvarks and the Paenungulata (elephants, hyraxes, and sea cows) (Springer et al. 1997). The three species of giant elephant-shrews (*Rhynchocyon*) are found in different forests in central and eastern Africa. The golden-rumped elephant-shrew is only found in Kenya and has the most restricted distribution of any elephant-shrew (Corbet & Hanks 1968).

R. *chrysopygus* is the largest of the elephant-shrews, being the size of a small cat. The body length is about 280 mm, tail length is 240 mm, and weight is about 540 g (Rathbun 1979a). Its body shape is unique, resembling a cross between a miniature antelope and an anteater. The legs are long and spindly, and a long nose, large eyes, and moderately large ears dominate the face. Unlike most small mammals, golden-rumped elephant-shrews are very colorful with a dark amber body highlighted by a bright yellow rump patch, black legs and tail, and a grizzled gold forehead (Corbet & Hanks 1968).

NATURAL HISTORY

Golden-rumped elephant-shrews spend much of the day slowly walking about on the forest floor searching with their long noses for invertebrates in the dense leaf litter. Prey includes earthworms, millipedes, spiders, and insects (Rathbun 1979b). They neither climb nor burrow, but when disturbed they take rapid flight across the forest floor. They sometimes take refuge in the hollow bases of large trees, if available. This elephant-shrew spends the night alone in one of several widely scattered nests that it builds on the ground with dead leaves (Rathbun 1979b).

Although much of the older literature describes *Rhynchocyon* as solitary, it actually exhibits a relatively rare social organization called facultative monogamy, whereby male/female pairs defend a joint territory by chasing away individuals of the same sex. Except for mating, however, members of a pair spend little time together (Rathbun 1979b; FitzGibbon 1997).

After a gestation of about 40 days, a single precocial young (independent, needing little parental care) is born in a leaf nest, where it stays for about 2 weeks. Females can produce up to six litters per year, but the males do not assist in raising the young (Rathbun 1979b). In primary forest, golden-rumped elephant-shrews may reach densities of 68/km², but in poorer habitats their densities are usually below 25/km² (FitzGibbon 1994).

CONFLICTING ISSUES

The coastal dry forests of eastern Africa are relatively small, isolated, and highly threatened (Burgess et al. 1996). In these forests live numerous species of amphibians, reptiles, birds, and mammals that are found nowhere else. For example, the Arabuko-Sokoke Forest, located between Mombasa and Malindi in Kenya, supports three endemic mammals, including the golden-rumped elephant-shrew. The fate of these animals rests in the fate of their forest habitat (Nicoll & Rathbun 1990), which is under increasing pressure from an expanding human population.

Although the Arabuko-Sokoke Forest is protected and managed by the Kenya Department of Forestry, there is increasing pressure to clear parts of the forest for urban, agricultural, and commercial uses. In addition, a legal and illegal selective tree harvest is changing the composition and structure of the forest, which in turn is threatening the golden-rumped elephant-shrew.

The coastal forest north of Mombasa has a long history of human use, as shown by the 13th- through 17th-century Swahili/Arab ruins at Gedi, near Malindi. The indigenous Sanya people were hunter-gatherers in the forest and led a mobile existence, but about 100 years ago the Mijikenda people arrived in the area, resulting in more permanent settlements and an expanding population (FitzGibbon et al. 1995).

As the number of people living near the coast has increased (by 3.8% per year), so has the need for forest products, agricultural land, and building sites for homes (Burgess et al. 1996). Over the last 100 years the structure and composition of the forest have changed greatly as poles, hardwoods, and firewood have been selectively extracted (Mogaka 1991). In addition, two significant and obvious changes have occurred in the distribution of the coastal forest. First, its area has been greatly reduced. Second, the remaining forest has become fragmented.

In 1943, 418 km² of forest were officially protected as the Arabuko-Sokoke Forest Reserve. Today, this area plus a few additional square kilometers of unprotected forest is all that remains of the estimated 1,000 km² of original coastal forest between Mombasa and Malindi. The loss has been owing to clear-cutting for various activities, including cashew nut and coconut plantations, charcoal production, exotic timber plantations, and cash crop and subsistence farming (Mogaka 1991).

The reason the forest has become fragmented is unique. Scattered throughout the region are about 45 sacred Mijikenda sites called *Kayas*, which are often associated with a cave or hilltop. Each *Kaya* usually includes from 10 to 300 hectares of forest that has been protected because of traditional beliefs (Hawthorne 1993). However, as tribal traditions are lost, so are the practices associated with protecting the forests on the *Kayas* (FitzGibbon 1994). In addition, as the unprotected forest between these sites has disappeared they have become isolated, and the wildlife in the forest patches, including the golden-rumped elephant-shrew, is now prone to extirpation.

As tourism has developed in Kenya, the Kamba people have developed a very successful and lucrative woodcarving industry. For example, it is estimated that 60,000 carvers produce rhinos, impala, giraffes, and the like with a yearly export value of about U.S. $20 million. Much of this success can be attributed to the effectiveness of several cooperatives, such as the Akamba Handicraft Cooperative in Mombasa, and another in Malindi. Over the years, however, the carvers have depleted many of the favored species of trees in Kenya's forests, including the mahogany, or muhugu (*Brachylaena huillensis*), from the Arabuko-Sokoke Forest (Marshall & Jenkins 1994), which the elephant-shrews use for shelter.

To better understand the ecological impacts of logging on wildlife, Kyalo (1997) tallied how many of 515 muhugu trees that were felled had hollow trunks and harbored wildlife. He found 91 of the 515 trees actually had *Rhynchocyon* in their hollow trunks. Kyalo estimated that 20,800 muhugu trees are harvested per year from the Arabuko-Sokoke, and that about 9,360 of these would be hollow. He further estimated that 4,200 of the hollow trees would be used by golden-rumped elephant-shrews for shelter. How important are these hollow trunks to the elephant-shrews?

Gedi Historical Monument is a 44-hectare isolated patch of forest that once was connected to the main Arabuko-Sokoke. In this regard it resembles the many isolated *Kayas* in the region. Rathbun (1979b) did his ecological studies of *R. chrysopygus* at Gedi in the early 1970s. When FitzGibbon (1994) began her fieldwork in the area 20 years later, she found that marauding dogs belonging to people living around the forest at Gedi had decimated the elephant-shrew population. Because virtually all of the old-growth trees with economic value have been harvested from this isolated forest, few hollow trees remain. Without these shelters, fleeing elephant-shrews have no secure escape from dogs. Is this to be the fate of the golden-rumped elephant-shrew, as the coastal forests become increasingly fragmented, surrounded, and used by more and more people? Coastal elephant-shrew populations, similar to other vertebrates that inhabit isolated fragments of forest, are very susceptible to this type of human-related extirpation, as well as to extinction from unpredictable or random natural events such as severe storms.

Although wildlife trapping and hunting in the Arabuko-Sokoke Forest is regulated by the Kenya Wildlife Service and the Kenya Department of Forestry, the illegal subsistence harvest of all mammals, including the golden-rumped elephant-shrew, has increased in recent years as the forest has shrunk and the human population has expanded (FitzGibbon et al. 1995). Although there are over 1,000 households hunting and trapping in the Arabuko-Sokoke, trapping only occurs in about 40% of the forest, mostly around the edges. The result is that the edges are depleted of elephant-shrews while the interior serves as a source. The densities of *R. chrysopygus* in the Arabuko-Sokoke average about 59 individuals/km^2, whereas the harvest is 8/km^2 and the maximum sustainable harvest is estimated at 20/km^2. Apparently the current level of illicit trapping is sustainable (FitzGibbon et al. 1996).

However, there are other issues. Elephant-shrews, being diurnal and showy animals, could attract tourists, which would mean income for forest management as well as local families. But animals must become tolerant of people and their densities must be high if they are to be viewed and photographed. This will not happen if trapping is allowed. Subsistence harvest is an important source of protein for the local people. In addition, when trapping and hunting are allowed, local support for forest and wildlife management is garnered. Without this support, future forest conservation efforts may not be successful. On the other hand, should a species that is endangered with extinction be harvested? If a regulated and sustainable harvest is allowed, will it encourage more trappers to venture deeper into the forest, thereby disrupting the delicate balance that now exists? FitzGibbon et al. (1996) suggest that the current "loose arrangement," with trapping around the perimeter, may be a workable compromise, but it needs a monitoring program to ensure that the elephant-shrew population does not decline further.

FUTURE AND PROGNOSIS

As pointed out by Turner and Corlett (1996), small and isolated fragments of forest are probably not as effective in preserving biodiversity as are large blocks of forest, but small patches are definitely more useful as sources of plants and animals for future conservation efforts than are urban and agricultural areas. For this reason, we commend the National Museums of Kenya for protecting the cultural and biological value of the 30 *Kayas* that have been designated as National Monuments, and we hope that the remaining *Kayas* will also be protected.

The depletion of hardwoods in Kenya's forests has not gone unnoticed by the woodcarvers and their cooperatives. They are being assisted in developing solutions to the problem by national and international conservation organizations. In association with the Mennonite "Ten Thousand Villages" project and the "People and Plants Initiative" of the World Wide Fund for Nature (WWF), UNESCO, and the Royal Botanic Gardens at Kew, woodcarvers are trying to (1) diversify the types of woods favored by tourists, and (2) develop agroforestry systems and plantations as an alternative source of hardwoods (Cunningham 1998). These programs offer real hope that in time one of the threats to the habitat of the golden-rumped elephant-shrew will decrease.

If some form of legal elephant-shrew harvest is allowed, we believe it will require not only a monitoring program to ensure that the trapping is indeed sustainable (FitzGibbon et al. 1996) but also some form of efficient regulation. However, regulating wildlife trapping will be tricky. In areas where the standard of living is low, funding of programs is difficult and there is great temptation for corruption by administrative and field staff with regulatory roles. Unless these problems are satisfactorily addressed, it is likely that elephant-shrew trapping cannot be managed adequately and thus not sustained.

Total protection is very attractive, given the difficulty of implementing a sustainable harvest, the highly restricted and fragmented distribution of *R. chyrsopygus*, and its vulnerable natural history. "But conservation is a human problem, not a biological problem. Conservation will not succeed unless human needs are catered for and adequate alternative resources provided" (Rodgers 1993:318). Perhaps a conservation effort similar to the grass roots effort that is evolving with the woodcarvers (Cunningham 1998) will develop for wildlife itself and thus generate alternative resources and support for conservation.

Gorilla

Amy L. Vedder and William Weber

Common Name: gorilla
Scientific Name: *Gorilla gorilla*
Order: Primates
Family: Pongidae
Status: Endangered under the Endangered Species Act of the United States; Listed in Appendix I of CITES.
Threats: Loss and fragmentation of habitat (especially for *G. g. graueri*); hunting: direct (*G. g. gorilla*) and indirect (*G. g. beringei*).
Habitat: Restricted to African tropical rain forest, from lowland to montane (up to 4,000 m). Present at greater densities in open forest (secondary and montane), though also existing in primary forests.
Distribution: Found in two separate regions of central Africa, separated by approximately 750 km: (1) forest fragments on the border of eastern Nigeria/western Cameroon, and larger forest blocks throughout southern Cameroon, most of Gabon, Congo, Cabinda, and Equatorial Guinea, and into the southwestern Central African Republic (CAR); and (2) forests of eastern Democratic Republic of the Congo (formerly Zaire), into southwestern Uganda and northwestern Rwanda.

DESCRIPTION

The largest of the primates, the gorilla varies significantly in size according to sex. Born at about 2 kg, adult males grow to 140–180 kg (1.8 m in height) and females to an average of 90 kg (1.5 m in height). The adult gorilla has a large head; a broad, flattened nose with flared nostrils; long arms with massive forearms; short and stocky legs; and a protruding abdomen. Its eyes, ears, length of fingers, and genitals, however, are small relative to overall body size. The gorilla is richly endowed with hair, especially long on the arms and legs. The coat is generally black, with the lower back region turning white or gray in adult males (approximately 12 years old). In some geographic areas, the crown of the head may be rusty red or brown instead of black. Skin is black as well, uncovered at the face, and only partially covered on hands, feet, and upper chest.

NATURAL HISTORY

Gorillas live in stable family groups with whom they travel, eat, play, and sleep. Family groups range from several individuals to as many as 40 (3–4 adult males, 9–12 females, and offspring), and averaging 5 to 11 (Tutin & Fernandez 1984). During the first 3 or 4 years of life, an infant is extremely dependent on its mother. It becomes more independent over time and normally leaves its birth group as it approaches sexual maturity. Males most frequently leave the group to initially travel alone. Females normally depart to join an existing family or a lone adult male, with the latter forming a new group.

Female gorillas become sexually mature at about 9 years of age, when 26 to 30-day estrus cycles begin. Birth of an infant (normally one, rarely twins) occurs following an 8.5 to 9-month gestation period and recurs every 4 years, on average. Given a maximum life span of 45 to 50 years, females normally produce 3 to 8 offspring in a lifetime. Mortality rates are greatest in the first 2 years of life (25%) but are very low subsequently. Although a small number of deaths are known to be inflicted by other gorillas (and there are anecdotal reports of gorillas killed by leopards), most known deaths are owing to disease or poor health. The resultant potential population growth rates are slow but very strong, which bodes well for conservation (Weber & Vedder 1983).

Gorillas are herbivorous, feeding largely on herbaceous leaves, stems, and shoots in open forest (secondary and montane), and significantly on fruits where they are abundant (in closed primary forest) (Vedder 1984, 1989; Watts 1984). Gorillas consume many species of foods in all habitats. They are known to be important seed dispersers in lowland rain forest, and for some tree species they are the only known agents for seed distribution (White & Tutin in press). As such, they play key ecological roles in the dynamics of the forest.

High food requirements—both caloric and nutritional—cause gorillas to travel widely to satisfy their needs. Daily movements range from several hundred meters to several kilometers per day and result in yearly home ranges of 5 to 30 km². At dusk each (noninfant) gorilla constructs its own nest on the ground or in a tree, bending plants to form a springy platform on which to sleep through the night. Counting of night nests is the most accurate method of estimating population densities of gorillas, which range from 0.18 km² (primary forest) to 0.76 km² (open forest) (Tutin & Fernandez 1984; Weber & Vedder 1983).

CONFLICTING ISSUES

Central African forests represent regions where great human poverty, pressures for development, and weak governments collide with conservation goals. Given strong population growth in the region, the recent influx of transnational logging and mining companies, and the continued instability of governments, the home of the gorilla is under great threat. The biggest challenge for conservation, therefore, is to address these conflicting interests and to seek recognition for the values of gorilla and forest protection.

In much of equatorial African forest, local people are without adequate investment or transportation systems, educational opportunities, or service sectors. As a result, they are necessarily tied to using on-site natural resources (most obviously agricultural), especially on forest fringes, and/or hunting, particularly in deeper forest regions. Lack of alternatives creates complete dependency on these resources, and increases in human population and access to technology drive deforestation and overhunting.

Gorillas are most threatened by forest clearing and fragmentation at forest fringes. As described by Weber (1989), in the eastern range (Rwanda, Uganda, and eastern Democratic Republic of the Congo) the human population density is the highest in Africa, reaching as many as 400 people/ km² and growing rapidly (by a rate as high as 3.7% annually). People in these areas are largely small-plot, permanent farmers. With family farms too small to subdivide among descendants, and few alternative livelihoods possible, there is tremendous pressure on the undeveloped land—and therefore on the habitat of gorillas. Each of the national parks created to protect mountain gorillas is surrounded by intensively cultivated agricultural lands (Volcanoes National Park, Rwanda; Virunga National Park, DRC; Mgahinga National Park, Uganda; and Kahuzi-Biega National Park, DRC). The borders of these parks have been compromised by incremental agricultural encroachment, and in certain cases (e.g., the Volcanoes National Park during 1969–1972) governmental decrees have converted major blocks of forest from park to cash crop production (Weber & Vedder 1983). Similar, though less intensive, pressure exists in the westernmost extensions of gorilla distributions of Nigeria and Cameroon (Harcourt et al. 1989).

Deeper in gorilla forest habitat, direct hunting is a greater threat from local people. Although agriculture may be practiced, it is frequently conducted on a subsistence, itinerant basis and does not result in permanent clearing of forest. Associated with these practices, however, is the tradition of bushmeat hunting, for which gorilla meat is often prized. Particularly with the introduction of shotguns in eastern DRC (Hall et al. 1998) and the promotion of bushmeat marketing by logging companies in west-central Africa (Cameroon, Gabon, and Congo), local hunting is now thought to have a major negative impact on gorilla populations. People in much of this region are only marginally involved in cash economies, but they see the potential benefits that cash earned from more intensive hunting can bring. Access to better health and education services, purchase of manufactured goods, and a possible buffer to purchase basic goods during "hard times" are all seen as desirable outcomes of bushmeat sales—which now drive hunting to greater exploitation of, and negative impact on, gorilla populations.

The large commercial enterprises of logging, mining, and oil extraction, which have accelerated rapidly in the last 5 years in central Africa, open up previously inaccessible gorilla habitat through the construction of roads and creation of villages. This results in direct loss of habitat. Although it is generally believed that tropical forest logging threatens wildlife, for the large terrestrial mammals of central Africa the case is more complicated. The few studies that have been done (White and Tutin in press) seem to indicate that gorilla and elephant populations may not be seriously affected by selective logging per se, as practiced in the region. These species may actually benefit from a more open forest canopy and abundant herbaceous growth that takes place beneath. However, these industries often promote hunting of large mammals by providing guns, ammunition, and transport for workers and families, and by creating cash economies, middlemen, and markets for such products (Wilkie et al. 1992). Some companies directly organize and hire hunters. Although gorillas may not be among the most common prey, they are subject to this hunting wherever it exists—and it exists in more and more remote regions owing to expansion of these industries.

It is logical to look to government to balance these demanding forces, all of which conflict with gorilla conservation. One might expect government to be mandated to provide basic services and opportunities for its citizens, regulate the activities and impacts of commercial industries, and manage natural resources and wildlife for the common good and for future generations. In fact, central African governments are technically, politically, and financially weak with regard to forest management and wildlife conservation. They do not focus on the needs of people located far from centers of power, they do not have sufficiently developed expertise in wildlife management, and they depend significantly on fees and taxes derived from extractive industries. Consequently there are few incentives or means to act effectively for wildlife conservation, and there are many disincentives that work against it.

Finally, a very different perspective is represented by traditional conservation forces. Most often, interest in wildlife conservation stems from deep, personal beliefs in the moral or esthetic values of wild species. Such conservationists base their actions on noneconomic, frequently nonutilitarian, values. They are motivated by the survival value of a species, in particular one as intelligent, long-lived, social, and human-like as the gorilla. Because of gorillas' similarity to our own species, many people recognize the right of existence of individual gorillas as second only (or in some cases, equal) to that of humans. This creates the potential for great conflict between those who want to save all gorillas from unnecessary death and those who see these animals as purely economic resources.

Perhaps it is not surprising that actors interested in gorilla conservation have looked to these conflicts to craft possible solutions. Some of the earliest efforts to integrate economic and political concerns into conservation programs were initiated in Rwanda, with the inception of the Mountain Gorilla Project (Vedder & Weber 1990). In this program, short-term protection from poaching and agricultural encroachment was combined with development of a high-profile ecotourism program and a concomitant public

awareness campaign (Weber 1987). Protection was identified as an immediate priority to safeguard a small and severely threatened population of
gorillas. Tourism was seen as an alternative land use option, with the potential to generate significant revenue and employment both locally and
nationally. Public education was initiated to provide information on unrec-
ognized values of the gorillas and their habitat: the uniqueness of the gorillas, but especially the importance of the forest for nationwide ecological
services (e.g., water catchment, soil stabilization, flood prevention). Embed-
ded in all of the project's activities were efforts to build the capacity of
national institutions to assume responsibilities and encourage the realization
of benefits by local people. The Mountain Gorilla Project became extremely
successful, providing a major source of foreign exchange, employing numerous and more effective park staff, and catalyzing a source of pride within
Rwanda. Within 7 years of its inception, gorilla population numbers began
to rise significantly (Vedder & Weber 1990).

This general approach (protection, revenue generation, and public awareness) has been adapted and applied in various forms in neighboring DRC
and Uganda, as well as in Gabon, Congo, and CAR. Underlying its conception is the need to confront those factors that conflict with gorilla conservation: local poverty, lack of economic alternatives, low priority accorded
by national governments, competing land use and economic pressures, and
lack of recognition of nonconsumptive values of the forest or its wildlife
(e.g., ecotourism, ecological, and/or heritage values). Some more recent
programs have generated major innovations. Staff and associates of the
Bwindi Impenetrable National Park in Uganda have created a trust fund
from international donor support, with proceeds that directly benefit local
communities surrounding the park (Macfie personal communication). Using
a different approach, personnel working at the Nouabalé-Ndoki National
Park of Congo have encouraged participation in conservation by the CIB/
Feldmeyer logging company, which operates in zones surrounding the park.
The company agreed to change road-building practices to minimize direct
habitat loss, set aside sensitive areas for wildlife protection, prohibit the
hunting of all endangered species (including gorillas), control other bushmeat hunting on its concession, and support a public education campaign
to reinforce these actions.

FUTURE AND PROGNOSIS

Despite integrated, well-targeted actions to address fundamental conflicts
in gorilla conservation, threats remain and the long-term prospect of success
is uncertain. There are many regions in which conservation activities are
absent and local people feel mired in poverty with little choice but to exploit
forest resources maximally. The vast majority of logging, mining, and oil
companies in the region have yet to recognize a responsibility for safeguard-

ing the environment in which they are operating, or even respecting the future harvest potential of the forest. In several cases where ecotourism is generating substantial revenue, corruption is appearing and may threaten the well-being of the gorillas themselves. Governments in the region are at best preoccupied with developing the economies of their nations and at worst remain uncaring about present or future resource management. Finally, gorillas frequently face another type of conflict: that of civil strife and war. Political instability in the region has rendered the tasks of gorilla conservation even more complicated than would otherwise be the case.

Yet in the midst of these difficulties, preliminary success has been demonstrated. Conservationists and governments have begun to recognize opportunities to reconcile some types of economic development with improved gorilla conservation practice. These types of programs must be expanded and innovation encouraged. It is extremely important to monitor these initiatives closely because pressures for development continue to grow and change, and they will leave gorilla populations and their habitats increasingly vulnerable to loss. As long as the values recognized by different stakeholders are not congruent, great challenges will remain in gorilla conservation.

Grevy's Zebra

Mary Rowen and Stuart Williams

Common Name: Grevy's zebra
Scientific Name: *Equus grevyi*
Order: Perissodactyla
Family: Equidae
Status: Endangered on the 1996 IUCN Red List; Threatened under the Endangered Species Act of the United States; listed in CITES Appendix I. Legally protected in Ethiopia; protected in Kenya by a 1977 hunting ban.
Threats: Primarily, competition with pastoralists and their livestock for water and forage; habitat reduction and fragmentation through loss of access to suitable watering areas; loss of habitat owing to heavy livestock grazing and drought.
Habitat: Semi-arid scrub and grasslands with permanent water.
Distribution: Historically, in northern Kenya, Ethiopia, and western Somalia. Currently, in northern Kenya and southern Ethiopia. The main concentration of animals exists in the southernmost part of their range near Buffalo Springs/Samburu National Reserves and the Laikipia Plateau in Kenya.

DESCRIPTION

The Grevy's zebra is the largest of the wild equids; adult males weigh up to 450 kg, about 10% more than females. Grevy's have a thick black dorsal stripe. A white margin on either side of the dorsal stripe is punctuated by characteristic thin black stripes that radiate vertically toward the belly. Over the hips, the stripes curve toward the rump. The lower belly is white, with a less uniform black/brown ventral stripe. Except for the upper, inside portions, the legs are covered with thin horizontal stripes. Grevy's zebras have a long upright mane, large ovoid ears, and, unlike the other species of zebra, a brown (not black) nose. Infants are born with mostly brown stripes that turn black during their first year. It is possible to determine infants' age according to their brown/black stripe pattern (Rowen 1992).

NATURAL HISTORY

The Grevy's zebra range covers the semi-arid mixed scrub and grasslands in northern Kenya, Ethiopia, and western Somalia. In the southern portion of their range, where the arid scrub turns into wetter grasslands, the Grevy's zebra range overlaps with that of the common, or Plains, zebra (*Equus bur-chelli*). To the north and east of their range, in the arid Ogaden and Danakil regions of the horn of Africa, the Grevy's zebra historic range abuts the historic range of the now extremely rare African wild ass (*E. africanus*).

The semi-arid area occupied by the Grevy's zebra is characterized by two dry and two rainy seasons. Failure of one or more rainy seasons is not uncommon, and droughts occur periodically. Grevy's zebra are principally grazers, although during dry periods they browse for up to 30% of their diet (Ginsberg 1988). Grevy's zebra, like all equids, are hindgut (caecal) digesters and can eat and process large amounts of food, which enables them to ingest large volumes of poor-quality forage.

Individual Grevy's zebra utilize their habitat independently of each other according to their water and forage needs. Drinking requirements dictate movement patterns. Adults need to drink every 2 to 5 days, although they will drink daily if possible. Lactating females must drink every 1 to 2 days. During the rainy season, when ephemeral water sources are common, Grevy's zebra can utilize their whole range. However, during dry seasons or drought, their movement patterns are constrained by the distance to available permanent water sources (springs, lake, river). Adult males and nonlactating females can range farther from water than can females with young foals. This is advantageous, as forage is generally more abundant farther from the heavily grazed areas surrounding water (Ginsberg 1988).

Grevy's zebra do not form stable social units (harems) like horses and common zebra. Rather, Grevy's zebra exhibit a territorial, resource defense social system (Klingel 1974; Ginsberg 1988). Territorial stallions maintain large (up to 12 km²) territories in which they have almost exclusive breeding rights to estrous females. The male's reproductive success is related to both the quality of forage on his territory and the distance to safe water (Ginsberg 1988). The best territories are those that attract mothers with young infants. When possible, females with young foals tend to graze within a few kilometers of a water source (Becker & Ginsberg 1990; Ginsberg 1989; Rowen 1992). Females experience a postpartum estrous 7 to 10 days after birth, and then regular cycles (about 27 days) continue until impregnation (unless a female loses condition and stops cycling). Hence if a female gives birth and remains on a particular stallion's territory, there is a high likelihood that the stallion will have access to her during several estrous cycles. Females with young foals form mother-infant groups that may remain relatively stable and stay on one to three territories for up to 6 months (Becker & Ginsberg 1990; Rowen 1992). At 6 months of age the foals are very mobile and nurse less frequently, thereby enabling both mother and foal to move farther away from water sources. The group dynamics of nonlactating females is much more fluid (fusion-fission; see Rubenstein 1989), with groups changing on a continuous basis as groups of animals meet, generally while moving toward watering points, and then separate.

Grevy's zebra females can give birth at 3.5 years of age, and gestation is about 13 months. Males can breed at 3 years but generally are not able to capture and defend territories until 6 to 7 years of age. There is no distinct birth season, although birth peak periods do occur according to drought and good rain cycles.

CONFLICTING ISSUES

The range and habitat of Grevy's zebra, sandwiched between that of the common zebra and the African wild ass, has probably never been able to sustain large numbers of animals. Although early population estimates do not exist, it is clear that Grevy's zebra numbers declined drastically in the latter half of the 20th century. During this period the Grevy's zebra became extinct in Somalia, with the last sighting in 1973. Klingel (1980) recorded a population decline of 90% during 1967–1977 on the Il Bonyeki plains, Kenya. Prior to and during this period, Grevy's were extensively hunted for their skins. With their thinner stripes, Grevy's zebra were more sought after than common zebra for their fine skins. Grevy's zebra pelts were used for rugs and fashion accessories such as handbags and wallets. With numbers of many species declining, the Kenyan government imposed a total hunting ban in 1977. At about the same time, Grevy's were listed on CITES Appendix I and the U.S. Endangered Species List. The combination of the hunting ban and international trade limits effectively curtailed trophy hunting of Grevy's zebra skins.

Despite the hunting ban, the numbers of Grevy's zebra have continued to decline. Using aerial census methods, the Kenyan Department of Remote Sensing and Resource Surveys documented a 70% decline in the Kenyan Grevy's zebra population (est. 13,718 in 1977 and 4,276 in 1988) (Dirschl & Wetmore 1978; Grunblatt et al. 1989). Given this continuing rate of decline, population extinction within 50 years was seen as a definite possibility (Rowen & Ginsberg 1992). Similar rates of decline have occurred in the Ethiopian population (Thouless 1995a, 1995b).

By 1988 the range of Grevy's zebra had shrunk to approximately 25% of its former size, and sightings of groups of animals (not just single males) were common in less than 10% of their former range. The majority of animals were found in or surrounding protected areas. Less than 0.5% of the Grevy's zebra's historic range is under some form of protected status (Williams 1998).

During the period 1977–1988, the Grevy's zebra population increased slightly in the southernmost part of their range only, in the area of overlap with the common zebra. This includes the main protected area for the species, the Buffalo Springs/Samburu/Shaba National Reserves (contiguous reserves) and several wildlife/livestock ranches. The ecology of the population in the reserves has been well documented (Ginsberg 1988; Rubenstein 1989; Rowen 1992; Williams 1998). Data from these studies indicate that the population of Grevy's zebra using the reserves has remained stable at about 1,500. Except for a very few animals, most Grevy's zebra move in and out of the reserve and onto adjacent pastoralist lands throughout the year; residence in the reserve is temporary, with concentrations of animals occurring in the dry season and at all times by lactating females. The reserves have a series of permanent springs and are bisected by the Ewaso Ng'iro River. Historically the river flowed all year except during severe droughts. Currently the river dries up more frequently owing to upstream off-take.

Today the outlook for Grevy's zebra is dependent on the status of "permanent" water sources and range quality throughout their habitat. The overlying threat to Grevy's zebra is competition for resources with pastoralists and their livestock and, more recently, with farmers. Although Grevy's zebra habitat is not being converted to other land uses, it is becoming increasingly fragmented by a reduction in the number of water sources available to the zebra and by diminished habitat quality from increased livestock grazing pressure.

In 1977–1988, the same period that Grevy's zebra population declined by 70%, livestock numbers in northern Kenya rose. Although the number of cattle declined by 18%, the number of sheep and goats increased by 53%. Unlike cattle, which need to drink every 1 to 2 days, sheep and goats need water only every 2 to 5 days—about as frequently as nonlactating Grevy's zebra. Hence the shift in livestock herd composition caused an increase in the number of animals competing directly for resources with the zebra. In northern Kenya, sustained grazing pressure by domestic livestock in conjunction with an increasingly sedentary lifestyle of local pastoralists has resulted in vegetation changes and widespread erosion, particularly around water sources (Herlocker 1992, 1993; Williams 1998). The most recent study of Grevy's zebra confirms that Grevy's zebra compete directly with pastoralists and their livestock in northern Kenya (Williams 1998). One such form of competition can be seen where Grevy's zebra and livestock use the same water sources: Grevy's must drink at night rather than during midday as they do in reserves. This shift alone results in a much greater risk of predation (Williams 1998).

The vast area available to Grevy's zebra is becoming less productive. Increased erosion and loss of perennial grasses decrease land fertility and lead to desertification. Desertification is further exacerbated by increasing agricultural use of upstream lands. Headwater off-take is increasing in the more fertile and populous areas of Mt. Kenya and the Aberdares. Recent proposals for upstream off-take threaten to completely stop downstream flow of the Ewaso Ng'iro, the river that currently flows through the nucleus of the remaining Grevy's zebra habitat.

Protected areas are also cause for concern. In semi-arid habitats, off-road driving can cause rapid and permanent loss of vegetation cover and erosion.

Currently, Buffalo Springs and Samburu National Reserves are a main focus of the tourism trade in Kenya. Tourist numbers are low in Kenya at present. However, the amount of damage caused by illegal off-road driving is increasing, in part owing to the additive effect of years of poor patroling and continued lack of enforcement of off-road driving restrictions.

Data from a ground survey in Kenya in 1995 (Wisbey 1995) and from Williams (in press) indicate that the remaining population of Grevy's zebra is becoming increasingly fragmented. Current groups are becoming more isolated, and recruitment into and between any population is low. These groups therefore become more susceptible to localized extinction owing to drought and other stochastic events.

Although there is serious cause for concern, the outlook is not entirely bleak. Grevy's zebra numbers are healthy and increasing in the southernmost part of their range on the Laikipia Plateau. Landowners have increasingly utilized their land for nature-based tourism in addition to livestock production. This trend is very good for wildlife populations, but only under the condition that animals are not fenced and can move freely between ranches, pastoral areas, and nature reserves. Without free movement in and out of protected areas the populations become essentially captive, and it will require translocation of individuals to maintain gene flow.

In a few areas near Buffalo Springs/Samburu National Reserves and on Laikipia Plateau there are efforts to work with pastoral people to encourage co-existence with wildlife populations. Community-based natural resource management (CBNRM) activities are increasingly being used to bring nature-based tourism and its revenue to communities that have traditionally shared their land with wildlife and, more recently, have competed with wildlife for resources.

FUTURE AND PROGNOSIS

Although it is relatively stable today, the future health of the remaining Grevy's zebra population is precarious. Over one-fourth of the total population of Grevy's zebra uses the Buffalo Springs/Samburu/Shaba complex of nature reserves during a portion of the year. The viability of this population is partially dependent on strategies for upriver water usage. Without the river, competition would increase between Grevy's zebra and pastoral people and their livestock for scarce water sources and a declining vegetation base. Unless more ranch and pastoralist land is converted to wildlife-friendly practices, it is doubtful whether the Laikipia region can sustain many more Grevy's zebra than exist there today. There is already an increasing lobby to permit culling of some animals on the ranches.

The continued viability of Grevy's zebra is dependent on the commitment of both Kenya and Ethiopia to protect and maintain the health of their protected areas. Expanded CBNRM activities with local communities can increase the amount of land available to wildlife. However, the success of these programs is dependent on growing levels of tourism to the region, particularly Kenya. Even though the numbers of tourists visiting Kenya have been very high in the past, and at one point tourism was the leading earner of foreign exchange, the numbers are currently low. More marketing by other countries and regional unrest have severely hampered the tourism industry in Kenya. It should be noted that increased tourism should go hand in hand with an increase in the amount of area for wildlife viewing (i.e., CBNRM activities) and better management of tourist activities to lessen the impact of tourist vehicles. If the problems of habitat destruction and water loss can be curtailed or mitigated, then the outlook for Grevy's zebra might turn around dramatically. Although the land is depressed, much of it is recoverable. The ecology of Grevy's zebra is such that given a chance, the population could expand into much of its former range with little direct intervention.

Hawaiian Goose

J. Michael Scott and Paul C. Banko

Common Name: Hawaiian goose, nene
Scientific Name: *Branta sandvicensis*
Order: Anseriformes
Family: Anatidae
Status: Endangered under the U.S. Fish and Wildlife Service.
Threats: Introduced predators (mongoose, feral dogs, feral cats, and rats); elimination of habitat, especially in lowlands; modification of habitat by introduced ungulates (cattle and goats). Historically, hunting for commercial and recreational purposes and possibly the collection of eggs and young were major factors in the decline.
Habitat: Native and exotic shrub and grassland areas.
Distribution: By 1900 the Hawaiian goose was found only on Hawaii. Originally it occurred on all high-elevation Hawaiian islands from sea level to 3000 m.

DESCRIPTION

The Hawaiian goose is a medium-sized goose standing 41 cm high. Males weigh 1,695–3,050 g, 11% more than females, which weigh 1,525–2,560 g (Kear & Berger 1980). It has longer legs and a more erect posture than other geese and is marked by a black head and light cheek, dark and light contrasting neck pattern, and black and white barring on its back. Its deeply furrowed neck plumage is unique among waterfowl. The Hawaiian goose is one of the most terrestrial species of waterfowl and does not use freshwater and ocean habitats in the same way as most other geese and ducks do. Its relatively long, strong legs and partially webbed toes allow it to walk and run steadily over rough ground and to reach high while browsing on vegetation and fruit.

NATURAL HISTORY

When Polynesians arrived in Hawaii about A.D. 400 (Kirch 1985), they encountered at least eight species of geese (Olson & James 1982, 1991). The Hawaiian goose, which once numbered in the thousands, was possibly the only species capable of flight, and the only one to survive to 1778 when Captain James Cook found Hawaii (Olson & James 1982, 1991). By about 1900 the Nene was restricted to the island of Hawaii (Henshaw 1902; Perkins 1903), with ambiguous records for the island of Maui as late as 1914 (Baldwin 1945).

Fossil records are known for all the high islands (Olson & James 1982), but historical records exist only for Hawaii and Maui (Henshaw 1902; Perkins 1903; Baldwin 1945; Banko, Black, and Banko 1999). A reintroduced population has been established on Kauai (Telfer personal communication 1996).

The Hawaiian goose is a grazer and browser that feeds on grasses, berries, seeds, flowers, and herbs. Historically it migrated from higher elevations to lower elevations in response to changes in food availability. At least 50 different native and alien plant species have been identified in the Hawaiian goose diet (Baldwin 1947; Black et al. 1994). The Hawaiian goose's ability to utilize alien species and highly altered habitats enhances its prospects for persisting in the wild (Banko, Black, and Banko 1999).

The Hawaiian goose is nonmigratory. Although there are no longer regular movements between high and low elevations (Henshaw 1902), seasonal movements between patches of habitat in the uplands on the island of Hawaii have been documented (Elder & Woodside 1958; Stone, Hoshide, & Banko 1983).

Vocalizations are similar to, but quieter than, those of the Canada goose (*Branta canadensis*) (Johnsgard 1965). Kear and Berger (1980) described two calls; the first, a low murmuring *"Nay"* or *"Nay nay,"* and the second, a shrill, disyllabic trumpet that ends with a moan. Calls are louder and more strident in the breeding season.

Hawaiian geese walk and run well over lava and through dense vegetation. They swim readily where water is available. They occur singly, in pairs, and as members of flocks of up to 30 individuals. Males defend territory centered on their nests or families, and they are social within their family units.

In the wild, the record for longevity of a Hawaiian goose is 28 years (Banko, Black, and Banko 1999), with annual mortality rates of 5 to 16% in some habitats (Black & Banko 1994). Nests have been recorded from August to April, but most eggs are laid from November to January (Henshaw 1902; Kear & Berger 1980; Banko 1988; Banko et al. 1999). Pairs are established after the first year, with first breeding usually occurring at 2 to 3 years of age. Pairs do not always breed every year but may lay multiple clutches during a single season (Banko 1992; Kear & Berger 1980). Females select the nest site and build the nest. Typically, three eggs are laid (range 1–5) on Hawaii and Maui, but four are typical on Kauai. Incubation period is 29 to 32 days (Kear & Berger 1980; Banko, Black, and Banko 1999). The female incubates, and male guards the female on the nest. Goslings are precocial and feed on their own. Young remain with their parents until the next breeding season (Kear & Berger 1980).

CONFLICTING ISSUES

The Hawaiian goose has been the focus of multinational, multiagency recovery efforts for more than 50 years (Elder & Woodside 1958; Berger

Endangered Animals

1978; Black 1998). Prior to non-Polynesian occupation of the Hawaiian Islands, threats to the Hawaiian goose included hunting, habitat loss in the lowlands, and introduced predators such as dog (*Canis familaris*), Polynesian rat (*Rattus exulans*), and Polynesian pig (*Sus scrofa*). Since 1778 additional predators, such as the cat (*Felis catus*), mongoose (*Herpestes auropunctatus*), Norway rat (*Rattus norvegicas*), and roof rat (*Rattus rattus*), have been introduced. The Polynesian pig has been replaced by the larger and far more wide ranging European strain (Tomich 1969). The introduction of firearms made hunting of geese much more effective, and the impact of hunting increased dramatically because the hunting season was nearly identical to the Hawaiian goose's breeding season (Henshaw 1902).

A recovery plan for the Hawaiian goose was published in 1983 (U.S. Fish and Wildlife Service 1983) and is currently being revised. Hawaiian geese were first reared in captivity in England at Knowsley Hall as early as 1834 (Wilson & Evans 1890–1899; Kear & Berger 1980). Hawaiians maintained Hawaiian geese in a semi-domesticated state, and some ranchers on the island of Hawaii raised Hawaiian geese. The state's captive rearing project at Pohakaloa, under the leadership of the Hawaii Department of Land and Natural Resources, has been successful. More than 2,300 birds were released to the wild between 1960 and 1997 (Banko, Black, & Banko 1999) to supplement the remaining wild population. These efforts also included the U.S. Fish and Wildlife Service, U.S. National Park Service, the Wildfowl and Wetlands Trust in Slimbridge, England, and private landowners. However, the ultimate goal of a self-sustaining wild population has not been achieved.

Release methods have varied. Hawaiian geese survived better after release if food was abundant, predators were absent, and the geese could learn adaptive social skills from their parents (Marshall & Black 1992; Black et al. 1997). Food supplementation and water have not been provided on a large scale in nesting areas, although small-scale efforts suggest that productivity could be enhanced if this were done. Hawaiian geese frequently forage and nest in areas that are recovering from fire or that are regularly mowed (Banko, Black, & Banko 1999). However, burning and mowing may be difficult to reconcile with other management objectives, such as preserving and promoting native biodiversity and restoring native ecosystems.

Efforts to control predators or supplement food have been inadequate and somewhat impractical. Mongooses and especially cats maintain relatively large home ranges, immigrate rapidly into vacant territories, and are difficult to control economically. Predator control methods involve labor-intensive trapping or distribution of poison bait stations that restrict nontarget species.

Population modeling suggests that status quo management of the Hawaiian goose would lead to extinction within 200 years in 98% (Maui), 59% (Hawaii), and 9% (Kauai) of 1,000 simulations (Black & Banko 1994). Thus more active management programs are necessary, especially on Maui and

Hawaii, where predators are more common and high-quality food is less available.

Stone and colleagues (1983) identified nine hypotheses regarding the ecology, distribution, and abundance of Nene; and Black et al. (1994) found evidence to support them all. To recover the Hawaiian goose in self-sustaining, wild populations, the most basic of issues must be addressed. The factors that have caused the decline of the Hawaiian goose (e.g., predation, loss of habitat) must be eliminated or significantly reduced. Sufficient knowledge of the ecology, biology, and release methods exists. It remains to control predators and provide quality habitat over areas large enough to be meaningful to a viable population in the wild. Conservation efforts on behalf of the Hawaiian goose constitute the most comprehensive ever conducted for a duck or goose (Black 1998). Despite this, Black (1998) judged that defensible re-introduction criteria were not met. He concluded that introduced predators are still having an impact and current unmanaged habitats do not contain sufficient quantity or quality of food.

The Hawaiian public has mixed feelings about current practices or the need to control introduced mongooses, feral cats, and rats. To some, it is a philosophical question: Why destroy individuals of one species in order to save another? Others primarily object to methods of capturing and disposing of these alien predators. For example, some advocate adoption of feral cats, not understanding or accepting the fact that feral cats generally make unsafe, unmanageable pets and that animal shelters already provide a seemingly endless supply of feline pets.

FUTURE AND PROGNOSIS

Private landowners have expressed concerns that if they participate in efforts to restore historic, currently unoccupied habitat, they will be vulnerable to penalties under the Endangered Species Act. It is uncertain if use of Habitat Conservation Plans (Beatly 1994) with a "no surprise" clause would alleviate their concerns. Such public-private partnerships could prove useful to the Hawaiian goose, as much of the historical lowland habitat is privately owned. There is a long tradition of cooperation between private landowners, conservationists, and government agencies to recover the Hawaiian goose (Kear & Berger 1980). Cooperative agreements to release and manage Hawaiian goose populations on private lands were established between the State of Hawaii and landowners of three widely dispersed sites on Hawaii Island prior to releasing the first captive-reared Hawaiian geese in Keauhou Sanctuary in 1960. Landowners agreed to allow state biologists access to their property to conduct research, construct pens for the release of Hawaiian geese, control predators, and manage feral ungulates. The State agreed not to restrict ranching or hunting activities.

Today the State of Hawaii and the U.S. Fish and Wildlife Service are developing Safe Harbor Agreements with landowners on Molokai and Lanai Islands and additional locations on Hawaii Island under a new state law (Act 380) intended to promote endangered species restoration on private lands (Terry personal communication). Predator control and habitat management are the key issues being developed as part of the agreements. Increased interest in Safe Harbor Agreements and Habitat Conservation Plans in Hawaii is the result of greater awareness by all concerned that endangered species recovery is not the sole responsibility of government agencies. Government organizations, nongovernment agencies, and individuals must work together more closely if endangered species are to be recovered. Because endangered species ignore ownership boundaries, recovery actions must work across property lines. Doing so will ensure more effective use of funds, trained personnel, and other limited resources.

Perhaps the single biggest challenge of removing the Hawaiian goose from the ESA is the large area required for recovered populations on Hawaii and Maui, and the failure of past management activities to apply remedial actions over such an area. Future management must be applied over areas large enough to influence the productivity and survival of a population. Actions limited to single breeding pairs or small habitat patches are insufficient for population recovery. Current populations are maintained on all but Kauai only through continued re-introduction. If recovery and delisting from the ESA are to be achieved, a higher level of commitment to restoration activities in the field will be necessary.

ACKNOWLEDGMENTS

Thanks to Tom Telfer for sharing unpublished data, Carol Terry for sharing information on the State of Hawaii's initiative to reintroduce Hawaiian geese to Molokai and Lanai, and to Ron Walker for providing information on cooperative agreements between the State of Hawaii, private landowners, and the National Park Service.

Indiana Bat

Nina Fascione

Common Name: Indiana bat
Scientific Name: *Myotis sodalis*
Order: Chiroptera
Family: Vespertilionidae
Status: Endangered under the Endangered Species Act (ESA) of the United States and the 1996 IUCN Red List.
Threats: Human disturbance; habitat loss; adverse habitat modification.
Habitat: Caves and mines in winter; riparian, bottomland and upland forests in summer.
Distribution: Eastern United States, with populations or individuals recorded in more than 25 states—from Oklahoma, Iowa, and Wisconsin eastward to Vermont and southward to northwestern Florida (Barbour & Davis 1969; U.S. Fish and Wildlife Service 1983).

DESCRIPTION

The Indiana bat is a medium-sized species of the genus *Myotis*, which means "mouse-eared." Head and body length of the Indiana bat averages 41–49 mm (Hall 1981), with the tail adding another 26–40 mm (Mumford & Whitaker 1982). The average weight of the Indiana bat is 6 g but can range from 3.3 to 10.5 g (Mumford & Whitaker 1982). In size and general appearance the Indiana bat is remarkably similar to the more common little brown bat (*M. lucifugus*). However, there are distinguishing characteristics. Whereas the pelage of the little brown bat is cinnamon-colored to dark brown and somewhat glossy, the hair of the Indiana bat is a dull grayish-chestnut (Hall 1981). In addition, the hair of the Indiana bat varies in color from base to tip, with lighter coloring at the base (Hall 1981). According to Fenton (1983), during the fall months little brown and Indiana bats can also be distinguished by their characteristically different odors. The toe hairs extend past the end of the toes in little brown bats, but not in Indiana bats.

NATURAL HISTORY

The Indiana bat was first described as a species in 1928 by Miller and Allen in Leavenworth, Indiana, hence the name (Barbour & Davis 1969). Undoubtedly the species was collected prior to that date but incorrectly identified as the little brown bat (Mumford & Whitaker 1982). Like most insect-eating bats, the Indiana bat is nocturnal, leaving roost sites at dusk and spending much of the night hunting flies, moths, beetles, mosquitoes, and other insects (Brack & LaVal 1985; Kurta & Whitaker 1998). Its role in controlling insect pests makes it very valuable to humans.

From mid-September to mid-April, Indiana bats hibernate in large, tightly packed colonies in caves and mines. Before their populations declined, Indiana bats could be found in extremely large numbers in their winter hibernacula (i.e., hibernation sites). Several caves in Kentucky, Missouri, and Indiana were known to each house more than 100,000 individuals (Barbour & Davis 1969; Mumford & Whitaker 1982). Because they maintain denser hibernating clusters than other bats, the Indiana bat has been nicknamed the social bat and the cluster bat. Hibernating Indiana bats prefer cave temperatures between 4 and 8°C, with a relative humidity above 74% (Barbour & Davis 1969; Humphrey 1978). However, recent re-evaluation of what constitutes "good" hibernation temperatures for the species indicates that their requirements may be more complex than previously thought (B. Currie personal communication). Indiana, Missouri, and Kentucky contain the largest populations of hibernating Indiana bats, although the latter two states have seen significant declines in recent decades (U.S. Fish and Wildlife Service 1983).

Beginning around mid-March, Indiana bats leave their hibernacula and migrate to cooler climates, traveling as far north as southern Michigan and Vermont. They remain in their summer habitats until late summer, foraging for insects in riparian areas and floodplains of small to medium-size streams, in bottomland hardwood forests and in upland forests, and roosting under loose tree bark in upland forests (U.S. Fish and Wildlife Service 1983). Oak, sycamore, cottonwood, black walnut, black willow, and hickory all appear to be important species for roosting sites (Mumford & Whitaker 1982).

Indiana bats begin their fall migration back to their winter hibernacula in mid-August or early September, and breeding occurs during a 2-week period in the fall (Hill & Smith 1984). During this time large numbers of bats visit caves that will become hibernating sites in what has been termed "swarming behavior" (Fenton 1992). Most mating occurs during this period (Cope & Humphrey 1977). Like some other hibernating bat species, Indiana bats demonstrate delayed fertilization in which sperm is stored by females until gestation begins in the spring. A single offspring is born in late June or early July (U.S. Fish and Wildlife Service 1983). Unlike little brown bats and other species that form large maternity colonies to raise their young, Indiana bat females raise their young in small maternity colonies under loose bark or in tree cavities (Kurta et al. 1993). Banded individuals have been recorded living up to 15 years (Humphrey et al. 1977, in Mumford & Whitaker 1982); however, it is possible that they live much longer, as little brown bats have been recorded to live for 33 years (Wilson 1997).

The Indiana bat's few natural threats include flooding of caves and ceiling collapse, particularly in mines (U.S. Fish and Wildlife Service 1983). Human activity poses a much greater threat to the species. Disturbance of hibernating bats in winter can cause mortality, as waking bats suffer depleted energy resources and starve. Vandalism can also be a significant problem. In one case 10,000 bats were killed in Carter Cave State Park, Kentucky, by three boys who stoned and trampled all of them to death in a single afternoon (U.S. Fish and Wildlife Service 1983). But undoubtedly the ultimate cause of the decline of many bat species, including the Indiana bat, is the expanding human population and increased human exploitation of natural resources (Pierson 1998).

The decrease in Indiana bat numbers in the last half-century is alarming (Barbour & Davis 1969). When the Indiana bat recovery plan was written in 1983, the species numbered approximately 550,000 (U.S. Fish and Wildlife Service 1983), down from approximately 500,000 in 1980 and 800,000 in 1960 (Barbour & Davis 1969; B. Currie personal communication). According to Barbour and Davis (1969), both these figures were much lower than in earlier decades, although the first systematic counts were not made until the 1950s. Despite completion of the recovery plan, the number of Indiana bats has continued to drop and was only an estimated 350,000 in 1997 (B. Currie, personal communication). Based on winter surveys, caves in New England, New York, and Pennsylvania that once harbored hundreds, if not thousands, of individuals now contain few or no bats. During the 1950s, Indiana bats nearly disappeared from West Virginia, Indiana, and Illinois (Barbour & Davis 1969). Missouri has experienced an 80% decline over the last 13 years (Clawson 1987; Indiana Bat Recovery Team 1996). Indiana bat populations are stable to increasing in Alabama (300), Illinois (4,530), Indiana (182,510), New York (14,990), and West Virginia (11,660), but continue to decline in most other states (B. Currie personal communication).

Biologists initially attributed the decline primarily to disturbance of the Indiana bat's winter hibernacula. The Recovery Plan developed by the Indiana Bat recovery team and approved by the U.S. Fish and Wildlife Service (FWS) is now putting greater emphasis on protecting caves and mines. Hibernacula with a documented population greater than 30,000 since 1960 are identified as Priority 1 and protected by a variety of state and federal agencies in cooperation with the FWS (U.S. Fish and Wildlife Service 1983). These include caves in Missouri, Indiana, and Kentucky. Because nearly 90% of the population winters in a relatively small number of hibernating sites (U.S. Fish and Wildlife Service 1983), prioritizing the protection of important caves is appropriate and permits the most efficient utilization of limited funds for protection and recovery.

Despite cave protections the Indiana bat population has continued to decline, forcing conservationists to look for alternative explanations beyond disturbance of winter hibernacula (Kurta & Whitaker 1998). Its migratory behavior exposes the Indiana bat to threats in both its winter and summer ranges. Because in summer Indiana bats require upland forest for roosting and quality riparian habitat for foraging, the threat may well come from disturbance to their summer range. It is during the summer months that females bear and raise young, so loss of habitat as a result of deforestation and urban and suburban sprawl probably has caused reduced bat numbers. Indeed, an alarming trend was noted by Mumford and Whitaker (1982: 158), who reported that in their Indiana study area "no foraging was detected in forest, open pasture, cornfield, upland hedgerow, or along creeks from which riparian trees had been removed." Since this paper was published it has been demonstrated that Indiana bats will forage in nonriparian forests (B. Currie, personal communication).

Complicating matters is the lack of information about Indiana bats, particularly their summer food and habitat requirements. Bats have been difficult for biologists to study because of their small size and elusiveness. This has been especially true of the Indiana bat. In addition, most state and federal conservation programs have focused traditionally on species with a narrow distribution, and the wide distribution of bats coupled with their migratory nature may impede recovery (Pierson 1998).

CONFLICTING ISSUES

Biologists first considered the Indiana bat imperiled in 1967, and it was listed as Endangered under the ESA soon after that law's passage in 1973. The most pertinent sections of the act in terms of Indiana bat conservation are Section 7, which requires review of federal actions, and Section 9, which prohibits taking bats or their habitat. Section 7 in particular affects the numerous National Forests that support Indiana bat populations. ESA protection requires federal agencies to consult with the FWS to ensure that any action on federal lands, such as logging, any sale of lands that support Indiana bats, or any project that is federally funded or permitted does not jeopardize the survival of endangered species. This places a procedural responsibility on the U.S. Forest Service (USFS), for it cannot move forward with projects without FWS consultation. The ESA also imposes a substantive requirement, that is, examination of whether the proposed action will indeed trigger jeopardy. Section 9 pertains to take. Will removal of trees or other actions directly or indirectly destroy the endangered species or its habitat (i.e., "take" bat)? For the Indiana bat, there is an automatic "may affect" determination for a 5-mile radius around every hibernaculum for tree-removal purposes. This "5 mile rule" was developed to ensure that when a federal action is planned in the vicinity of a hibernation site, the needs of the species over the long term are addressed. It was not intended to be an absolute prohibition against timber management or other habitat modification around winter roost sites (B. Currie personal communication).

Protection of the Indiana bat is a classic case of economic growth versus biodiversity conservation. Although scientists continue to garner additional information about the species' behavior and habitat, questions remain about the sort of restrictions that should be placed on forestry practices where Indiana bats are known or thought to occur. Legal protection of the Indiana bat can affect logging and mining operations, road building, and other forms of development; and it has already pitted this small bat against federal agencies, private landowners, and large corporations. A recent example in Pennsylvania involves a complaint filed by environmentalists to stop logging in the 510,000-acre Allegheny National Forest to prevent potential damage to Indiana bat habitat. In January 1999, U.S. District Judge Donetta Ambrose ruled that logging could continue and that the USFS had developed adequate standards to ensure protection of bat habitat. The USFS intends to stop logging operations on April 1 each year, when bats cease hibernating and begin looking for roosting sites in the Allegheny National Forest. Environmentalists may appeal the ruling. Ironically, although the Indiana bat is known from five caves in Pennsylvania (at least one of these sites is a mine), it wasn't known to exist in the Allegheny National Forest until one specimen was discovered there in August 1998.

In the Northeast, environmentalists are challenging timber harvests in both the White Mountain National Forest in New Hampshire and the Green Mountain National Forest in Vermont. Indiana bats have been identified in Vermont, but whether the species inhabits New Hampshire is uncertain. Accordingly, logging has been suspended as of November 1998 in the Green Mountains. Logging in White Mountain National Forest has been halted until at least the summer of 2000 while the USFS does an ESA consultation with the FWS on the Indiana bat. Surveys for the bats, as well as assessments of the likely impacts of logging to bat populations, will continue during this time.

Federal protection and litigation require the USFS to more closely evaluate its forest management practices. It is now monitoring and surveying bats and developing guidelines to protect them, even if their presence is speculative. In some cases, tree removal has been limited to winter months to avert direct take. However, it is important to note that this remediation technique protects the bat, but not the bat's habitat. It is critical to study the impacts of these actions to gauge their effectiveness. The goal should be to provide and maintain suitable roosting and foraging habitat for the species over the long term. Other federal agencies and organizations have taken additional actions. For example, the FWS, the National Park Service, most state wildlife agencies, non-profit organizations, and other private groups and individuals are installing gates at cave and mine entrances to reduce the impacts of human disturbance.

FUTURE AND PROGNOSIS

The Endangered status of the Indiana bat and increasing awareness of its plight have environmentalists and wildlife agency planners struggling to protect the species throughout its range. A combination of conservation strategies is required to save this species. More research is needed to determine the cause of decline and to answer questions regarding habitat utilization and foraging needs. Technological advances are facilitating study through miniature radio transmitters and other devices (Pierson 1998). This increased knowledge will facilitate better conservation decisions. More stringent protection is needed to protect winter roost sites, including increasing the number of protected caves and mines and more liberal use of gate devices to prevent human access. Perhaps most important, critical riparian and upland forest used for summer foraging and for roosting habitat should be defined and safeguarded. Finally, government agencies and nonprofit conservation organizations should continue public education to dispel existing myths and improve public support for bats. Economic burdens imposed by endangered species protections must be viewed in light of the ecological and potential economic benefits gained from protecting the species. Although the overall downward trend of the species is disturbing, the population of Indiana bats has grown in its namesake state from 177,000 to 182,000 in the last few years. With the combined efforts of conservationists, federal agencies, and private interests on behalf of the species, there may be hope for the Indiana bat's future.

Ivory-billed Woodpecker

Jerome A. Jackson

Common Name: Ivory-billed woodpecker
Scientific Name: *Campephilus principalis*
Order: Piciformes
Family: Picidae
Status: Endangered under the Endangered Species Act of the United States; listed in CITES Category I; possibly extinct in the United States; a few may survive in eastern Cuba.
Threats: Destruction and fragmentation of old-growth forests; overhunting by collectors, Native Americans, and others.
Habitat: Limited to old-growth bottomland hardwood forests of the southeastern United States and old-growth montane pine forests of Cuba during the past 150 years; evidence suggests earlier more strict use of all old-growth forests in the United States and Cuba.
Distribution: No known U.S. populations; possible remnant population in montane forests of eastern Cuba; formerly ranging from eastern Texas to South Carolina and southern Florida.

DESCRIPTION

The ivory-billed woodpecker is the largest woodpecker known from the United States, and in the New World it is second in size only to the equally endangered and possibly extinct imperial woodpecker (*Campephilus imperialis*) of the old-growth montane pine forests of western Mexico. Primarily a black bird, the ivory-bill is best identified by the extensive white "shield" formed on its lower back by white secondaries and white-tipped primaries. Although it has an ivory-white bill, the bill color is not diagnostic of the species. The pileated woodpecker (*Dryocopus pileatus*) has often been mistakenly identified as the ivory-bill, but it differs in several ways: (1) the ivory-bill has the white shield on the folded wings; the pileated does not; (2) the ivory-bill male has a black forehead and red only on the back of the crest; the pileated male has red extending to the bill; the female pileated has red only on the back of the crest but generally has a light-colored forehead; (3) the ivory-bill in flight is very long-necked and long-tailed in appearance, reminiscent of a pintail; the pileated in flight is stockier, shorter-necked, and more blunt-tailed.

NATURAL HISTORY

The ivory-billed woodpecker has been extremely rare since the late 1800s. Most knowledge about the behavior and ecology of the species comes from anecdotal accounts and from the extensive work done by James Tanner in Louisiana (Tanner 1941, 1942a, 1942b), as well as from efforts by Dennis (1948) and Lamb (1957, 1958) in Cuba. Reliance on these major studies probably led to the conclusion that the ivory-billed woodpecker in the United States is a bottomland hardwood forest species whereas the population in Cuba is a montane pine forest species. It may be more accurate to say that in both areas the species requires forests that are warm temperate/tropical, old growth, and humid. In contrast to 20th-century accounts, through the 1800s there were several reports of ivory-bill use of pines for nesting and foraging in the southeastern United States as well as reports of ivory-bills in lowland swamp forest of Cuba. These 19th-century habitats just happened to be the ones from which the birds were first extirpated.

As a primary cavity-nesting and roosting species (i.e., first to use cavity site), the ivory-bill, like all woodpeckers, is limited by the availability of trees suitably large for cavity excavation. Most described nests were in bottomland hardwoods, but some Florida nests were in pines near wetland areas (see review in Jackson 1996). It seems likely that as with other woodpeckers, the ivory-bill was dependent on fungal decay to facilitate cavity excavation, and the high-humidity environment of swamps and riverine forests would promote such decay. Fungal decay also creates habitats required by the 3 to 5-inch wood-boring beetle (Cerambycidae) larvae on which the ivory-bill feeds (Tanner 1942a; Jackson 1991). Factors that limited the ivory-bill's range probably included climatic limits to decay, other factors limiting the geographic range and availability of its prey, and possibly the distribution of potential predators such as the northern goshawk (*Accipiter gentilis*).

Behaviorally the ivory-bill differs somewhat from other woodpeckers. It has not been reported to drum as other woodpeckers do, although it (and other *Campephilus* woodpeckers; Short 1982) does communicate via a characteristic, sharp, double rap on a resonant surface—a mechanical sound that may function as drumming. The ivory-bill's voice is also rather unwoodpecker-like: a single, double, or triple note sounding like the toot of a tin horn and easily imitated by blowing on the mouthpiece of a clarinet. In searching for food, the ivory-bill characteristically knocks large slabs of bark from dead trees, thereby exposing the tunnels of its larval prey. Anecdotal accounts suggest that the ivory-bill may be more social than most woodpeckers in that it has been observed foraging or traveling in pairs or family groups (Tanner 1942a).

CONFLICTING ISSUES

Because of the ivory-bill's large size and habitat requirements, the home range of individuals and family groups is probably several square miles (Tanner 1942a), thus naturally limiting populations. A long, slow decline in numbers probably began with the arrival of aboriginal humans in the New World as many as 40,000 years ago. Because of its daily return to its roost cavity, the ivory-bill is vulnerable to human hunting. Ivory-bills have been considered good eating—"better than duck" (Wayne 1893:338)—and because of their large size might have been a prime target of early hunters.

The development of Amerindian cultures resulted in the development of extensive trade in ivory-billed woodpecker bills and scalps. This posed a great threat to the species. Trade extended far outside the natural range of the ivory-bill. In the early 1700s, the colonial naturalist Mark Catesby (1731) wrote that the bills of ivory-billed woodpeckers were valued "by the Canada Indians, who make coronets of them for their princes and great warriors, by fixing them round a wreath, with their points outward." He also noted that the northern Indians had to purchase these bills from the southern tribes at the price of two to three buckskins per bill. In the early 1800s, Audubon (Audubon & Chevalier 1840–1844: 216) related that "its rich scalp attached to the upper mandible forms an ornament for the war-dress of most of our Indians, or for the shot-pouch of our squatters and hunters, by all of whom the bird is shot merely for that purpose."

Anthropologists (e.g., Skinner 1926) suggest that the Native American warrior sought to gain the power of the woodpecker. Just as the woodpecker used its bill to find and seize the wood-boring grubs that it ate, so the spirit of the woodpecker could bring warriors success as they attacked their enemies. Although this is not mentioned by anthropologists, I suspect that the symbolism went even further. One Ioway Indian war pipe I examined had the upper bill and scalp of seven male ivory-bills on it. It seems to me that the flaming red crest of the male might easily have been symbolic of success at war, the red symbolizing blood—successful scalping.

European cultures brought their own reasons to kill ivory-billed wood-peckers. Sometimes they were killed as food, at other times simply as curiosities. Audubon (Audubon & Chevalier 1840–1844: 216) commented that "Travellers of all nations are also fond of possessing the upper part of the head and bill of the male, and I have frequently remarked that on a steam-boat's reaching what we call a wooding place, the strangers were very apt to pay a quarter of a dollar for two or three heads of this Woodpecker." By the late 1800s leisure activities among Americans and Europeans included maintaining a "natural history cabinet" including collections of birds and their eggs. Just as with baseball cards and postage stamps, the collected items were traded and sold. Ivory-bills, because of their size, limited range, and inhospitable habitat, had a high price on their head.

At first, human killing of the birds might have been a minor problem, but as weapons became more sophisticated and killing became easier, the impact would have become greater. As with all desired commodities, the rarer the commodity, the higher the price placed on it—supply and demand. The greater the price, the greater the pressure on the species.

In spite of the serious toll of direct killing of ivory-billed woodpeckers, the greatest past and current threat to the species is habitat destruction. Even the killing of birds was greatly facilitated by habitat destruction. Following the Civil War, vast acreages of southern lands reverted to the federal government. Recognizing that no local taxes were being paid on these lands, Alabama Congressman Goldsmith Hewitt was successful in getting a bill passed that allowed their sale (Lillard 1947). They were quickly bought up—mostly by northern forest industries—at prices often averaging $1.25/ acre. The forest industry moved in and railroads were built to remove the virgin timber (Croker 1979).

One forestry practice that might have temporarily benefitted the ivory-billed woodpecker was the girdling of large trees to kill them during dry weather and returning to cut them down only when river water was high enough to float the logs out. Because ivory-bills are thought to favor the larvae of beetles that live in recently dead trees (Tanner 1942a), this practice may have provided birds with additional food. However, it may also have lured them to their death, because bird collectors typically took advantage of the transportation and expertise of lumbermen.

Some virgin forest survived the turn of the century, and conservation efforts saved some swamplands early in the 20th century (Dennis 1988). But World War I brought increased demands for lumber for the war effort. Following these pressures came flood control projects on the Mississippi that allowed drainage of the Mississippi Delta and cutting of much of the remaining bottomland forest. World War II brought further "patriotic cutting," and clearing of bottomland forests continued for crop lands. Very little mature forest survived the 1940s.

One of the last remaining large tracts of virgin hardwood bottomland forest was the Singer Tract along the Tensas River near Tallulah, Louisiana. Owned by the Singer Sewing Machine Company, it gained protection as a state wildlife refuge in 1926 although the state did not acquire title to the land. Local citizens knew ivory-billed woodpeckers were in the Singer Tract, but it wasn't until 1932 that ornithologists confirmed their presence (Bird 1932) and not until 1935 that they were seriously studied by Allen and Kellogg (1937). However, under contract to Singer, the Chicago Mill and Lumber Company logged the forest in the late 1930s and early 1940s, claiming it had to do so to avoid layoffs and corporate losses and, further, that it was in the national interest because the timber was needed for the war effort.

While James Tanner (1942a) was making his classic study of the ivory-billed woodpecker there from 1937 to 1939, the birds may well have been making their last stand. Serious pleas from the National Audubon Society and support from the National Parks Association and the American Forestry Association did not slow the cutting (Baker 1942). Nor did efforts within the Louisiana Conservation Department, the U.S. Fish and Wildlife Service, and the National Park Service. Why didn't the National Audubon Society do more? There were at least some who questioned whether the National Audubon Society was doing its best (Edge 1943).

Tanner (1942a, 1942b) concluded that ivory-bills disappeared from most areas of their range shortly after logging began within their habitat, and that all other factors were inconsiderable in comparison. He also outlined a careful plan for selective cutting that might have saved the species had the plan been cautiously implemented. It was not. Baker (1942) suggested that in the case of the ivory-bill, selective cutting could be as devastating as clearcutting. Thomas Barbour's book, *Vanishing Eden* (1944:71–72), gives a clear death knell for the Singer Tract ivory-bills: in the spring of 1944 Ludlow Griscom went to Louisiana to see what the chances were of saving the last great stand of cypress and of preserving what he thought were the last existing pairs of the magnificent ivory-billed woodpecker. "I awaited his report fearfully, and was not surprised when he said: 'The whole area is full of portable sawmills. You can't get away from the sound of tractors hauling out logs. It is too late. It can't be done.' " It could have been done earlier.

Ivory-bills disappeared from their last South Carolina stronghold along the Santee River coincident with the upstream construction of the Santee-Cooper Hydroelectric Project in the late 1930s and impoundment in 1942 (Cahalane et al. 1941; Sprunt & Chamberlain 1970).

When ivory-bills were discovered along the Chipola River in northern Florida (Eastman 1958), National Audubon entered into an agreement along with the State of Florida, the St. Joe Paper Company, and the Neal Lumber and Manufacturing Company to establish a 1,300-acre refuge for the birds (Andrews 1951; Baker 1950). Following an apparent conflict between the Audubon president and the refuge manager—as well as lack of substantiation of the birds' presence—Audubon withdrew its support and no more was heard regarding the birds.

FUTURE AND PROGNOSIS

The future for the ivory-billed woodpecker is bleak. It is quite possible that the species is extinct in North America and is only holding on by a thread in Cuba. The U.S. Fish and Wildlife Service would like to declare the species extinct in North America, and in 1986 it gave me a grant to search the swamps of the Southeast for evidence of the survival of the species (Anonymous 1985; Jackson 1989). I found good habitat in several areas, habitats much improved as a result of growth since the 1940s. I also had one possible encounter with the species near Vicksburg, Mississippi, where a bird responded to a tape-recording for several minutes. However, I obtained no sound recordings or photographs to document the existence of the species. Yet with the quality and extent of habitat I found, and with the number of unverified reports that keep coming in, one cannot "prove" extinction.

The ivory-bill was rediscovered in Cuba in 1985 (Short 1985; Short & Horne 1986), and I was fortunate to hear the bird on 8 different days and get a glimpse of one in the same area of Cuba in 1988 (Jackson 1991). Later efforts to find the birds in Cuba (Lammertink 1992, 1995; Lammertink & Estrada 1995) were unsuccessful and resulted in the declaration that the bird was extinct in Cuba. The folly of such a declaration was proven 1998 when once again the bird was discovered in another area of eastern Cuba (Wege and Lee personal communication; see also Diamond 1987).

Any hope for the ivory-billed woodpecker rests in providing the extensive old-growth forest habitats that it requires. Political differences between the United States and Cuba further complicate conservation efforts. Even though the U.S. Fish and Wildlife Service has been anxious to declare the species extinct, it has made minimal effort to search for it. More efforts are needed, and habitats for the species should be protected as if the birds are there. What species of lesser glamour that share these old-growth habitats are on the brink of extinction yet not recognized? The scientific community may also share the blame for lack of efforts.

Jaguar

Rodrigo Núñez, Carlos López-González, and Brian Miller

Common Name: jaguar
Scientific Name: *Panthera onca*
Order: Carnivora
Family: Felidae
Status: Federally listed as Endangered in the United States (Austin & Palmer 1997) and Mexico (NOM 069 SEDESOL 1994); listed in Appendix I of CITES; Near Threatened on IUCN 1996 Red List.
Threats: Habitat destruction; direct persecution; declining prey.
Habitat: The jaguar is mainly associated with tropical forests and wet areas, but it is also found in temperate forests and desert thornscrub.
Distribution: Historically, jaguars occurred in the southern parts of California, Arizona, New Mexico, and Texas to northern Argentina (Seymour 1989). The jaguar has been eliminated from the United States and is thought to exist in 38% of its original range in Mexico and Central America and in 67% of its original range in South America (Swank & Teer 1989). In many areas where jaguars remain, their densities are reduced (Swank & Teer 1989).

DESCRIPTION

The jaguar has a large head and stout body with relatively short legs; the skin color is yellowish-brown with a spotted pattern of large, broken-edge, black rosettes that surround a small block dot (Seymour 1989). Black jaguars occur in South America (Nowell & Jackson 1996). The weight varies considerably, with jaguars from forested areas being smaller than those from more open habitats. In Central America, the males average 57 kg and females 42 kg, but in South America males average 100 kg and females 76 kg (Nowell & Jackson 1996). In taxonomic classification Nowell and Jackson (1996) list nine subspecies, but a recent review lists only one (Larson 1997).

NATURAL HISTORY

Jaguars are known to prey on more than 85 different species, and they take prey according to availability (Seymour 1989). When large prey become scarce, jaguars switch to domestic livestock, which can become a major component of their diet (Hoogesteijn et al. 1993).

Jaguars play a key role in maintaining the ecological integrity of the landscape. For example, jaguar prey are also predators of seeds and plants, so jaguar predatory activities affect the structure of plant and animal assemblages that are far removed from the original sources of predation.

Home range size varies greatly and is related to age and sex of the animal and productivity of the area. The land tenure system is characterized by a male home range that overlaps several female ranges. The female range is related to prey availability, whereas the male home range reflects behavioral spacing and access to females. Female ranges vary between 10 and 11 km² in Belize (Rabinowitz & Nottingham 1986), 25 and 65 km² in western Mexico (authors' unpublished data), and 97 to 168 km² in Brazil (Crawshaw & Quigley 1991).

Likewise population densities vary greatly, and productivity can also affect degree of spatial overlap within a sex. Densities range from 1.4 residents per 100 km² in Brazil, to 2 per 100 km² in western Mexico, to 6.6 per 100 km² in Belize (same references as above).

Jaguars probably can breed at any time of year, but there may be a birth peak that relates to prey availability (Rabinowitz & Nottingham 1986). Gestation lasts 90 to 111 days, and average litter size is two (Seymour 1989; Nowell & Jackson 1996). The kittens are hidden in caves, fallen trees, rocks, brushy areas, and the like, and they remain with the mother for 1.5 to 2 years (Nowell & Jackson 1996). Young animals are then forced out of their natal areas, with females moving short distances while males travel far from their original location. This often places young animals, particularly males, in inferior habitat and in areas of high risk.

CONFLICTING ISSUES

Humans persecute jaguars because they are perceived as a threat to livestock, and many people do not know or understand the jaguar's role in maintaining ecosystem processes. Historically, large ranches kept a full-time cat hunter on staff, and some modern ranches continue to contract professional trappers. Many rural people shoot jaguars opportunistically; in fact, illegal shooting has been listed as the principal cause of death for jaguars in the United States (Austin & Palmer 1997). Shooting has caused five of six known jaguar deaths in and around the Chamela-Cuixmala Biosphere Reserve in western Mexico.

Persecution of jaguars in any form is usually not penalized under the law. Typically, there are few individuals assigned to enforcement, salaries are barely livable, and the level of training may not be sufficient. Additionally, in the rural areas of developing countries the legal system does not yet play a strong role in society.

African Wild Dog *(Lycaon pictus)*. Photo by M.G.L. Mills. Courtesy of M.G.L. Mills.

Altai Argali Sheep *(Ovis ammon ammon)*. Photo by H.M. Mix. Courtesy of H.M. Mix.

Top: Anegada Iguana
  (Cyclura pinguis).
  Photo by G. Mitchell.
  Courtesy of G. Mitchell.

Left: Aruba Island Rattlesnake
  (Crotalus unicolor).
  Photo by R.A. Odum.
  Courtesy of R.A. Odum.

Black-footed Ferret *(Mustela nigripes)*. Photo by T.W. Clark. Courtesy of T.W. Clark.

Boreal Toad *(Bufo boreas boreas)*. Photo by E. Muths. Courtesy of E. Muths.

Adult Female Chinese Alligator (*Alligator sinensis*). Photo by J. Thorbjarnarson.
Courtesy of J. Thorbjarnarson.

Dalmatian Pelican *(Pelecanus crispus)*. Photo by H.M. Mix. Courtesy of H.M. Mix.

Right: Douc Langur *(Pygathrix nemaeus nemaeus)*. Photo by B. Nichols. Courtesy of B. Nichols.

Bottom: Eastern Barred Bandicoot *(Perameles gunnii)*. Photo by J. Seebeck. Courtesy of J. Seebeck.Courtesy of B. Nichols.

Ethiopian Wolf *(Canis simensis)*. Photo by C. Sillero-Zubiri. Courtesy of C. Sillero/BFF.

European Mink *(Mustela lutreola)*. Photo by T. Maran. Courtesy of T. Maran.

Top: Florida Manatee
(Trichechus manatus latirostris).
Photo by G. Rathbun.
Courtey of G. Rathbun.

Right: Giant Panda
(Ailuropoda melanoleuca).
Photo by R.P. Reading.
Courtesy of R.P. Reading.

Golden Lion Tamarin *(Leontopithecus rosalia)*. Photo by J.M. Dietz. Courtesy of J.M. Dietz.

Golden-rumped Elephant-shrew *(Rhynchocyon chrysopygus)*. Photo by G. Rathbun. Courtesy of G. Rathbun.

Mountain Gorilla *(Gorilla gorilla)*.
Photo by A. Vedder and W. Weber. © Vedder/Weber.

Adult Male Hawaiian Goose *(Branta sandvicensis)*. Photo by P.C. Banko.
Courtesy of P.C. Banko.

Three Asian Elephants *(Elephus maximus)*. Photo by J. Plungy.
Courtesy of Denver Zoological Foundation.

Fiamanga *(Paratilapia bleekeri)*. Photo by R. Haeffner. Courtesy of R. Haeffner.

Jaguar *(Panthera onca)*. Photo by M. Kinsey. Courtesy of Denver Zoological Foundation.

Marbled Murrelet *(Brachyramphus marmoratus)*. Photo by U.S. Fish and Wildlife Service. Courtesy of U.S. Fish and Wildlife Service.

Mexican Prairie Dog *(Cynomys mexicanus)*. Photo by J. Treviño-Villarreal. Courtesy of J. Treviño-Villarreal.

Mountain Plover *(Charadrius montanus)*. Photo by Stephen J. Dinsmore.
Courtesy of Stephen J. Dinsmore.

Northern Rocky Mountain Wolf *(Canis lupus nubilus)*. Photo by W. Campbell.
Courtesy of W. Campbell.

Left: Po'o-uli (Black-faced
Honeycreeper)
(Melamprosops pheaosoma).
Photo by J. Kowalsky.
Courtesy of J. Kowalsky.

Bottom: Scarlet Macaw
(Ara macao cyanoptera).
Photo by K. Renton.
Courtesy of K. Renton.

Top: The Stellate Sturgeon or
Sevruga *(Acipenser stellatus).*
Photo by P. Vecsei.
Courtesy of P. Vecsei.

Right: Aye-aye *(Daubentonia
madagascariensis).*
Photo by J. Edwards.
Courtesy of Denver
Zoological Foundation.

Grevy's Zebra *(Equus grevyi)*. Photo by J. Edwards.
Courtesy of Denver Zoological Foudation.

Male Asiatic Lion *(Panthera leo persica)*. Photo by R.S. Chundawat.
Courtesy of R.S. Chundawat.

Loss of habitat is a major cause of decline. Deforestation rates in Latin America vary from country to country, but they are generally among the highest in the world. Demographic expansion, poverty, devaluing economies, and social conflicts in Latin America have pushed people farther into the tropical forests, and their main activity is subsistence agriculture. Indeed, 90% of the human population growth occurs in tropical countries (Conway 1992).

Sport hunting is permitted in Bolivia, and there is no protection in Ecuador or Guyana. A professionally guided jaguar hunt can cost as much as U.S. $10,000, but sport hunting does not have as much impact as opportunistic shooting by ranchers and farmers. CITES protection has drastically curbed the harvest and trade in jaguar skins, which was very high until the 1970s (Nowell & Jackson 1996).

Many Latin American countries also have large international debts, and in some cases the present interest on debts now exceeds the original loan (Middleton et al. 1993). Roughly 45% of the economy for all developing countries relies on resource extraction (Middleton et al. 1993). That situation leaves little alternative but to destroy the environment for short-term economic survival (Frazier 1990).

As quality habitat of the jaguar is fragmented, it means that each cat must cover more ground to fulfill its needs, and there is a higher proportion of movement in disturbed areas where the jaguar is more vulnerable. Even in protected ares, jaguars are vulnerable near the boundaries; and the smaller the protected area, the higher the proportion of jaguars that will travel near those boundaries (Woodroffe & Ginsberg 1998).

As forests are converted by logging or subsistence agriculture, settlers hunt game for food. When large prey are reduced, jaguars expand their home range size and their movements, increasing the risks mentioned in the previous paragraph. In addition, low prey numbers enhance any jaguar's chances of killing livestock, and this is particularly true for young dispersers that have relatively little experience in hunting (Hoogesteijn et al. 1993).

Our study in Mexico indicates that the ecological balance between jaguars and pumas shifts in favor of the puma when habitat is fragmented. Pumas, like jaguars, prefer to eat large prey, but in altered landscapes pumas can switch to smaller prey that are not readily utilized by the jaguar.

Another problem is a general lack of biological knowledge, and there have been few published studies on jaguar ecology. Indeed, there is even a lack of knowledge on basic distribution and densities, and there is no common method by which to census jaguars.

Across the range, decisions that affect jaguars are made by a large number of different administrative units. The more governmental, administrative, and agency borders that are crossed by a species, the more difficult it becomes to forge partnerships for recovery. Different organizations have different goals and different definitions of problems. Administrative boundaries drawn on maps may not reflect informal levels of power in the region (Frazier 1990). For example, large areas of land may be controlled by factors that have nothing to do with official government policy (e.g., drug cartels or political rebellions).

Wildlife agencies are often understaffed, which causes delays in permitting and planning (Mares 1991). Economic devaluation is a constant threat, and unstable economies make it difficult for governments to commit funds to conservation projects (Mares 1991). The nongovernmental conservation movement in Latin America is growing, but it still lacks a strong power and financial base.

In general, Latin American political agencies are organized as top-down bureaucracies. Rules and procedures are complicated, and the slow process can impede science (Mares 1991). In addition, when there is political change, entire institutions are often replaced.

Political power in Latin America is often centralized in the capital, and many rural people thus feel that their needs are not represented. Historically, the politically powerful have used their advantage to accumulate wealth, so there is sound basis for that skepticism (Middleton et al. 1993). A conservation plan for jaguars must move through those political circles of the capital while still including local people in the process.

There is an incredible array of biodiversity in Latin America, but there are not many professional biologists (Mares 1991). This situation allows a few people to dominate the conservation arena, and their influence often extends outside their area of expertise (Mares 1991). The creation of such "empires" can result in fighting between different factions, which impedes the science needed for conservation plans (Mares 1991).

The costs of maintaining an "empire" often mean that a significant portion of funding does not reach the field project. In addition, institutional procedures may delay the arrival of funding to the project, and if the currency is devaluing, that can impede, or end, a project (Mares 1991). Gathering data on jaguars is both labor intensive and expensive, so secure funding is important to success.

FUTURE AND PROGNOSIS

Even though issues of population growth and poverty in tropical areas (coupled with excessive consumption in other areas), unstable economies, and debt seem too large to address, they have a negative impact on jaguar habitat. The U.N. Conference on the Environment and Development, held in Rio de Janeiro in 1992, offered a forum in which to discuss such global issues; but as a species, humans still place short-term personal advance over the needs of nature or other people. That attitude remains at the root of many environmental problems.

The effects of fragmentation are most visible over an entire region, and one must start planning across landscapes. The piecemeal approach (fighting brushfires) has failed. Even if jaguars presently exist in isolated populations, surrounding development will limit dispersal and increase vulnerability. The smaller the isolated population, the more vulnerable it becomes over time.

If fragmentation isolates and shrinks jaguar populations, then the logical method of countering this problem is to expand the size of core areas and reconnect them. Plans such as Paseo Pantera in Central America are a bold step in this direction, and nature will benefit from such ventures.

A recent effort by the Wildlife Conservation Society to cooperatively prioritize geographical areas for action will also help tremendously. This should push the scale of thinking past administrative boundaries and foster a more unified definition of the problem. It could also ensure that methods are coordinated so that results are more comparable across different regions.

Plans to reduce livestock predation are essential. In reality, jaguars are not predisposed to kill livestock, and they often do so only after their natural prey have been reduced by human competitors (Hoogesteijn et al. 1993). Simply moving "problem animals" often results in death or homing, and alone it does not address the conditions that created a livestock killer. We also do not advocate the sport hunting of problem animals for the same reasons. It would be far more productive to increase prey and work with local ranchers to alter their management programs (see Hoogesteijn et al. 1993).

Sport hunting has been proposed as a way to increase the value of jaguars. However, there are three counterarguments. First, there can be no regulated hunting without effective enforcement and reduced corruption. Second, jaguars did not evolve as a prey item. They therefore have a low reproductive rate. For such animals the discount rate can exceed the sustainable harvest rate, so there is economic pressure to quickly convert the population to cash rather than sustainably harvest over the long term; the pressure is increased when money is in short supply (Caughley 1993). Third, much of the money generated from hunting jaguars has historically been concentrated in the hands of a few, not with the people of the region.

Finally, protected areas, even with strong enforcement, will not be sufficient. If local people do not support the reserve, it will simply become an island with all the associated problems of isolated fragments. How local people are reached will depend on the situation, but buffer zones around reserves can provide the linkages between cores for plans such as Paseo Pantera.

Kemp's Ridley Sea Turtle

Jack Frazier

Common Name: Kemp's ridley sea turtle, bastard turtle
Scientific Name: *Lepidochelys kempii* (Garman 1880)
Order: Testudines (stem reptiles, turtles and tortoises)
Family: Cheloniidae (hard-shelled sea turtles)
Status: IUCN—one of the 12 Most Endangered animals in the world (1986), Critically Endangered (1996); CITES—Endangered, Appendix I (1973); Mexico—Totally Protected (1973, 1990); United States—Totally Protected (1970), Endangered (1973—Endangered Species Act); Florida—Totally Protected (1974); Texas—Protected (1963) (Márquez 1994; TEWG 1998).
Threats: Historically local egg harvests were intense, as was commercial exploitation of adults and immatures for food and leather. Contemporary threats include incidental capture in fisheries, notably shrimp trawls; disturbance of nesting habitat; and marine and coastal contamination.
Habitat: Nests on sandy beaches, mainly in Tamaulipas, Mexico; newly hatched turtles live in pelagic (open ocean) gyres in the Gulf of Mexico; adults live in near-shore waters of the Gulf; juveniles live in near-shore areas of the Gulf and western North Atlantic, often in shallow, protected inshore bays and lagoons in the east, or along beachfronts in the west.
Distribution: The Gulf of Mexico and Atlantic coast from Florida to Maine, occasionally Nova Scotia and Newfoundland; uncommonly found in European Atlantic waters; dubious records from the Caribbean (Brongersma 1972; Márquez 1994; TEWG 1998).

DESCRIPTION

The carapace (top shell) of Kemp's ridley sea turtle has 39 scutes: 5 vertebrals, 5 pairs of costals, and 12 pairs of marginals. In juveniles it is dark green, contrasting strongly from the white plastron (bottom shell) and other ventral surfaces; the posterior of each vertebral scute in the central dorsal keel develops a prominent spine or knob. The smallest living sea turtle, adult females average 38 kg (25–54 kg), and straight carapace length varies from 55 to 78 cm (usually 60–70 cm). The plastron is joined to the carapace on both sides by four inframarginal scutes, each scute with a conspicuous pore. Front limbs are flippers and hind limbs are paddles, each with two claws—at the first and second digits. Adult shells are nearly as wide as long, smooth with no dorsal keel, drab gray-green dorsally, pale cream or yellow ventrally, and with no sexual difference. Adult males have muscular, prehensile tails (i.e., grasping), as long as a hind leg, with a claw-like tip, a well-developed, curved claw on the first digit of each limb (Márquez 1994), and a softened plastron (Owens personal communication).

The difference between Kemp's and the congeneric (same genus) olive ridley (*L. olivacea*) is deceptive (some earlier authors considered *kempii* to be a subspecies of *olivacea*). Genetic analyses (Bowen & Karl 1997) indicate that Kemp's ridley diverged 3 to 6 million years ago, following the separation of the Atlantic and Pacific Oceans by the formation of the Isthmus of Panama. The two species rarely occur in the same area and may be identified by geographic origin (see Márquez 1994).

NATURAL HISTORY

After nesting in Tamaulipas, Mexico, females migrate northward or southward along the coast; nonbreeders are abundant in Florida Bay, the Mississippi Delta, and Campeche Banks. Adults and juveniles are found mostly in the Gulf of Mexico; juveniles also occur in the western Atlantic from Florida to Maine. Kemp's ridleys are often found in shallow waters with white shrimp and portunid crabs; they feed mainly on crustaceans associated with clay, sandy clay, or sandy bottoms (Bjorndal 1997). Benthic (bottom-feeding) juveniles along the eastern seaboard of the United States make seasonal north-south migrations to avoid cold water; smaller animals are recorded farther north (Musick & Limpus 1997). Juveniles have strong site fidelity in the western Gulf. Kemp's may burrow into bottom mud, becoming torpid when water temperature drops to about 11°C (Márquez 1994).

Nesting, from April to July or August, peaks in May and June. Clutch size ranges from about 20 to 190, averaging between 97 and 112 eggs; since 1966 there has been a trend for decreasing average annual clutch size. Eggs, usually white and virtually spherical, range from 35 to 46 mm in diameter and from 24 to 41 g in weight (Márquez 1994). Within a season, a female nests an average of three times, with 24-day intervals (range 7–60 days) between nests. Massed nestings (called *arribadas* in Spanish) are usual, and often occur with a third-quarter moon and onshore winds. Tens of thousands used to nest in each *arribada*, but now only a few hundred nest annually. Unlike other sea turtles, *kempii* regularly nests during daylight hours; most females nest every 2 years (Márquez 1994; TEWG 1998).

In natural nests incubation (together with hatchling emergence) takes from 44 to 52 days, occurring without parental care. Hatchlings usually emerge from the nest (as much as 40 cm deep) during the night, especially between midnight and dawn, and run directly to the sea. Little is known of the first few years of life; hatchlings are only known from the Gulf of Mexico. The distinctly counter-shaded coloration and dorsal armor of "dinner-plate sized" turtles is evidence for a pelagic existence. At about 20–25 cm carapace length they change from pelagic-netonic (open-ocean, surface feeders) to benthic-neritic (near-shore, bottom feeders) (Márquez 1994; Musick & Limpus 1997).

Age at first reproduction may be 10 to 15 years in wild turtles (Zug et al. 1996) and as early as 5 years in captivity (Wood & Wood 1988). Nesting females have continued nesting for at least 11 years, so the potential reproductive contribution of a single female is thousands of eggs (see Márquez 1994; Burchfield et al. 1997; TEWG 1998).

CONFLICTING ISSUES

Kemp's ridley conservation has been extraordinarily contentious. With a long tradition of direct exploitation throughout their range, *kempii* were openly sold and eaten in Mexico and the United States until the 1970s; commercial operations centered in Texas and Florida, lasting into the 1950s in the latter. Intense exploitation of eggs at Rancho Nuevo, Tamaulipas, the primary nesting beach, occurred until 1966; in 1970 the Mexican government issued a permit for capture and processing of 5,000 *kempii* (although none were caught). Focused conservation actions began after 1963, when a 1947 film of an *arribada* was discovered; an estimated 40,000 females were on the beach in the film. By the end of the 1960s a few thousand were estimated to nest each year, and by the 1980s fewer than 800 nests were counted in each of six sequential nesting seasons (Márquez 1994).

Incidental mortality in fisheries went undetected until the 1970s, when, after decades of mechanization and intensification, shrimp trawling was identified as significant. Turtles are caught in various fisheries, but a study commissioned by the National Research Council (NRC 1990) estimated annual mortality from shrimp trawling in the United States at 500 to 5,000 Kemp's ridleys, far above all other forms of fisheries mortality (50 to 500 annually), as well as other sources of human-induced mortality. Although Millett (1998) disputed these figures and defamed the authors, these are the best estimations available. About half the *kempii* stranded dead in the United States were recorded from Texas (Márquez 1994).

The catastrophic decline in nesters and increased mortality led to dramatic changes in the policies of both countries, from uncontrolled exploitation to total protection, with various governmental listings by Mexico and the United States (as well as two individual states). The IUCN even dubbed this one of the 12 most endangered species in the world.

Nest protection began in 1966 at Rancho Nuevo, and in 1978 a binational, multi-institutional project began, providing intensified support (material, financial, and personnel) for beach patrolling, nest and egg protection, basic research, and community education. In addition to fieldwork focused at Rancho Nuevo, experimental beach imprinting and captive rearing ("headstarting") programs were carried out, mostly in Texas. Between 1978 and 1988 more than 22,500 eggs (about 2,000 eggs annually) were sent to a hatchery on North Padre Island, Texas, using the island's sand for imprinting during incubation. After hatching, the young were headstarted (reared) for about a year in Galveston, Texas, then marked and released into the Gulf of Mexico. From 1989 to 1992 headstarting and release continued, but without "imprinting" on North Padre. During these 13 years more than 20,000 yearling, captive-reared *kempii* were released into the coastal waters of Texas and Florida (Márquez 1994), ostensibly to provide rapid recruitment and rehabilitation of the population, and to start/restart a nesting colony in south Texas—an area of high intensity shrimp trawling. Concern about headstarting's experimental nature, operational costs, and no clear results led to an evaluation, which concluded that captive rearing should be stopped, while clear objectives should be established—including intensified monitoring to gather information on survival and fecundity (Eckert et al. 1994). Between 1995 and 1998 a total of 32 Kemp's nests were detected on North Padre and nearby islands, at least 6 of them from headstarted animals (Shaver & Caillouet 1998).

Also in 1978, after 5 years of growing concern about sea turtle mortality caused by shrimping, the National Marine Fisheries Service (NMFS) started a program to develop turtle excluder devices (TEDs). The Sea Turtle Stranding and Salvage Network was established in 1980 in the United States by NMFS to record turtles, live and dead, found on beaches. During the 1980s scores of studies identified shrimp trawling as an important source of mortality for sea turtles in U.S. waters of the Gulf and Atlantic. Simultaneously conflict in the United States escalated, as conservationists promoted voluntary (1981–1985), then mandatory, use of TEDs. Shrimpers, shrimp industry representatives, conservationists, U.S. senators and congressmen, U.S. government (especially NMFS) employees, a White House chief of staff, and even a president were involved in the "TED wars." There were bodily threats and even a port blockade. Consequently several politicians, a secretary of commerce, other civil servants, and some representatives of shrimping interests are viewed by conservationists as caricatures of inept leadership (Weber et al. 1995).

Major changes in fisheries occurred during this period. Several social scientists have written about the "TED wars" in the United States, describing complex problems, many of which derived from actions beyond the shrimpers' control. U.S. shrimpers were phased out of Mexican waters beginning in 1976; 2 years later the Exclusive Economic Zone (EEZ) was implemented with drastic reduction of U.S. fishing in Mexico. Hence, during the "TED wars" an overcapitalized U.S. shrimp fleet lost access to rich shrimping grounds in Mexico, while simultaneously operating costs soared. Moreover, refugee Vietnamese immigrants entered the fishery industry with low-interest loans and modern vessels that are unattainable for most long-established shrimpers. Perhaps most significant was that low-cost, pond-reared shrimp from third world countries began to flood the U.S.

market, driving down prices. All this happened when anti-government feelings proliferated during and after the Reagan years. A common criticism was the way in which TED technology was developed and implemented—with inadequate participation by the shrimping community. Suspicions of ulterior motives for TEDs compounded the issue, helping to escalate the rhetoric and further polarize parties.

The history of this mêlée belies a rational policy for renewable, common property resources. After nearly two decades of denial, court cases, administrative delays and reversals, broken promises by the U.S. government, escalating conflict, and no clear advancement in resource protection or conflict resolution, three states enacted their own TED regulations: South Carolina in 1988; Florida, 1989; and Georgia, 1990. TEDs were federally implemented, although sporadically, in U.S. offshore waters in May 1990; and in 1993 TEDs were implemented from North Carolina to Texas in both state and federal waters (Weber et al. 1995). A debate continues about the economic effect of TEDs on the U.S. shrimp industry, yet South Carolina shrimpers maintained that because the gear is useful in excluding unwanted catch, they would employ TEDs whether or not they were required. Federal TED regulations have been expanded in the United States since 1995.

The history of Kemp's ridley conservation in Mexico is less detailed and contentious. Exploitation of eggs did not stop in 1966 when they were legally protected; there has been an ongoing, major effort at community education in the area of the main nesting beach at Rancho Nuevo. The response has been enthusiastic, and egg harvest is now only a minor problem (Burchfield et al. 1997). The exclusion of U.S. shrimpers from Mexican waters in 1978 had major consequences, for the Mexican fleet in the Gulf of Mexico is relatively small and much of it has been inactive. The first TED trials in Mexico were in 1984, where implementation of TEDs has met resistance from the shrimping community, but with less spectacle than in the United States. This is partly because the conservation community is smaller, weaker, and less confrontational in Mexico, and partly because the government has greater financial and political control over much of the fishery. In 1990 Mexico banned all sea turtle exploitation, and in 1993 TEDs were required on all Mexican shrimp trawlers in the Gulf and Caribbean. However, systematic information on sea turtle strandings and incidental captures is lacking.

Enforcement of TEDs, as well as closed areas to trawling—notably Rancho Nuevo—has been loose in Mexico. There has probably been significant trawl-related mortality in northern Tamaulipas, just south of Texas, where the problem is well documented. Although many questions are still unanswered, data indicate that in recent years incidental capture in Mexican waters has been less important than in the northern Gulf. Of enormous concern is the threat of massive coastal development and expansion of fisheries activities adjacent to the main nesting area, as well as plans for dredging an intercoastal waterway in Tamaulipas. Nearly all Kemp's nest in this state, and over 60% of nests recorded in 1997 were from Rancho Nuevo (Burchfield et al. 1997); hence the main breeding population and area are vulnerable to focused, small-scale catastrophes (e.g., development, hurricanes, and oil spills).

Because of habitat and dietary preferences throughout their range, Kemp's ridleys regularly occur in the same places where shrimp are trawled (moreover, these turtles eat by-catch from shrimp trawlers). Thus their habits make these turtles liable to incidental capture—and drowning—in one of the most intense fisheries that targets one of the most valuable products: shrimp. Destruction of coastal areas, as well as watershed and marine contamination, deteriorate the turtles' habitat with poorly understood, but clearly negative, effects. The history of Kemp's ridley epitomizes the conflict between intensive, industrialized extraction of marine resources (shrimp trawling), modern forms of coastal development, and agricultural practices on the one hand, and protection of species with decimated breeding populations on the other. Balancing the lure of short-term financial and political gain with long-term environmental and social needs is a serious challenge, resulting in intense, complex, poorly understood conflicts (Frazier 1997).

Its highly endangered status has resulted in Kemp's ridley being the focus of diverse reports and publications (e.g., Caillouet & Landry 1989; Ross et al. 1989; Eckert et al. 1994; USFWS & NMFS 1992; Weber et al. 1995; TEWG 1998). Yet essential information and basic understanding about many fundamental biological and conservation issues are still inadequate. This fact is employed by some organizations (e.g., Millett 1998) to invalidate attempts by experienced professionals at developing conservation and management recommendations using available information. The size, dispersion, and educational level of some of the major stakeholder groups exacerbate the problem. For example, of the tens of thousands of shrimpers spread along the shores of the Gulf, historically few have had a high school education, much less an appreciation for complex fisheries management issues. When combined with intentional misinformation (e.g., Millett 1998), the results are devastating (Pulliam 1998).

Although efforts in monitoring nesting beaches have been increasing over the last few years, the Turtle Expert Working Group (convened by the U.S. government) recently concluded that the Kemp's ridley population now seems to be in the early phases of recovery (TEWG 1998). This has taken more than three decades of strict control of nesting beaches, a U.S. federal program to develop and implement selective fishing techniques, and a binational, multi-institutional enterprise over the past two decades—all involving hundreds of thousands of person-hours and millions of dollars. The costs of achieving this initial phase of recovery have been immense for most parties in the conflict, not just financially but also environmentally and socially. Expenditures on preventive measures are unlikely to have totaled even a small part of these costs; recovery efforts are paying the accumulated damages of delaying for nearly two decades to address major threats to the species.

FUTURE AND PROGNOSIS

Substantial species recovery will require a sustained commitment by agribusiness, coastal developers, commercial and sport fin-fishers, conservationists, educators, fisheries industry financiers and investors, politicians, researchers, shrimpers, and other members of both Mexican and U.S. societies. Sincere efforts to identify and resolve major issues in a complex, changing world, without locking into one paradigm, require the nurturing of ongoing public education, awareness, and involvement in the issues. The history of polarized interest groups, fueled by short-term political and financial gain and operating on inadequate information, must become an object lesson on how *not* to proceed. "Saber rattling" and "blood-letting" persist: some environmentalists still make strident claims, taking intransigent stands, while industry contractors (e.g., Millett 1998) make scathing—if not libelous—attacks on scientists and national committees responsible for findings inconsistent with the desired image of fishing and seafood enterprises. The fishing industry, often viewed as a monolithic enemy of turtles, is a complex, multisectorial enterprise (with its own internal conflicts), but it must support conservation programs if they are to function. This is now happening with some sectors: for example, the National Fisheries Institute, Inc., and Texas Shrimp Association, Inc., have provided material and labor for beach protection facilities in Mexico.

The root of the problem is the way that modern fisheries and coastal land use have been developed: with unlimited access to resources and massive investments of capital, involving continual and profound environmental impacts. Bringing together members of different—often antagonistic—interest groups to foster communication, collaboration, and respect is essential (e.g., Caillouet & Landry 1989; Owens and Evans 1999), because this lays the foundation for common understandings and plans for access, use, and protection of resources shared by diverse groups. This is not to say that simply convening meetings will solve basic problems, but it must be recognized that fundamental issues are social and political—not biological (Bookchin 1994; Brulle 1996).

ACKNOWLEDGMENTS

Valuable comments on earlier drafts of this entry were made by S. Branstetter, P. Burchfield, J. C. Cantú, M. Coyne, M. Donnelly, B. Gallaway, A. Landry Jr., R. Márquez M., B. Miller, D. Owens, D. Shaver, and M. Weber.

Leatherback Turtle

Jack Frazier

Common Name: Leatherback, leathery, or trunkback turtle
Scientific Name: *Dermochelys coriacea*
Order: Testudines (stem reptiles, turtles, and tortoises)
Family: Dermochelyidae (soft-shelled sea turtles)
Status: IUCN—Endangered (1996); CITES—Endangered, Appendix I (1996); United States—Totally Protected (1970), Endangered (1973—Endangered Species Act).
Threats: Extensive harvests of eggs and nesting females; loss and disturbance to nesting habitat; incidental capture in oceanic fisheries; marine contamination—especially plastics.
Habitat: Nests on sandy beaches; feeds and lives in the open ocean.
Distribution: Occurs from tropical to subpolar oceans; nests on tropical, rarely subtropical, beaches.

DESCRIPTION

The carapace (top shell) is composed of hundreds of interlocking osteoderms, irregular polygonal bones that are rarely more than 3 cm long. Seven longitudinal keels and a prolongation at the posterior distinguish the carapace. A narrow oval of osteoderms outlines the plastron (bottom shell). One tooth-like projection is located on either side of the upper beak. The front flippers are well over half the length of the carapace; a fleshy fold joins the hind flippers and tail. Adult body color is black with irregular white spots, and the plastron is off-white with black blotches. There is a pink rosette on the crown of adults. The largest of living chelonians (sea turtles), adults frequently surpass 2 m in total length and straight carapace length may exceed 1.5 m. Nesting females average about 300 kg; one male weighed 916 kg (Deraniyagala 1939; Morgan 1989). Bead-like scales on hatchlings are lost during the first months of life, but a thin beak is present throughout the turtle's life. Claws are absent, except in the odd embryo. This species is highly specialized (Frazier 1987; Rhodin et al. 1996).

Several internal and physiological characteristics of leatherbacks are also distinctive. Lung ventilation rate is extraordinarily high; blood-oxygen carrying capacity is twice that of other sea turtles; myoglobin (iron-containing pigment in the muscles) stores are very large; the limbs have vascular counter-current heat exchangers, vascular beds where arteries lie next to veins to allow the warm blood from the heart to heat up the cool blood from the extremities; the ends of long bones, or epiphysis, do not ossify and fuse but remain cartilaginous and are perforated with an extensive vascular system. These features, typical of marine mammals, are unknown in other living chelonians (i.e., turtles & tortoises) (Paladino et al. 1996; Rhodin et al. 1996; Lutcavage & Lutz 1997; Spotila et al. 1997).

NATURAL HISTORY

Leatherbacks nest mainly in the tropics on high energy (i.e., beaches exposed to wave action), sandy beaches that have unobstructed approaches to deep water. Some individuals lay up to 12 nests a season, and 4 to 7 is usual; the interval between nests is usually 9 or 10 days. These sea turtles generally have the largest number of nests per individual, per season, and the shortest inter-nesting interval. Leatherbacks lay the largest eggs of any sea turtle, often more than 5 cm in diameter and 70 g in weight. Clutch size may exceed 100, but about one-third of the eggs are small and "yolkless," with no known function. As in other sea turtles, incubation temperature determines the sex of leatherbacks, and 29.5°C is the "pivotal temperature" (where hatchling sex ratio is even). Incubation and emergence from the nest take about 2 months.

Nest destruction from tidal inundation and erosion is common, and more than 50% of the nests can be lost (Boulon et al. 1996). Often hatchlings pip the eggshell but do not emerge and die in the nest. Important nest predators are feral dogs, along with wild mammals, vultures, crabs, and ants; orcas prey on adults (Sarti et al. 1994). Hatching success and hatchling recruitment vary greatly between localities and seasons, from 5% to 80%. Hatchling survival during the first day at sea may be as low as 25%; fishes and birds can cause heavy at-sea mortality. The interval between nesting seasons is often 2 years, although it can be longer or as short as 1 year. Despite a 50% loss of conventional flipper tags between nesting seasons, certain females are known to have nested in seven different seasons, and some are known to have been nesting for at least 18 years (Boulon et al. 1996). Passive integrated transponder (PIT) tags have good retention and are now being employed (McDonald & Dutton 1996). Next to nothing is known of juveniles or adult males (Spotila et al. 1996; Miller 1997).

Leatherbacks eat soft-bodied invertebrates such as jellyfish (cnidarians), combjellies (ctenophores) and pelagic tunicates (salps), which sometimes appear in tremendous concentrations. Both turtles and prey occur regularly in temperate waters, and turtles evidently migrate across oceans in search of abundant food. Some females have moved thousands of kilometers from their nesting beaches in periods of a few months, averaging 70 km per day.

One of the deepest diving of air-breathing vertebrates, leatherbacks can reach depths of more than 1,000 m and can spend up to 42 minutes submerged, yet they seem to stay within the Aerobic Dive Limit (i.e., never become anaerobic and build up lactic acid) (Eckert et al. 1996). Leatherbacks can breach out of the water and have attacked swimmers and small boats. Adults can maintain relatively constant deep body temperatures, up to 18°C above water temperature. Thermal regulation is through "gigantism" and depends on large body size, efficient insulation, and low metabolism. Hatchling metabolism is three times that of other sea turtle species. Respiratory and blood physiology shows adaptations for an active, diving, air-breathing vertebrate (Paladino et al. 1996; Lutcavage & Lutz 1997; Spotila et al. 1997). Unlike other sea turtles, leatherbacks grow rapidly with unique skeletal adaptations (Rhodin et al. 1996). They mature in 6 to 13 years (Zug & Parham 1996). Nesting females from the east Pacific average smaller than those from Atlantic or Indian Ocean beaches. Genetic differentiation within a nesting population, as well as between females from distant nesting beaches, is less marked in leatherbacks than in other sea turtles (Dutton et al. 1999).

CONFLICTING ISSUES

The slaughter of leatherback turtles for meat and oil occurs widely, and in many coastal communities the takes have increased during recent years. Leatherback hunts, having lost traditional customs and controls, are serious threats; former sources of food for community rituals have become trade items for expanding markets (Suarez & Starbird 1996).

Leatherbacks are routinely subjected to heavy egg predation, especially by coastal peoples. For example, the population at Rantau Abang, Terengganu, Malaysia, once thought to be the largest nesting population in the world, now has been reduced to less than 1% of what was recorded during the 1950s. Other nesting populations in Asia, also subjected to intense egg takes, have shown similar declines (Chan & Liew 1996; Spotila et al. 1996).

Heavy egg predation and nest destruction have led to hatchery programs around the world. Often, these have been poorly conceived and executed (Chan & Liew 1996) and may have resulted in 100% "masculinization" of "protected" hatchlings (Dutton et al. 1985) or hatchling release sites that feed offshore fishes. Hence some attractive, hands-on conservation activities have been counterproductive.

Nesting populations on Pacific Costa Rica and Mexico have also declined dramatically. Intensive egg takes impacted the population at Playa Grande, Costa Rica (Spotila et al. 1996), which dropped from 1,372 to 109 annual nesters in one decade (Paladino et al. 1996). There has been consistent protection of the nesting beach at Mexiquillo, Mexico, for more than a decade, yet annual nesting has plummeted from thousands to just one turtle (Sarti et al. in press).

Sustained, intensive egg/hatchling mortality can alone decimate a population, but egg harvests and inadequate hatchery practices have been compounded with (1) the mechanization of offshore fisheries using bottom trawls and gill nets, and (2) the high seas squid driftnet fisheries (Chan & Liew 1996; Spotila et al. 1996). Egg takes and direct slaughter of nesting females are apparent, but less obvious sources of mortality are easily overlooked or even discounted (e.g., incidental catch in fisheries, especially long lines and driftnets). Known since the 1950s, evidence of incidental capture and death has burgeoned during the past decade (e.g., Wetherall et al. 1993; Lutcavage et al. 1997; Witzell 1999). Recently some authors have identified high seas fisheries as the greatest culprit (e.g., Eckert & Sarti 1997; Sarti et al. in press).

Mortality from pollution is also easily overlooked because documentation depends on chance observations. Since the 1970s, leatherbacks have been known to be especially liable to mortality from ingestion of plastics and entanglement (Lutcavage et al. 1997). For example, phthalate esters (compounds liberated from plastics) have been found in leatherback egg yolks (Juarez-Cerón et al. in press). Pollution effects are typically chronic, lasting for many years, and marine pollution is intensifying in many areas (NRC 1995).

Researchers concerned about decreases in nesting numbers have reported that certain populations are in a critical state (e.g., Chan & Liew 1996; Sarti et al. in press) and even that leatherbacks may become extinct (Spotila et al. 1996), but one person disputes that conclusion (Pritchard 1996). This controversy merits discussion, for countless turtle enthusiasts have looked to Pritchard for leadership, especially regarding leatherbacks. Although several of his points regarding weaknesses in demographic modeling are convincing, his overall conclusion (Pritchard 1996:303) "that the leatherback is a vigorous and dynamic species" and, hence, is in no danger requires thoughtful consideration.

Some nesting populations have increased. Over the past decades, nesting at both Tongaland, South Africa (Hughes 1996), and St. Croix, U.S. Virgin Islands (Dutton et al. in press), has risen steadily to about 100 nesters per year. In addition, important nesting populations have recently been documented on the western coast of Africa, although it is unknown what trends these have experienced. The nesting population at Ya:lima:po Beach, French Guiana, seemed to be holding its own, with some 50,000 nests (about 10,000 turtles) reported in 1988 and again in 1992 (Girondot & Fretey 1996). This was reassuring, as it is by far the largest known leatherback population. However, annual numbers of nests there have dropped rapidly, to 7,800 nests (perhaps 1,500 turtles) in 1998. Beaches are changing and populations moving, so the intensity of the trend is unclear, but there is no doubt that a serious decline has occurred at Ya:lima:po Beach (Girondot personal communication). Hence the "vigorous" nesting populations of the Caribbean Sea and Indian Ocean are, when taken together, nowhere near as large as any one of the decimated populations from the eastern Pacific, Malaysia, or South America. Loss of populations is the first step toward extinction of a species.

The complaint by Pritchard (1996) that demographic models of leatherbacks (Spotila et al. 1996) are incomplete and imperfect, and depend on "inputs" from species other than the one being modeled, is indisputable. Nevertheless, demographic simulations for all sea turtles have consistently emphasized how important juvenile and adult survival are to the status of the respective populations (Spotila et al. 1996 and references therein). These are models and thus are perfectible. However, insisting that a model be perfect before trying to make the best possible conservation and management decisions is a luxury that heavily impacted species, particularly if endangered, cannot afford.

Two of Pritchard's recent declarations are remarkable and symptomatic of a fundamental problem. He stated, "Man has had no known influence upon any of the juvenile and subadult stages of the leatherback, and thus the survival of hatchlings as they pass through these stages, lacking data to the contrary, may be assumed to be essentially unchanged from pre-human conditions" (Pritchard 1996:304); and "The leatherback does not feature in international commerce, and its juvenile stages (indeed, all stages between hatchling and adult) remain so cryptic that it is unlikely that humans have any effect on them" (Pritchard 1997:24).

Scientific literature on both sea turtles and oceanic fisheries shows that sources of at-sea mortality on sea turtles, although poorly understood, are innumerable, diverse, and increasing. Although known 30 years ago, the problem was underrated. Today burgeoning proof of incidental capture and death in oceanic long-line and driftnet fisheries cannot be denied. Thousands of records show that leatherbacks, from 43 cm long to adult size, are caught and killed in oceanic fisheries (e.g., Zug & Parham 1996; Lutcavage et al. 1997; Witzell 1999 and references therein). This is not to mention the complex challenge of documenting impacts of ever-increasing marine pollution on every age group of leatherbacks. Hence contentions (e.g., Pritchard 1996, 1997) that human effects on leatherbacks on the high seas (at any part of their life cycle) are insignificant, or that juvenile and subadult mortality is not critical to the status of a population, are untenable.

Moreover, stating that there is no problem because scientific studies do not prove otherwise is consistent with the philosophy of unlimited exploitation, not rational management and conservation. Such arguments were fashionable during the first half of the 20th century and earlier, when resources were considered boundless and technology was regarded as the panacea. Because these assumptions have led to serious environmental disasters, there have been intensive international efforts to develop and implement a "Precautionary Approach" for fisheries. A milestone is the *Code of Conduct for Responsible Fisheries*, adopted by the Twenty-eighth Session of the Food and Agriculture Organization of the United Nations Conference on 31 October 1995, for this establishes several basic principles (FAO 1995:12):

The absence of adequate scientific information should not be used as a reason for postponing or failing to take conservation and management measures. . . .

In implementing the precautionary approach, States should take into account, *inter alia*, uncertainties relating to the size and productivity of the stocks, reference points, stock condition in relation to reference points, levels and distribution of fishing mortality and the impact of fishing activities, including discards, on non-target and associated or dependent species as well as environmental and socio-economic conditions.

Siding vigorously against the "precautionary approach" manifests a philosophy regarding global fisheries that undermines contemporary efforts to turn the tide on one of the greatest debacles of modern society, the depletion and destruction of marine resources and habitats. Slogans of "scientific rigor" and "sustainable use," when ultimately promoting laissez faire extraction, will only further degenerate an already delicate and complex situation (Hunter 1997).

FUTURE AND PROGNOSIS

Basic information on status, demography, migration routes, and other life history characteristics is needed. Nonetheless, despite the incompleteness and imperfection of contemporary knowledge, several conservation priorities are clear.

Protection of nesting beaches is fundamental. Without a safe place to reproduce, any species is doomed. Necessary actions include the reduction, or even prevention, of coastal development, egg predation (especially by people, whether or not it is legal or customary), and contamination (especially light pollution). This is much easier said than done, for around the world there are socially accepted customs, as well as strong economic incentives, for coastal peoples to harvest turtle eggs, and controls are rare. Promoting processes in which these resource-users become responsible stewards necessitates community-based conservation; this implies long-term commitments by conservationists and community workers (Frazier 1999).

Generally, the nations involved also suffer economic, political, and social problems, which add to the enormity of the dilemma. Even so, the first step is to recognize that leatherback conservation depends intimately on the participation of peoples, often referred to as "poachers," who exploit turtles and their eggs. More complex is controlling coastal development. There is a massive global move to develop tourist and residential facilities throughout tropical seashores. Because these initiatives are regularly driven by powerful transnational corporations, promulgated by third world elites and their governments, and packaged as part of the "sustainable development" fantasy

(Frazier 1997), they are almost impossible to control. The beaches around Rantau Abang, Malaysia (Chan & Liew 1996), or even Las Baulas Marine Park, Costa Rica (Steyermark et al. 1996), are proof of how difficult it is to control coastal development, even directly adjacent to areas legally protected for leatherback nesting.

Nonetheless the best marine park in Costa Rica, the most wonderful laws and turtle field camps in Mexico, and the most dedicated efforts of beach protection and hatcheries in Malaysia will not protect leatherbacks adequately; once the turtles set off on their migrations through the Pacific Ocean they will be vulnerable to intensive fisheries and chronic pollution. Leatherbacks, more than any other turtles, provide the example that highlights the complexities of biological conservation in this modern world. In the end they are global resources and therefore require global cooperation to be conserved. The yet-to-be-proven Inter-American Convention for the Protection and Conservation of Sea Turtles is an attempt to meet this goal in the Western Hemisphere (Frazier 1999).

High seas fisheries are notorious for their rates of extracting living marine resources and are often characterized as "the tragedy of the commons," in which resource users exploit as rapidly as they can with little concern for the future costs of their actions or for others who rely on the same resource. Because these activities occur outside the jurisdiction of any single nation, they are the concern and responsibility of *all* nations (e.g., McGoodwin 1990; Fairlie 1995).

The devastating impacts that this mining of potentially renewable resources has on leatherback turtles alone are manifest. The future of this ancient animal depends on far more than protecting nests from "egg poachers." Rather, it is a litmus test for modern society. How will today's governments respond, when they are intoxicated with the illusion of "sustainable development" (Frazier 1997) and virtually controlled by the interests of transnational corporations (Korten 1995), particularly the fishing industry (McGoodwin 1990; Fairlie 1995)? Much more than the future of leatherback turtles is at stake.

ACKNOWLEDGMENTS

Valuable comments on earlier drafts of this entry were made by A. Barragán, P. Dutton, J. Fretey, M. Girondot, H.-C. Liew, C. Limpus, B. Miller, F. V. Paladino, P. Plotkin, A. L. Sarti M., and J. Spotila.

Malagasy Freshwater Fishes

Rick Haeffner

Common Name: (see Table 1)
Scientific Name: (see Table 1)
Order: (see Table 1)
Family: (see Table 1)
Status: Protected (as is all wildlife in Madagascar) by the Malagasy government.
Threats: Introduced exotic fish species; deforestation with resultant increase in river silt loads; loss of historic food items.
Habitat: Freshwater lakes and rivers.
Distribution: All restricted to the island of Madagascar. Many species formerly widespread are now restricted in range or extinct.

DESCRIPTION

The exact number of freshwater fish species on Madagascar is in constant revision and requires extensive survey work (Stiassny & Raminosoa 1994). Most species lists contain approximately 38 species within 11 families of freshwater fishes; however, each time field collectors and researchers visit the island, new species are discovered or revived from extinction lists. From surveys over the last 3 years a compilation of 53 species, within 11 families, has been determined (Haeffner 1998). This appears to be a conservative number (Table 1), with a slight projected increase in new species within the next 5 years.

Of the 53 listed species, all but 4 are endemic, with 2 endemic families and 13 endemic genera. The family Cichlidae represents the largest family, with species occurring in diverse aquatic ecosystems throughout the island. Killifish, members of the Cyprinodontidae and Aplochelidae, are restricted to localized habitats in the northern third of the island, and the eastern coast contains many species of rainbowfish (Bedotiidae).

NATURAL HISTORY

After Madagascar split from Gondwana over 65 million years ago, the fauna of the island became isolated, resulting in a high degree of endemism. Current estimates list about 80% of all plant and animal species as endemic (Reinthal & Stiassny 1991). These unique species are highly diverse. Madagascar, along with five other countries, accounts for 50 to 80% of the world's biodiversity (Mittermeier 1988). The freshwater fishes are no dif-

Table 1. Provisional List of Endemic Freshwater Fishes of Madagascar

Scientific Name (Latin)	Common Name (English)

Order: CLUPEIDAE
Clupeidae — Herrings
Sauvagella madagascarensis Sauvage 1883 — —

Order: SILURIFORMES
Anchariidae — Malagasy Catfishes
Ancharius brevibarbus Boulenger 1911 — —
Ancharius fuscus Steindachner 1881 — —

Order: ATHERINIFORMES
Atherinidae — Silversides
Teramulus waterloti Pellegrin 1932 — —
Teramulus kieneri Smith 1965 — —

Bedotiidae — Madagascar Rainbowfishes
Bedotia gaeyi Pellegrin 1907 — Zono Rainbowfish
Bedotia tricolor Pellegrin 1933 — Tricolored Rainbowfish
Bedotia sp. (Nosivolo River) — Steel Blue Rainbowfish
Bedotia sp. (Namorona River) — Ranomofana Rainbowfish
Bedotia longianalis Pellegrin 1914 — Yellow Rainbowfish
Bedotia madagascarensis Regan 1903 — Madagascar Rainbowfish
Bedotia sp. (Lazana River) — White-finned Rainbowfish
Rheocles alaotrensis Pellegrin 1904 — Katrana
Rheocles lateralis Stiassny & Reinthal 1992 — Zono Ala
Rheocles sikorae Sauvage 1891 — Spotted Zono
Rheocles wrightae Stiassny 1990 — Wright's Zono
Rheocles sp. nov. (Anjingo) — Anjingo Threadfin Zono
Rheocles sp. nov. (Ambomboa River) — Threadfin Zono
Rheocles sp. ("Andapa") de Rham 1993 — Andapa Zono

Order: CYPRINODONTIFORMES
Apolochelidae — Killifishes
Pachypanchax omalonotus Dumeril 1861 — Madagascar Killifish
Pachypanchax sp. (Betsiboka River) — Betsiboka Killifish
Pachypanchax sakaramyi Holly 1928 — Sakaramy Killifish
Pachypanchax sp. (Anjingo River) — Anjingo Killifish

Cyprinodontidae — Lampeyes
Pantanodon madagascariensis Arnoult 1963 — Malagasy Lampeye

Table 1. (*continued*)

Scientific Name (Latin)	Common Name (English)
Order: PERCIFORMES	
Ambassidae	**Glassfishes**
Ambassis fontoynoti Pellegrin 1932	Madagascar Glassfish
Cichlidae	**Cichlids**
Paratilapia bleekeri Sauvage 1882	Fiamanga
Paratilapia polleni Bleeker 1868	Marakely
Paratilapia sp. nov. (Lac Ihotry)	—
Ptychochromis oligaacanthus Bleeker 1868	Juba
Ptychochromis nossibeensis (Nosy Be)	
Bleeker 1868	Tsipoy
Ptychochromis sp. (Mahanoro)	Golden Saroy
Ptychochromis sp. (Ambila)	Black Saroy
Ptychochromoides betsileanus Boulenger 1899	Trondro Mainty
Ptychochromoides sp. (Mangoro)	Fia potsy
Ptychochromoides katria Reinthal & Stiassny 1997	Katria
Paretroplus polyactis Bleeker 1878	Red-eyed Damba
Paretroplus dami Bleeker 1878	Bicolor Damba
Paretroplus kieneri Arnoult 1960	Kotsovato
Paretroplus maculatus Kiener & Mauge 1966	Spotted Damba
Paretroplus petiti Pellegrin 1929	Kotso
Paretroplus menarambo Alleghayer 1996	Pin Stripe Damba
Paretroplus nourissati Alleghayer 1998	Lamena
Paretroplus sp. (ac Androongy)	Striped damba
Paretroplus sp. (Akalimolitra River "new Lamena")	Tsimoly
Oxylapia polli Kiener and Mauge 1966	Songatana
Eleotridae	**Sleeper Gobies**
Eleotris vomerodentata Mauge 1984	—
Ratsirakia legendrei Pellegrin 1919	—
Typhleotris madagascarensis Petit 1933	—
Typhleotris pauliani Arnoult 1959	—
Hypseleotris tohizonae Steindachner 1881	Tohovily
Gobiidae	**True Gobies**
Acentrogobius therezieni Steindachner 1881	—
Chonophorus macrorhynchus Bleeker 1867	—
Teraponidae	**Target Perches**
Mesopristis elongatus Gulchenot 1866	—

Note: Names in parentheses indicate localities of undescribed species. Many species do not have common names.

ferent, with as many as 60% of the genera and 84% of the species being endemic (Stiassny & Raminosoa 1994).

Most of the endemic freshwater fish species (61%) are found exclusively in the forested rivers and streams of eastern and northeastern Madagascar (Stiassny & Raminosoa 1994). The detailed, historical range of many species has not been well documented; thus only recently has field research revealed a snapshot of the ichthyofaunal distribution (i.e., distribution of fish species). A few, such as Marakely (*Paratilapia polleni*), were once widespread but have been severely reduced in range (Reinthal & Stiassny 1991).

The inland waters of Madagascar, which constitute about 1% of the land area, have a diverse hydrographic profile. Most of the major rivers are subject to severe flooding. In portions of the northeast and northwest, the rise and fall of water levels can be so radical that large streambeds in the south are left dry with no surface water or are reduced to mere streams 25 cm deep. The chemistry of the rivers and lakes is correspondingly highly diverse. There is a significant east-west asymmetry in the relief of the island, with the central plateau and eastern regions being slightly acidic to very acidic and the western plain waters being neutral or slightly basic. The infusion of marine waters from tides in the large coastal drainage of the western coast plays an important role in influencing the ichthyofauna, as marine species migrate up into the rivers and salt-tolerant freshwater fishes utilize the lower mangrove areas.

Little is known of native fish diets; however, the stomach contents of two genera of Bedotiidae revealed a predominance of terrestrial adult invertebrates and their associated aquatic stages (Reinthal & Stiassny 1991). Members of the Cichlidae have been observed feeding in the soft bottoms of slow-moving streams by head-shaking the substrate aside to stir up aquatic invertebrates. One entire genus of Cichlidae have specialized teeth to facilitate invertebrate predation. From observed captive behavior of *Pachypanchax* species, it is likely these fish feed on surface insects.

CONFLICTING ISSUES

Without the conservation of habitat, conservation cannot occur. This is nowhere more evident than with fishes that are dependent on seasonal rainfall, correct water chemistry, food, and appropriate aquatic vegetation for shelter and spawning arenas. In Madagascar, introduced exotic species have severely impacted the native fish fauna. The number of exotic species is over 27 (Stiassny & Raminosoa 1994), and in recent surveys every body of water sampled contained exotic fish (Haeffner 1998). These two factors cause freshwater fishes to be the most severely threatened Malagasy vertebrates. Although Stiassny (1990) listed three of the five *Rheocles* species as extinct, this may not yet be the case; but as they survive in such low numbers and with such restricted ranges, extinction is imminent. The same is true for many other fishes from the other families.

Cutting forests for firewood and charcoal began with the arrival of humans approximately 1,500 to 2,000 years ago. This process, along with clearing forests for rice paddy and slash-and-burn agriculture, has caused the loss of over 80% of the island's forests. Although the practice of slash-and-burn agriculture has been in place for centuries, a human population increase from 5.4 million in 1960 to 12 million in 1990 tremendously accelerated deforestation. The original primary rain forest of Madagascar was reduced from approximately 11.2 million hectares to less than 3.8 million hectares by the late 1980s. At the current annual rate of 1,110 km^2/yr in loss, little forest will remain into the 21st century.

The exact effects of forest reduction on fishes are not clearly understood. The presence of abundant populations of native fish in locations where rain forest remains provides some empirical support for the importance of forests to fish survivability. Stomach analysis of fishes living within forested ecosystems further demonstrates the dependence of these fishes on food sources emanating from the trees (Reinthal & Stiassny 1991). During a survey in 1995 on the Onilahy River, it was evident that deforestation along the river's sloped banks caused high levels of soil run-off during the rainy season. Significant silt loads caused increased algae proliferation, which severely affected the Madagascar lace plant (*Aponogeton fenestralis*), an aquatic plant species important to this ecosystem. Fishermen on the river also spoke of diminished water clarity over the last 10 years, owing most likely to forest removal.

The presence of exotic fishes in large numbers has a direct impact on the native fishes. Although the actual detrimental mechanism of introduced fish in Madagascar is unknown, competition for food and spawning sites most likely contribute to native fish reduction. Direct predation from large, introduced fish piscivores (fish eating fish) also decimates local fauna. As one example, the introduction of redbreast tilapia (*Tilapia rendalli*) into Lake Kinkony, Madagascar, caused complete devastation of primary aquatic plant species and subsequent disappearance of the once locally common and native cichlid, Kotso (*Paretroplus petiti*) (Cann 1996).

About 1.8% of Madagascar's land area is set aside in 41 reserves. These reserves fall under the central government administration of the Direction des Eaux et Forêts. Conservation of this approximately 1 million hectares has been spotty at best, and because little money is allocated to administer these parcels, edge cutting and burning continue unabated. These protected areas have no buffer zones and are eroded at the edges. Without intensive monitoring and control they will further diminish in size. Because elsewhere the illegal cutting of forest and burning of grasslands go uncontrolled, it is vital that reserves be protected. Even with this protection it is hard to quantify how beneficial these small, original forest tracts will be to native fishes. Healthy fish populations rely on large, undisturbed portions of forest along the river and lake banks.

With many fish facing extinction, Dr. Paul Loiselle, curator of fishes at the New York Aquarium, began working to conserve representative specimens of all the freshwater species on the island. The initial goal was to collect and bring viable specimens back to U.S. zoological institutions to form the nucleus of captive breeding colonies. Collection and survey expeditions began in 1993. Since that time, annual trips with New York Aquarium and Denver Zoo staff have successfully surveyed and returned specimens to captive programs at the New York Aquarium, Denver Zoo, and Old World Exotic Fish, Inc. Currently 27 species (or around 50% of the total fish number of species) have been established in colonies, with approximately 50% of these successfully reproducing and being disseminated to other facilities.

It has proven impossible to spawn some species except in large outdoor ponds used in southern Florida. Some of the gobies (Eleotridae) have presented special challenges in fry rearing, and various catfish species (Anchariidae) are very difficult to spawn in captivity, but these hurdles will need to be surmounted soon. One species of cichlid (*Paretroplus menarambo*) appears to have gone extinct in the wild within the last 2 years (De Rham 1997). Fortunately, specimens of this species are reproducing in captive programs in the United States.

FUTURE AND PROGNOSIS

Given the importance of freshwater fishes to Malagasy people and aquatic ecosystems, a two-pronged series of efforts is needed. First, a series of in-country programs is necessary. This should consist of (1) more surveys to complete an understanding of the ichthyofauna and to collect founder specimens, (2) additional research into the limnology (i.e., study of lakes) and fisheries biology of the aquatic environments to assist the Malagasy government in preserving water resources for fishes and the local populations, (3) a halt to any attempts to expand introductions of nonindigenous species, and the pursuit of methods of restricting expansion of already established exotic species, (4) support of programs to establish natural refugia (like the island of Nosy Be) and other locations for artificially constructed ponds, and (5) training of Malagasy personnel to continue and expand the programs.

The second part of a conservation initiative is ex situ (in captivity) and involves the establishment of viable, captive, reproducing populations of as many species as possible. Currently no proper facilities or financial resources exist to enable this to occur within Madagascar. Without this captive reservoir, it will be impossible to preserve and re-establish extinct species into viable habitats.

Marbled Murrelet

Blair Csuti, Gary S. Miller, and Steven R. Beissinger

Common Name: marbled murrelet
Scientific Name: *Brachyramphus marmoratus*
Order: Charadriiformes
Family: Alcidae
Status: California, Oregon, and Washington populations listed as Threatened under the U.S. Endangered Species Act (1992); nationally Threatened in Canada (1990); Endangered in California (1992); Threatened in Oregon (1995); Threatened in Washington (1993).
Threats: Loss of nesting habitat, mainly from commercial timber harvest; adult mortality owing to gill net fisheries in near-shore marine habitat; potential increased nest predation owing to forest fragmentation; mortality and loss of large population segments from marine oil spills.
Habitat: Old-growth Douglas fir, western hemlock, western red cedar, and redwood forests; sometimes nests in younger forests with old-growth characteristics; forages in near-shore marine waters, usually within 2 km of the coast.
Distribution: Occurs along the coast of the eastern Pacific Ocean from the Aleutian Islands eastward and southward to central California (Figure 5); nests up to 80 km inland; in winter, occurs in offshore waters as far south as southern California.

DESCRIPTION

The marbled murrelet is a small (length = 25 cm), compact seabird about the size of a mourning dove (*Zenaida macroura*). It has a short neck and tail. The wings, used while swimming underwater, are about 11 cm long. Males and females have identical plumages; however, breeding and winter plumages differ. Breeding adults have sooty-brown upperparts with dark bars. Underparts are light, mottled brown. Winter adults have brownish-gray upperparts except for a white band below the nape extending up from white underparts and white scapulars. The plumage of fledged young is similar to that of winter adults, but the breast and sides have a slight brownish mottling.

NATURAL HISTORY

The biology of the marbled murrelet is similar to that of other species in the Alcidae family (Marshall 1988; Nelson & Sealy 1995; Ralph et al. 1995;

Figure 5. Range of the Marbled Murrelet and Location of Threatened Washington, Oregon, and California Populations (adapted from Ralph et al. 1995)

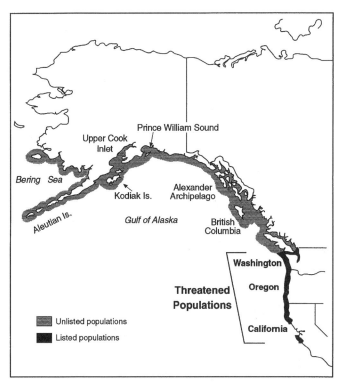

U.S. Fish and Wildlife Service [USFWS] 1997). During the nonbreeding season it spends most of its time in the near-shore marine environment. The marbled murrelet usually forages from 300 m to 2 km from shore, making repeated short (28–69 second) dives 20 to 80 m deep in search of prey. Its diet consists mostly of small fish, including Pacific sand lance (*Ammodytes hexapterus*), Pacific herring (*Culpea harengus*), northern anchovy (*Engraulis mordax*), smelts (family Osmeridae), and seaperch (*Cymatogaster aggregata*). Some small marine crustaceans (euphausiids, mysids, and gammarid amphipods) are also eaten, primarily during the nonbreeding season.

Along parts of the treeless coast of the Gulf of Alaska, some marbled murrelets are ground nesters on barren, inland slopes. The nesting habitat of the threatened populations was unknown until the discovery of a nest on the limb of a large tree in 1974 (Binford et al. 1975). The marbled murrelet is the only alcid known to nest in trees. The average nest tree diameter is 161 cm and the average age of forests in which nests occur (*N* = 16 nests) is 522 years. All 61 nest trees found in North America have been built in old-growth or mature forests (USFWS 1997). Most nests are within 40 km of the coast, although a nest has been found 84 km inland. Eggs are placed in shallow, bowl-like depressions in moss or drift on large, platform-like limbs, usually 30 to 70 m high.

Marbled murrelets have been detected at inland sites throughout the year, although detections are more frequent during the nesting season. The reason for the birds' visits to nest sites during the nonbreeding season is unknown. Marbled murrelets are relatively fast flyers, with an average ground speed of 77 km/hr (range 56–104 km/hr)(Hamer et al. 1995). Sociality of marbled murrelets at inland sites is not fully understood. No indications of colonial nesting have been observed; most nests occur singly or, at most, in the vicinity of a few others. Marbled murrelets lay only one egg per clutch and are thought to nest once a year. Not all adults may nest in a given year. Incubation lasts about 30 days, and chicks fledge in about 28 days. Both sexes incubate the egg, exchanging duties at dusk and dawn. The chick is fed a single fish an average of four times per day. Fledglings apparently fly directly from the nest to the sea. Marbled murrelets are unlikely to breed until their third year, and the estimated average life span is 10 years.

The historic size of the California, Oregon, and Washington marbled murrelet population is unknown, but the species has been described as plentiful during winter in offshore waters by many sources (USFWS 1997). Based on extrapolation from currently known population numbers in relation to remaining available nesting habitat, it was estimated that at least 60,000 marbled murrelets may have been found historically along the coast of California. In contrast, current population estimates for the same area range from 2,000 to 6,000 individuals. Evidence indicates that the main factors causing murrelet decline are loss of nesting habitat and poor reproductive success in remaining habitat (USFWS 1997).

CONFLICTING ISSUES

In the 20th century, most marbled murrelet habitat was lost owing to commercial timber harvest. The low elevation, older forests close to the coast, which marbled murrelets require for nesting, have been heavily harvested throughout the bird's range and are severely degraded as a result of fragmentation. At the time of listing, old-growth forests throughout western Oregon and Washington had been reduced by about 82% from pre-logging levels (Booth 1991). Additional habitat losses result from natural disturbances such as windthrow (i.e., trees toppled by strong winds), natural and human-caused fire, and development. Most suitable nesting habitat on private lands has been eliminated by timber harvest. In addition, many remaining stands of old-growth and mature forests occur as small fragments surrounded by second-growth forest. It has been hypothesized that logging activities heighten the susceptibility of marbled murrelets to nest predation owing to increased edge and fragmentation. From 1974 to 1993, of those murrelet nests in Washington, Oregon, and California where success/failure was documented, approximately 64% of the nests failed. Of failed nests, 57% failed because of predation (USFWS 1997). Corvids (the common raven, [*Corvus corax*], American crow [*Corvus brachyrhynchos*], and Steller's jay [*Cyanocitta stelleri*]) are suspected to be the major predators.

In Puget Sound and some other parts of the seabirds' range, murrelet adult and juvenile mortality from nearshore gill net and purse seine fisheries may have contributed to population decline. Relatively few studies have quantified the mortality of murrelets in net fisheries; however, it likely has had impacts on populations in Washington (USFWS 1997). Net fisheries in Puget Sound are operated by both native and non-native Americans, and they are managed by a variety of state and federal agencies, native nations, and tribal fishing groups. The amount of murrelet mortality from net fisheries is likely to be higher than reported owing to a reluctance on the part of net operators to jeopardize their livelihood by reporting murrelet bycatch (Carter et al. 1995). Because Native American and tribal fisheries operate under sovereign treaties, efforts to reduce murrelet bycatch in these fisheries depend on voluntary cooperation, and many options are currently being explored.

Because of their extensive use of near-shore waters, marbled murrelets are susceptible to the impacts of oil spills and have been given one of the highest oil spill vulnerability index values among seabirds. Large oil spills result periodically from oil tanker mishaps and similar mishaps by other large ocean-going vessels. Small oil spills are chronic in many areas owing to tank cleaning at sea, bilge pumping, and seeps. The thousands of marbled murrelets killed in 1989 during the Exxon *Valdez* spill (Piatt & Lensink 1989) have caused increased concern about the impacts of oil pollution on the species. It was estimated that the 1991 *Tenyo Maru* spill off the Olympic Peninsula, Washington, killed between 200 and 400 marbled murrelets. Although the location and frequency of oil spills are impossible to predict, measures to minimize potential impacts on murrelets include rerouting marine traffic, instituting double-hulled construction standards for oil tankers, and changing maritime maintenance practices, such as bilge pumping.

The marbled murrelet is the second of two birds listed as Threatened under the ESA that are dependent on older forests of the Pacific Northwest. The northern spotted owl (*Strix occidentalis caurina*) was listed as Threatened in 1990. Several major planning efforts culminated in the adoption of the President's Forest Plan (U.S. Department of Agriculture & U.S. Department of the Interior 1994), an ecosystem approach to management of late-successional forests and their associated species in a network of Late-Successional Reserves (LSRs). There are approximately 526,000 hectares of potential marbled murrelet nesting habitat protected as part of the mapped

LSR network. The marbled murrelet Recovery Plan builds on the Forest Plan in areas that were not considered or could not be considered (e.g., nonfederal lands) during development of the Forest Plan.

The commercial timber industry has suggested that natural climatic events (El Niño) and at-sea mortality, rather than loss of old-growth terrestrial habitat, may be responsible for the species' decline. In addition, the timber industry has produced reports suggesting that marbled murrelets successfully nest in younger forests. The Recovery Team found that although marbled murrelets occasionally display behavior associated with nesting in some younger forest stands, the vast majority of documented nesting and nesting behavior is associated with old-growth and late-successional forests.

FUTURE AND PROGNOSIS

As a part of the recovery planning process, a demographic model was developed to better understand marbled murrelet population dynamics and trends (Beissinger 1995; Beissinger & Nur 1997). Projections show that murrelet populations in Washington, Oregon, and California are apparently declining at a rapid rate (4–7% per year at most locations from 1990 to 1995). Current estimates of nesting success and recruitment in most years are well below levels required to sustain the threatened population. Furthermore, the naturally low reproductive potential of the murrelet means that populations will recover slowly from declines or disasters (growing at a rate of about 3% per year), even if the reproductive potential were to be fully realized over several years.

The effects of loss and degradation of old-growth forest nesting habitat are chronic and can persist for 100 to 200 years, until second-growth forests are old enough to provide structure that permits marbled murrelet nesting. Demographic analyses suggest that murrelet populations will probably continue to decline for at least the next 50 years, at which time some forests in LSRs will have matured sufficiently to provide new nesting habitat. Until that time, immediate conservation efforts that minimize and mitigate the loss of actual and potential nest sites, as well as increase adult survivorship, will be necessary.

A substantial portion of these efforts can take place on federal land; however, in parts of the range of the threatened population, including forests inland from the Humboldt County, California, and northern Oregon coasts, most remaining nesting habitat is privately or state owned and subject to harvest. Recognizing that recovery will involve nonfederal forests in these areas, some private and state land has been designated as Critical Habitat. Habitat documented to be occupied by nesting murrelets may not be harvested under provisions of the ESA. Incidental Take Permits may be issued under the provisions of a Habitat Conservation Plan (HCP), ESA, Section

10. An HCP has been proposed for Pacific Lumber Company (PALCO) lands in Humboldt County, California, which would protect a 1,215-hectare tract of old-growth redwoods (known as the Headwaters Grove) at a cost of $380 million in federal and state dollars. However, under provisions of the HCP, PALCO would be allowed to harvest a 200-hectare grove of old-growth redwoods and an additional 3,200 hectare of forest containing scattered stands of old-growth redwoods, including stands occupied by murrelets. Environmental groups have opposed this exchange because of PALCO's previous violations of timber harvest regulations. If the HCP allowing incidental take of some marbled murrelets is not approved, harvest of occupied habitat would be prohibited by other sections of the ESA. Enforcement of the ESA in this case, however, may result in PALCO bringing suit against the federal government for reducing the value of its land without compensation. This very real threat to the ESA could lead to a Supreme Court decision with broad ramifications for enforcement of federal environmental regulations on private land.

Recovery of the marbled murrelet is possible, especially because the listed populations are still relatively large compared to those of many other threatened or endangered species. However, these populations are likely to continue to decline for decades until they reach equilibrium with remaining nesting habitat. Additional loss of nesting habitat will reduce the probability of the species' survival in the Pacific Northwest. Conversion of forests from old growth to second growth represents a significant threat to the marbled murrelet and other old-growth dependent species (Noss & Cooperrider 1994). A moratorium on harvest of old-growth forests, whether publicly or privately owned, has been suggested by the Oregon Natural Resources Council but is strongly opposed by timber industry advocates, such as the Northwest Forestry Association. Norse (1990:277) suggests that a less extreme solution lies in a "shift from maximizing the cut in the short term to sustaining biological diversity and timber production far beyond the next cut." A measure banning clear-cutting was placed on the Oregon ballot in November 1998 but lost by a 3:1 margin in the face of a well-financed campaign in opposition, funded largely by forest products industry contributions. Conflict over timber harvest practices, once polarized in the public arena, is unlikely to be resolved in a manner acceptable to all parties involved. However, compromises that continue harvesting of murrelet habitat—for instance, through the development of HCPs—must be carefully evaluated for consequences that would reduce the probability of the marbled murrelet surviving its current population decline. Owing to the length of time needed to regrow murrelet nesting habitat (50–200 years), there is a chance that incremental habitat loss from timber harvest and direct mortality owing to oil spills and other factors could drive the threatened population to extinction before suitable habitat becomes available in LSRs. A conservative approach to marbled murrelet recovery argues for minimizing both habitat loss and mortality until data suggest the population is stabilized or growing.

ACKNOWLEDGMENTS

This entry is based largely on the marbled murrelet Recovery Plan (USFWS 1997). We thank Recovery Team members Harry R. Carter, Thomas E. Hamer, and David A. Perry for their contributions to the plan. We are grateful to Gus van Vliet for permission to use his photograph of the marbled murrelet in winter plumage and to C. J. Ralph for providing a copy of the map. We thank David Kato for adapting the marbled murrelet range map.

Mariana Crow

Robert E. Beck Jr. and Julie A. Savidge

Common Name: Mariana crow
Scientific Name: *Corvus kubaryi*
Order: Passeriformes
Family: Corvidae
Status: Endangered under the Endangered Species Act of the United States; CITES Appendix I; listed as Endangered by Guam and the Commonwealth of the Northern Mariana Islands; listed as Critically Endangered by BirdLife International (formerly International Council for Bird Preservation) (Collar et al. 1994).
Threats: Predation by the brown tree snake (*Boiga irregularis*) on Guam. Decline on Rota may be owing to habitat destruction and additional unknown factors.
Habitat: Forests.
Distribution: Endemic to Guam and Rota of the Mariana Islands. The current population on Guam is restricted to forest habitat in the extreme northern portion of the island. On Rota the crow is found islandwide in native forest.

DESCRIPTION

As described by Baker (1951), the Mariana crow is a small, black crow with a slight blue gloss on the back, wings, and tail and a slight green gloss on the head and underparts. The bases of its feathers range from a light grayish to white on the neck, giving the bird a ragged appearance. The nasal bristles extend over the nostrils from the base of the culmen (ridge on upper mandible of beak). The bill and feet are black and the irises brown. Females are smaller than males. Immature birds have less gloss and browner wings and tail than do adults.

NATURAL HISTORY

The Mariana crow (referred to as the *Aga* by the native Chamorros people) is the only member of the Corvidae family in Micronesia and is endemic to Guam and Rota in the Mariana Islands, a north-south chain of 15 islands halfway between Japan and New Guinea. Guam is the southernmost and largest (550 km²) island in the archipelago. Situated about 58 km north of Guam, Rota is considerably smaller (85 km²). Mariana crows have been recorded in a variety of forested habitats including mature forest, second growth, coastal strand, and even coconut plantations (Strophet 1946; Marshall 1949; Baker 1951; Jenkins 1983). Observations indicate an avoidance of human habitation (Baker 1951; Morton 1996).

The Mariana crow is gregarious and will form small foraging groups. Frequently, family groups of two to five individuals were seen on Guam (Jenkins 1983). Jenkins (1983) considered the crow as one of the least wary of the forest birds, noting it would perch and vocalize within 2 to 3 m of an observer.

The Mariana crow is an omnivore and has been recorded eating a variety of insects (grasshoppers, caterpillars, etc.), lizards, hermit crabs, and various plant parts, including fruits, flowers, and buds (Marshall 1949; Jenkins 1983; Tomback 1986).

Evidence of breeding has been observed nearly year-round (Jenkins 1983). Mariana crows on Guam build nests primarily in emergent trees such as joga (*Elaeocarpus*) and fig (*Ficus*). In contrast, crows on Rota nest well inside the canopy, perhaps in response to harassment by introduced black drongos (*Dicrurus macrocercus*). Whereas females do most of the incubation (Michael & Beck unpublished data), both mates appear to share in care of the young (Jenkins 1983). Clutch size is 1 to 4 eggs, but 1 to 3 young generally fledge (Morton personal communication). Renesting is a common occurrence after nest failure.

During a 1982 survey of crows on Rota, density varied from 5 to 23/ km^2 (Engbring et al. 1986). Although these densities were lower than those of most other native birds on Rota, the Mariana crow has a large territory and crows appeared, at that time, to be at or near carrying capacity (i.e., maximum number of animals the habitat can support) (Engbring et al. 1986).

CONFLICTING ISSUES

Recovery efforts for the Mariana crow have been fraught with both biological and nonbiological challenges. Various stakeholders have included the government of Guam, the Commonwealth of the Northern Mariana Islands (CNMI), the U.S. Fish and Wildlife Service (USFWS), the U.S. Air Force, various nonprofit conservation groups, scientific researchers, and local landowners.

An initial challenge was to discover why crows and other forest birds on Guam were declining. Although historic densities are not known, Mariana crows had been distributed throughout most parts of Guam and were considered numerous on Rota (Baker 1951). Mariana crows, as well as the other native forest birds, disappeared from Guam's southern ravine forests in the 1960s, and gradually their ranges contracted to the north of the island (Savidge 1987). Currently the crow population on Guam is 12, which includes 6 Rota crows translocated to Guam after being held in captivity for several years at various mainland zoos (DAWR 1997).

At first many biologists, including those from the USFWS, discounted the possibility that predators were responsible for the decline of Guam's birds. Federal funds were given primarily for research on disease because of a similar decline in native Hawaiian birds caused by avian malaria and pox. Only after collecting considerable data in support of predators and none in support of disease did the USFWS and scientific community begin to accept the hypothesis that brown tree snakes were indeed the primary cause of the decline of Guam's native forest birds (Savidge 1987). This exotic predator was first reported on Guam in the early 1950s and has since spread throughout the island.

The crow population on Rota is larger and still found islandwide, but a comparison of surveys done in 1995 with earlier surveys from 1982 suggests a 55% decline (Grout et al. 1996). Because brown tree snakes are not known from Rota, the more recent decline on that island is presumed to be primarily a result of habitat modification (National Research Council 1997). Other, unknown factors may also be involved.

Another challenge focused on Critical Habitat designation under the ESA. The governor of Guam first petitioned for federal listing of the Mariana crow in 1979. The USFWS conducted a survey of crows (and other forest birds) in 1981 and 1982 and estimated the Guam and Rota populations at 357 and 1,318, respectively (Engbring & Ramsey 1984; Engbring et al. 1986). Mariana crows were formally listed in 1984 (USFWS 1984), but the USFWS decided it was not prudent at that time to designate Critical Habitat. By 1985, surveys indicated that less than 100 crows remained on Guam (DAWR 1985; Michael 1987). In 1987, the U.S. Navy proposed the construction of a Relocatable Over-the-Horizon Radar (ROTHR) facility at Northwest field, Anderson Air Force Base. Northwest field had been designated as "essential" in the USFWS Recovery Plan for the native forest birds of Guam (Beck & Savidge 1990). The governor of Guam and the Sierra Club Legal Defense Fund, on behalf of the Marianas and National Audubon Societies, petitioned the USFWS to designate Critical Habitat to stop the ROTHR project. In 1988, the governor of Guam again requested that the USFWS designate Critical Habitat, this time on an emergency basis. By February 1989, the USFWS agreed that the petition to designate Critical Habitat had merit and announced that a final rule designating Critical Habitat would be published by "the summer of 1990." However, the USFWS issued a "nonjeopardy" opinion in December 1989 on the ROTHR project as part of the Section 7 consultation process. This paved the way for the Navy to issue the final environmental impact statement. If unchallenged, the project could proceed.

However, during 1990 the idea of a USFWS National Wildlife Refuge in northern Guam gained support as an alternative to Critical Habitat designation. Additionally, the Navy decided not to build the radar for "budget" reasons, a decision that also corresponded with the collapse of the Soviet

Union in 1991. In October 1993, 150 hectares at Ritidian Point in northern Guam were transferred from the U.S. Navy to the USFWS, thereby creating the Guam National Wildlife Refuge (GNWR), which officially opened in June 1997. The Navy and Air Force signed a cooperative agreement with the USFWS designating an additional 8,903 hectares of forest habitat on their bases in Guam as an "overlay" to be included in the GNWR. Biologists and conservation organizations saw the refuge as a haven for the Mariana crow and Guam's other remaining birds and also as a potential area for future reintroductions. However, by this time the government of Guam was opposed to the idea, arguing that the land should be returned to private landowners from whom it had been taken by condemnation during and after World War II. The conflict quickly erupted into a local "indigenous rights" issue.

Meanwhile, Mariana crow populations on both Rota and Guam were declining. In 1993, the crow population on Guam was estimated at 51 (DAWR 1993). Grout et al. (1996) estimated crow numbers on Rota to be 592 in 1995, and a DAWR (1995) survey on Guam found only 26 in the same year. No successful reproduction had been observed in the wild on Guam since 1986, and biologists were concerned that the low viability of eggs might be the result of an aging crow population (DAWR 1986–1995). Because of the extremely low numbers and possible senescence of crows on Guam, the DAWR requested a USFWS Endangered Species Permit in 1995 to translocate crows from Rota to Guam. Genetic analyses of the Guam and Rota crow populations suggested the Rota population may be a subset of the Guam population, and lack of distinction indicated its origin was relatively recent (Tarr & Fleischer 1999).

The USFWS was concerned that the Guam government would see the translocation (ironically proposed by Guam's own wildlife agency) as an attempt to prevent overlay refuge land from being returned to the people of Guam. Subsequently the USFWS asked the National Research Council of the National Academy of Sciences to review Mariana crow recovery efforts in general and DAWR's translocation request in particular. Thus the GNWR, which was created to enhance preservation and recovery of Guam's endangered forest birds, was now an obstacle preventing recovery actions. In 1996, the National Research Council's Board on Biology established the Committee on the Scientific Basis for Preservation of the Mariana Crow. As part of its final report issued in 1997, the Mariana Crow Recovery Team was established, 18 years after the initial petition to list the species. A permit to translocate crows from Rota to Guam was finally issued to DAWR in November 1998, and the first crow chick was transferred to Guam in December 1998. However, the Guam population was by then essentially genetically dead because all recent eggs were infertile. If the translocations from Rota succeed, the recovered crow population on Guam will be entirely descendent from the Rota population.

FUTURE AND PROGNOSIS

It appears that causes for the crow's decline on Rota are primarily habitat related, but additional studies are needed to investigate other potential contributing factors. A draft USFWS/CNMI Habitat Conservation Plan for Rota is being developed. If approved, this will preserve much of the best crow habitat left on Rota, which is believed to harbor 60 to 80% of Rota's remaining crows. Development would be allowed in other parts of the crow's range. Such compromises may be needed to ensure some habitat protection. It is also critical that efforts continue to keep the brown tree snake from spreading to Rota and other islands (Fritts 1988).

On Guam the current priority is to implement area snake control in endangered forest bird habitat. A 40-hectare snake-free area created on Northwest Field for introductions of another endangered bird, the Guam rail, is available for crows. Another adjacent area almost twice the size is presently being readied for translocation of Rota crows. Should translocated crows wander out of these areas, snake-proofing of individual crow nest sites by means of electrical barriers is feasible (Aguon et al. 1998).

Biological control has the potential to reduce snake populations islandwide and could be self-sustaining, but identification of appropriate control organisms is difficult and introduction of pathogens or parasites carries some risk to native reptiles on Guam. Research on various methods of control, including toxicants and snake attractants for traps, would be useful.

At least two viable populations of the Mariana crow are needed to reduce the risk of extinction of the species. Preservation of the Rota crow population and re-establishment of the population on Guam should secure the future of this island corvid. However, timely cooperation among stakeholders will be required. As is frequently the case, economic development has priority over preservation of habitat for endangered species on both Guam and Rota. Positive incentives for habitat conservation are needed, as is involvement of major stakeholders in the recovery process. Public information programs on the dangers of the brown tree snake and other exotics are essential, as well as programs that stress the value of native species to the forests of Guam and Rota, to tourism, and to local culture and heritage.

Mediterranean Monk Seal

Luis Mariano González

Common Name: Mediterranean monk seal
Scientific Name: *Monachus monachus*
Order: Pinnipedia
Family: Phocidae
Status: Critically Endangered under the 1996 IUCN Red List and in the CITES, Bonn, and Bern Conventions.
Threats: Historically, extensive killing for commercial purposes (oil, meat, and skin). Recently, persecution by fishermen who perceive the species as a competitor; loss of coastal habitat; reduction of food; pollution; cave collapse; toxic algal blooms.
Habitat: Coastal waters with sheltered beaches and caves along inaccessible rocky cliffs and small islands.
Distribution: Historically, the species occurred on suitable islands and coasts throughout the Mediterranean and Black seas, the Atlantic coast of North Africa as far as 20°N and in the Azores, Madeira, Canary, and Cape Verde Islands. Presently it is found in the Desertas Islands (Madeira), Cabo Blanco Peninsula (Western Sahara–Mauritania), the Mediterranean coast of Morocco and Algeria, and the Ionian and Aegean seas (Greece-Turkey).

DESCRIPTION

Monk seals are among the largest seal species (phocids), with an adult length of 300 cm and weight of 400 kg. They have primitive anatomical features (structure of the skull, skeleton, and vein system), and the claws on the hind flipper are small. The monk seal has the shortest hair of all pinnipeds (i.e., seals and sea lions). Females have four mammary teats, in contrast with other phocids, which have two. There are two, instead of three, upper incisors on each side (King 1955).

The pelage of the pups is uniformly black, with a white irregular patch on the belly. Juveniles and adult females are silver-grey or brownish overall, with slightly lighter bellies. Adult males are black overall with irregular white patches on the belly and throat, therefore displaying similar pelage coloration to pups.

NATURAL HISTORY

The genus *Monachus* is the most primitive of living seals (Muizon 1982), and they seem far more sensitive than other phocids to the intrusion of people into their environment. Monk seals are the only phocid seals living in tropical and subtropical waters. Cabo Blanco has the only remaining large aggregation of this species, and it is situated in one of the richest fishing grounds in the world. Up to 100 individuals have been seen on a beach at the same time (González, Aguilar, et al. 1997).

Adult males defend aquatic territories at the entrance to their caves and surrounding areas. Females and young are regularly found only in breeding caves, where some large males also occur. Females are sexually mature at 3 years of age and can breed at 4 years as subadults. Copulation takes place in the water. Mating and pupping occur throughout the year, although less frequently in winter. The interval between consecutive births averages 375 days. The Cabo Blanco colony produces an average of 50 pups per year. Females give birth inside caves on beaches, and fall and winter storms can wash pups away from caves into the strong surface currents and swells. Sometimes these currents move the pups a great distance. Rupture of the mother-pup bond and physical injury are the major causes of pup mortality. Pup survival rate is 0.55, lower than that of pinnipeds breeding on open beaches. It has been suggested that monk seals breed in caves because persecution by humans has modified important aspects of their biology.

During lactation, which lasts 120 days, females leave pups unattended during feeding trips. Lactation ends gradually after about 4 months of age, and young are dependent on the mother longer than the rest of the phocids (Universidad de Barcelona 1997).

Monk seals have poor underwater hearing because the structure of the ear region offers little resistance to high pressure. Indeed, the monk seal is a littoral species (coastal) and a shallow diver (Muizon 1982). They generally feed on fish, lobsters, and cephalopods, with daily intake estimated at 8 kg. They often parasitize and damage fishing nets, particularly static gear. The monk seal seems to be a rather sedentary species, but sporadic sightings of juveniles in areas far away from the breeding colonies suggest that juvenile dispersal over long distances can occur (Universidad de Las Palmas de Gran Canaria 1997).

CONFLICTING ISSUES

The world population of monk seals is small and rapidly decreasing. There are approximately 300 to 500 animals living in fragments that are isolated by distances too great to traverse (Reijnders 1998).

In the Mediterranean, conflicts with fishermen present the most serious threat to the seals. Fishermen continue to deliberately or incidentally kill seals, particularly off the coasts of Greece and Turkey, causing up to 50% of adult seal mortality (Aguilar 1998). Most deaths result from fishermen attempting to reduce damage to gear and competition for fish. However, loss of income in coastal fisheries is caused by amateur and illegal fishing in addition to competition from seals (Aguilar 1998). For example, the use of driftnets and illegal fishing methods, including dynamite and spear-fishing while scuba diving at night, continues to be widespread in many areas of the western Mediterranean: Morocco, Algeria, and Tunisia. To address these conflicts, public education and awareness programs are being carried out by several organizations in most locations where seal aggregations occur. In addition, fishermen from some areas receive compensation for collaborating with conservation initiatives; however, in other locations fishermen do not receive compensation, leading to friction.

The most important and perhaps only viable population is the Cabo Blanco colony in the Atlantic Ocean. Despite the fact that the Cabo Blanco seal colony had been known since about 1923, the first intensive biological studies were not carried out until 1984–1988 (Marchessaux 1989). Those studies were interrupted after the deaths of researchers owing to land mines. Warring factions and guerrillas deterred visitors and fishermen from using the area. In 1993 a Spanish Monk Seal Team (SMST) arrived at the colony, and after opening a safe passage they set up a base camp from which they have been carrying out surveillance work, monitoring, and research on the colony (González, Aguilar, et al. 1997). The team is made up of naturalists and scientists from the Spanish universities of Las Palmas and Barcelona, the Environmental Ministry of Spain (Madrid), and a nongovernmental organization (Isifer). The SMST has received financial support from the European Union, the Spanish authorities, and the nongovernmental organization Euronature. Scientific support and advice came from the IUCN Seal Specialist Group.

Until 1997 more than half the world population of Mediterranean monk seals lived in the Cabo Blanco colony, which numbered 350 to 400 individuals. However, in 1997 a toxic algae bloom killed over two-thirds of the colony in 1 month (Forcada et al. 1998). Two groups of scientists suggested different hypotheses about the cause of the crisis. One group thought mortality came from a morbillivirus infection (Osterhaus et al. 1997). The other thought it was the result of paralyzing shellfish poisons (PSP) produced by a bloom of marine dynoflagellates (Hernández et al. 1998). Necropsies (i.e., autopsies performed on animals) showed that seals died by drowning after paralysis, and immunological tests showed no evidence for an infection at the time of the die-off, so the virus hypothesis was rejected (Harwood et al. 1998). Oceanographic conditions in Cabo Blanco are favorable to the development of toxic algae blooms, particularly *Gymmodinium catenatum*. Between 1971 and 1994 this algae was responsible for seven outbreaks of algae toxins along the Atlantic coast of Morocco.

Conflict became particularly apparent during the disaster in the Cabo Blanco colony. During the crisis, scientists and the environmental organizations (World Wide Fund for Nature, World Conservation Union, United Nations Environment Programme/Bonn Convention, U.S. Marine Mammal Commission, and International Fund for Animal Welfare) met and devised an emergency plan to remedy the problem. But controversy disrupted action, impeded discussions, hindered fundraising, and stymied implementation of the plan. Indeed, some factions obstructed implementation of ideas they opposed, and only a few elements of the plan were ever enacted (Aguilar 1997; Harwood et al. 1998).

During the episode, the Seal Research and Rehabilitation Center (SRRC) informed Mauritanian authorities that it was convinced the mortality was owing to a virus (SRRC 1997a, 1997b). Meanwhile, the SMST warned authorities that mortality was owing to toxins and that there were possible health risks to local people (Aguilar 1997). The idea that the mortality was caused by a virus was initially embraced, as mortality from toxins would cause problems for the fishing industry, which is very important to the country. The pressure for SMST to accept the virus hypothesis included a smear campaign in the local press. Nevertheless the SMST had received funds from the European Union, and it informed the European Community that toxins may have been responsible for the die-off. The European Commission, in consultation with outside experts and in light of the evidence provided, ordered a careful examination of fish imports from the area, causing delays and financial damage to the fishing industry. However, 90% of fishery exports from the region are cephalopod species that are not very sensitive to the accumulation of toxic algae. Health inspections did not detect toxic levels, and after 2 months normal exports were resumed.

Even though there was agreement over plans for monk seals, disagreements and conflicts arose during implementation. The last monk seal conference, held in Monaco in 1998, listed inappropriate intervention owing to insufficient ecological knowledge and lack of consensus on the implementation of proposed conservation measures as major threats to the seals (Reijnders 1998).

An Emergency Action Plan included the capture of a limited number of seals to be held in semi-captivity and protected against PSP poisoning. An antidote to the algae bloom was provided by the U.S. Navy, but the group that thought mortality was owing to a viral infection rejected it (SRRC 1997a). Efforts to capture and maintain individuals in semi-captivity were paralyzed. As a consequence, at least 40 seals died even though the capture methodology and toxin antidote were available (Aguilar 1997; Harwood et al. 1998).

Another conflict surrounded the taking of abandoned wild-born pups to a seal rehabilitation station. This action has been one of the flagships for conservation of the monk seals in Greece, where at least nine abandoned pups were removed (Vedder et al. 1998). In accordance with the general life pattern of phocids, it had been assumed that Mediterranean monk seal mothers do not feed themselves when nursing young and always remain close to their pups while lactating. It had therefore been recommended that pups found alone in caves be removed. However, people working in the Cabo Blanco colony discovered that mothers do feed while nursing and do leave pups alone in the caves for quite some time without affecting pup survival.

The re-introduction of rescued pups was the object of another conflict between the same groups. One group proposed that pups be acclimatized in an ocean-beach enclosure before release, and it cited significant biological and behavioral shortcomings of rescued pups when compared with wild animals. Confinement of rescued pups in rehabilitation facilities probably precludes muscle development and attainment of the fitness necessary to survive in the ocean. Acclimatization had been routinely implemented with success in similar species, such as the Hawaiian monk seal (Ragen personal communication).

In accordance with recommendations of the IUCN Reintroduction Specialist Group (1996) animals should be radio-tagged to facilitate post-release monitoring, and releases should take place near the area of origin to make it easier for animals to integrate into the population. An opposing group considered the acclimatization phase to be unnecessary and suggested that radio-transmitters damage the seals' skin, affect the seals' locomotion and orientation, and compromise their survival (Vedder 1998).

In the end, only two animals were radio-tagged. Those animals were released without a pre-acclimatization period directly from an indoor pool into the sea and far from their colony (30 km south). This release area is exposed to a high level of fishing activity and is easily accessible to the public, but it was chosen because it was accessible to the media and the operation was organized primarily as a publicity event (González, López-Jurado, & Lopez 1997). A year later only one juvenile was observed alive.

FUTURE AND PROGNOSIS

Recommendations that might help improve the species' situation include the following:

1. Current plans do not outline conservation activities with sufficient detail, accuracy, or thoroughness because there is no information or previous experience relating to recommended actions. Therefore it seems advisable to review action plans thoroughly. In recent years, much biological information on the species and its problems has been generated. This new knowledge should be translated into conservation actions and incorporated into plans and strategies that are specific and justifiable.

2. Incomplete knowledge on the species represents a limitation to implementing activities; however, this should not halt activities already in progress that are based on current knowledge and are in accordance with the principles underlying the 1992 Convention on Biodiversity. The absence of scientific data should not be an obstacle to carrying out urgent actions that already possess a sufficient scientific basis.

3. It is recommended that the spirit of cooperation among countries, institutions, groups, and people working with the species be improved, giving a greater role in leadership, dissemination, and coordination to the relevant international conventions (Bonn, Bern, and Barcelona).

4. Management activities for this species are most appropriately carried out by wildlife and management professionals. No group should monopolize the conservation actions for the species, carrying out all efforts, research, and actions unilaterally. Previous experiences should be considered and expert advice followed.

5. Conflicts with local people must be addressed. Public awareness programs should be extended to all areas inhabited by monk seals and to countries in which monk seals may still exist or where recolonization may occur. Studies of local public values and attitudes should continue and, if possible, be strengthened in Morocco, Algeria, and Tunisia, where monk seals still thrive. The economic impact of seal damage to coastal fisheries should be precisely assessed, and compensation to collaborating fishermen should be standardized and coordinated through the relevant ministries. Finally, the program should make use of knowledge gained from experiments to protect other marine mammals, so as to develop methods of precluding Mediterranean monk seals' entanglement in nets.

Mexican Prairie Dog

Julián Treviño-Villarreal

Common Name: Mexican prairie dog
Scientific Name: *Cynomys mexicanus*
Order: Rodentia
Family: Sciuridae
Status: Classified as Endangered by the IUCN (1990), the U.S. Fish and Wildlife Service (USFWS 1991), and the Secretaría de Desarrollo Social (SEDESOL 1994); listed on Appendix I of CITES (1992).
Threats: Habitat fragmentation and destruction; competition from cattle grazing; agricultural pesticides; unauthorized eradication programs that make use of poisons.
Habitat: Valleys, prairies, and intermontane basins at elevations of 1,690 to 2,200 m and associated with loamy soils that are dominated by grasses, forbs, and bare soil (Treviño-Villarreal et al. 1997; Treviño-Villarreal & Grant 1998).
Distribution: Endemic, restricted, and relict distribution (Hall 1981) of less than 500 km² in the Mexican states of Coahuila, Nuevo León, and San Luis Potosí. The only prairie dog colony historically reported for the state of Zacatecas is now extinct (Treviño-Villarreal & Grant 1998).

DESCRIPTION

Five species of prairie dogs are found in North America: the black-tailed (*Cynomys ludovicianus*), white-tailed (*C. leucurus*), Gunninson's (*C. gunnisoni*), Utah (*C. parvidens*), and Mexican (*C. mexicanus*). White-tails, Utahs, and Gunnisons all have short tails (30–65 mm and less than 20% of total body length) with a variable amount of white or gray hair; Mexicans and black-tails have longer tails (60–110 mm and more than 20% of total body length) with a distinct black tip (Hoogland 1995).

The origin and isolation of the Mexican prairie dog presumably relate to post-Pleistocene climatic changes and the isolation of black-tailed prairie dog populations around 40,000 years ago (Baker 1956; McCullough & Chesser 1987). Hall (1981) noted that the Mexican prairie dog skull differs from that of black-tails by having inflated auditory bullae, triangular cheek-teeth, and broad and usually posteriorly truncate nasals. The dental formula for *C. mexicanus* is 1/1 incisors, 0/0 canines, 1/1 premolars, and 3/3 molars. Adult males average 1,011 g, whereas adult females average 955 g (Treviño-Villarreal 1990). Mexican prairie dog color is pinkish-buff above and streaked with numerous black hairs, producing a grizzled effect (Ceballos & Wilson 1985). Below, the hairs are dark basally and yellow distally. The tail has more individual black hairs extending proximally along the lateral margins of the tail as a dark border. There are two complete pelage (coat of hair) renewals (winter and summer) and two transition pelage periods (spring and autumn) (Treviño-Villarreal 1990).

NATURAL HISTORY

Treviño-Villarreal and Grant (1998) estimated that Mexican prairie dog current (1992–93), recent, and historical ranges are 478, 768, and 1,255 km², respectively. This indicates that the Mexican prairie dog has lost 62% of its historical, natural habitat as large and medium-size colonies have been fragmented into numerous, small, and isolated ones. Fully 74% of the current geographical range is in Nuevo León (354.7 km²), 24% in Coahuila (112.5 km²), and 2% in San Luis Potosí (9.5 km²) (Figure 6).

Mexican prairie dogs are gregarious and social animals that live in colonies. A typical colony consists of distinct groups of individuals that occupy and protect small areas within the colony (Treviño-Villarreal 1990). Hoogland (1981a) and King (1955) referred to such groups in black-tailed prairie dogs as coteries. Groups are structured by age and usually consist of one or two adult males, one to four females, and 16 to 20 young of the year. Intergroup movements in Mexican prairie dogs involve not only individuals but also groups (Treviño-Villarreal 1990). Group formation apparently is associated with availability of resources, especially food. Dispersal behavior has been described for several members of the Sciuridae family; however, dispersal of Mexican (Treviño-Villarreal 1990) as well as black-tailed (Hoogland 1981a) prairie dogs is poorly understood. Chesser (1983) suggests that increased agricultural use of the land and associated ranching practices, widespread poisoning programs, and natural decimation of populations owing to disease (e.g., sylvatic plague) have reduced the distribution of black-tailed prairie dog populations.

Mexican prairie dogs begin to reproduce either during the first or second year of life depending on body weight, current social structure of the colony, and environmental conditions, with litter size ranging from 4.0 to 4.5 (Treviño-Villarreal 1990). Mild climatic conditions throughout the range of the Mexican prairie dog permit a longer reproductive season than in other species of *Cynomys* in North America, which either hibernate or are inactive throughout most of the winter (King 1955). Reproduction usually occurs between early January and early April depending on food availability, which, in turn, depends primarily on rainfall patterns (Treviño-Villarreal 1990).

No figures on longevity for Mexican prairie dogs are available; however, in black-tailed prairie dogs, females may live 5 to 6 years whereas males usually do not survive longer than 4 years (Hoogland 1981b). Mortality factors affecting Mexican prairie dogs include plague and other diseases,

Figure 6. Distribution of Mexican Prairie Dogs

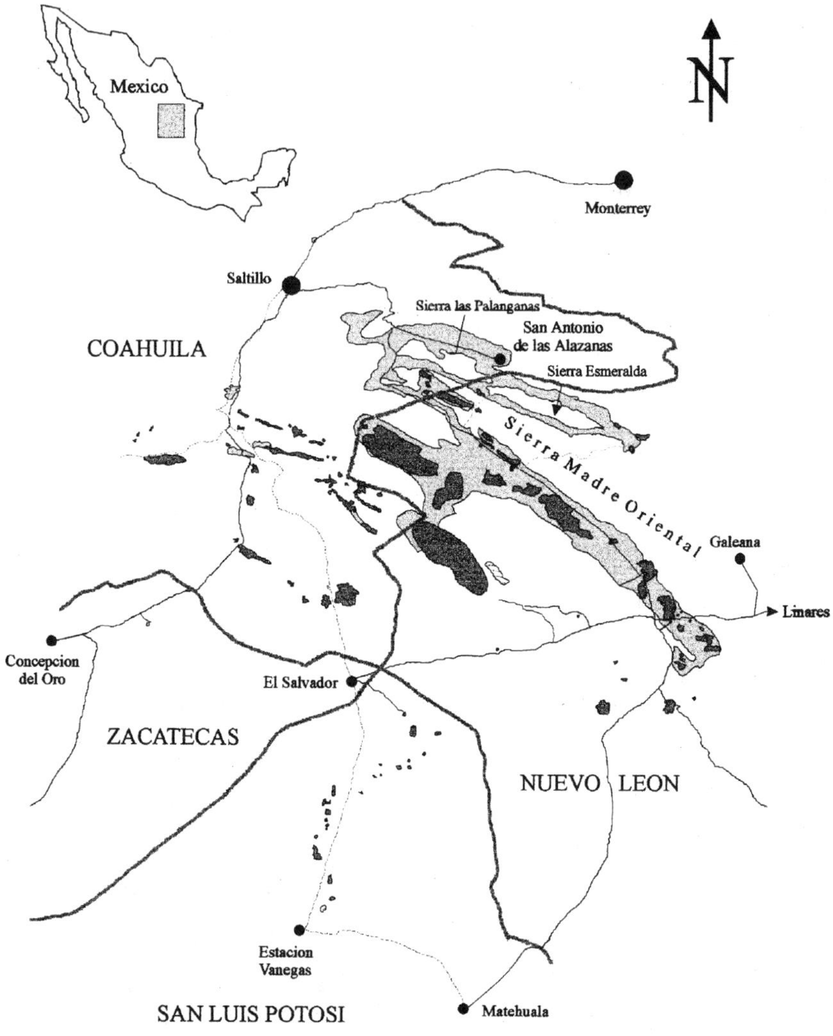

accidents, starvation, weather, parasites, and predation (King 1955; Hoogland 1995; Garret 1982), but human activities have caused the greatest loss
(Chesser 1983; Clark 1977; Treviño-Villarreal 1990).

　　A large and diverse number of plants and animals are present throughout
the grasslands inhabited by Mexican prairie dogs. A total of 84 species of
vertebrates and 163 species of plants have been reported within the geographical range of the Mexican prairie dog and surrounding habitats
(Jiménez-Guzmán 1966; Scott-M. 1984; Treviño-Villarreal 1988). Specially
protected species, such as rattlesnakes (*Crotalus scotulatus*) and red-tailed

hawks (*Buteo jamaicencis*); threatened species, such as badgers (*Taxidea taxus*), kit foxes (*Vulpes macrotis*), prairie falcons (*Falco mexicanus*), and burrowing owls (*Athene cunicularia*); and endangered species, such as golden eagles (*Aquila chrysateos*) (SEDESOL 1994), rely on Mexican prairie dogs as an important source of food or on their burrows for habitats.

CONFLICTING ISSUES

Prairie dogs in the United States and Canada have been considered economic pests because they clip and eat plants utilized by livestock. Similarly, Mexican farmers consider Mexican prairie dogs pests because they feed on commercial and subsistence crops (Treviño-Villarreal et al. 1996). Therefore the establishment of croplands led to extensive, unauthorized poisoning programs to eradicate Mexican prairie dogs. These efforts are local in planning and implementation, and they are not sponsored or coordinated by government. Despite widespread practice, the published literature has documented only one Mexican prairie dog poisoning eradication program (Medina & De La Cruz 1976). Poisoning is generally accepted because prairie dogs are perceived to conflict with human needs.

An estimated 62% of Mexican prairie dog habitat has been destroyed over the past four decades as a result of agricultural activities, and remaining habitat is under heavy grazing pressure from cattle (Treviño-Villarreal & Grant 1998). Densities of Mexican prairie dogs began declining in 1955, at the same time that high technology agriculture was established in the region (Treviño-Villarreal et al. 1996). With the opening of several new highways, the human population has increased, there is easier access for heavy agricultural equipment, and crop production has become the most intensive activity in Nuevo León and Coahuila. If agriculture continues advancing at its present rate in these areas, there is a strong possibility that 98% of remaining Mexican prairie dog habitat will disappear in the near future. Large farms also are being established in northern San Luis Potosí.

The fragmented distribution of remaining Mexican prairie dog habitat also may play an important role in determining the fate of colonies. The proximate cause of abandonment of the six colonies that were identified as inactive by Treviño-Villarreal and Grant (1998) remains speculative—none were close to croplands, and local residents did not observe any of the factors that have caused abandonment of colonies in other areas, such as disease (Treviño-Villarreal et al. 1998), poisoning eradication programs (Medina & De la Cruz 1976), direct or indirect effects of cattle grazing within the colonies and adjacent desert scrub communities (Mellink 1989), or catastrophic events (Oldemeyer et al. 1993). Chances of recolonization may be low owing to the colonies' relative isolation (Treviño-Villarreal & Grant 1998).

FUTURE AND PROGNOSIS

Mexican prairie dogs are Endangered and therefore protected by the Mexican Law for the Ecological Equilibrium and the Protection of the Environment (LEPEMA); however, there has been only one case in which the Mexican government, through LEPEMA, fined a private landowner in the state of San Luis Potosí because his land use practices were destroying and fragmenting Mexican prairie dog habitat (Treviño-Villarreal et al. 1992).

Studies on the effects of competition with cattle grazing, agricultural pesticides, and sylvatic plague, as well as pilot prairie dog reintroduction programs, are urgently needed to help future plans for the conservation and management of the Mexican prairie dog. However, such plans will require a strict regulation of agriculture; enforcement of the endangered species legislation by the Mexican Ministry of the Environment, Natural Resources, and Fisheries; and resolution of conflicts among the Mexican Agrarian Reform Law, the Forestry Law, and the Law for Ecological Equilibrium and the Protection of the Environment.

Mexican Wolf

Jane M. Packard and José F. Bernal-Stoopen

Common Name: Mexican wolf
Scientific Name: *Canis lupus baileyi*
Order: Carnivora
Family: Canidae
Status: Endangered in Mexico and the United States; a reintroduced population is designated as experimental and nonessential (east-central Arizona and west-central New Mexico).
Threats: Human-related mortality; depletion of prey populations owing to human activities.
Habitat: Dependent on adequate populations of large-bodied mammalian prey; primary habitat is in mid-elevation oak woodlands, with surrounding grasslands as secondary habitat.
Distribution: Formerly located throughout the western mountains and high plateau region of Mexico (from Michoacán and Puebla to the border regions of Arizona/Sonora, Chihuahua/New Mexico/Texas/Coahuila) (Leopold 1959; Young & Goldman 1944); presently there are infrequent signs of a small remnant population in the mountains on the border between Sonora and Chihuahua, where it has been logistically difficult to verify the information (Crane 1989). Captive-raised family groups have been reintroduced to the Blue Range Area of the Apache National Forest in Arizona (Dinon 1998; U.S. Fish and Wildlife Service 1997).

DESCRIPTION

The Mexican wolf is a relatively small-bodied subspecies of the gray wolf (25–42 kg), varying in color from grizzled-gray to tawny-cinnamon on the upper parts of the body, with lighter underparts (Brown 1983; Young & Goldman 1944). This subspecies is morphologically and genetically distinct from red wolves (*Canis rufus*) and coyotes (*Canis latrans*) (Hedrick et al. 1997).

In a revision of the taxonomic status of three subspecies of wolves described from Mexico and the southwestern United States, no convincing morphological evidence was found to separate the Mexican wolf (*C. l. baileyi*) from the extinct subspecies found in New Mexico (*C. l. mogollonensis*) and Texas (*C. l. monstrabilis*) (Brown 1983; U.S. Fish and Wildlife Service 1997). Body size and skull size overlap considerably with the variation observed in other subspecies of gray wolves (Brown 1983).

Genetically, Mexican wolves may be distinguished from other subspecies on the North American continent by one unique genetic marker (Hedrick et al. 1997; Savage 1995). This genetic marker also has been identified in the Italian population of wolves. Ron Nowak hypothesized that current southern wolf populations represent remnants of an ancestral form that spread from North America to Eurasia during an interglacial period; subsequently these small-bodied forms were restricted by glaciers to southern regions when a larger-bodied form evolved in Siberia and re-invaded the North American continent (Servín-Martínez 1993).

NATURAL HISTORY

Mexican wolves occupied mountain "islands" of Madrean oak woodland surrounded by "seas" of arid grasslands on the high plateau extending from the southern Rocky Mountains to the Trans Volcanic Range of central Mexico (Leopold 1959). Wolves primarily inhabited oak/pine forests of mid-elevation zones (above 1,400 m) where prey was more abundant compared with the surrounding semi-desert (below 900 m) or the mountain peaks that supported little more than alpine tundra.

Life history and behavioral characteristics of Mexican wolves are very similar to those of other subspecies of gray wolves (Brown 1983; Leopold 1959). After a gestation of about 2 months, litters are born in March–May, usually with three to eight pups (Bernal & Packard 1997; Servín-Martínez 1997). Similar to other wolf subspecies (Packard in press), each breeding pair of wolves travels together throughout the year, defending a territory from neighboring families and potential rivals. Males assist their mates in caring for pups during the first 2 to 3 months (Bernal & Packard 1997; Servín-Martínez 1991). Litters are born in dens and shallow scrapes (Young & Goldman 1944).

Some authors have suggested that the Mexican subspecies is less social than other subspecies of wolves (Brown 1983). Yet no systematic field studies of Mexican wolves have been conducted. Ranch workers and predator-control agents have reported wolves in small groups, ranging from one to seven (Brown 1983). However, wolf pack size tends to be smaller throughout the Northern Hemisphere wherever (1) wolves are trapped, (2) prey within the home range of each breeding pair is scarce, and (3) there are opportunities for dispersing offspring to find food and mates outside the parents' home range (Packard in press).

CONFLICTING ISSUES

The last reported capture of a Mexican wolf was in 1980, when a male that had been killing livestock in the Sierra del Nido of Chihuahua was brought into captivity to supplement the captive breeding program (Brown

1983). The main threats to Mexican wolves have been predator control activities, depletion of prey populations, and habitat alteration owing to logging and grazing (Brown 1983; U.S. Fish and Wildlife Service 1997). The oak woodlands previously occupied by deer and wolves are now the most productive summer range for livestock in this arid region.

Persecution of wolves caused rapid declines in the late 1800s; by 1960 wolves were exterminated from all but three remote sites in Mexico (Leopold 1959; Brown 1983). Extermination was the result of two federal policies in support of agricultural expansion: (1) predator eradication in the southwestern United States in the early 1900s, and (2) rabies control in northern Mexico from 1952 to 1960.

By the late 1800s bison and elk had been eliminated by market and subsistence hunters, and deer and pronghorn populations had declined as a result of landscape changes, subsistence hunting, and sport hunting (Leopold 1959). Wolves turned to livestock where native prey populations were drastically reduced (Brown 1983; Leopold 1959). The lure of an easy meal may have attracted the few remaining wolves to livestock areas, where they were more vulnerable to predator control. Those wolves that escaped capture or poisoning apparently learned not to eat twice from the same kill. This wariness caused the rate of livestock depredation to rise to the point at which it was intolerable to livestock producers.

Passage of the U.S. Endangered Species Act (ESA) in 1973 started a series of recovery events unmatched by Mexico (Brown 1983). Under Mexican laws, there are no mechanisms for mobilizing the resources for recovery efforts on a level comparable to what was available in the United States (Bernal-Stoopen 1999). The ESA provided the administrative tools and funding to transform previous predator control efforts into recovery efforts for Mexican wolves (Savage 1995).

Several key actors guided the Mexican wolf recovery effort through difficult situations (Bernal-Stoopen 1999). Mexican ranchers cooperated with a predator control agent in the live capture of seven wolves to start a captive population managed by the U.S. Fish and Wildlife Service (FWS) (Brown 1983). Owing to the small number of breeding founders and unclear subspecific status of the Mexican wolf, the initial request for development of a formal Species Survival Plan was rejected (Burbank 1990). Undaunted, leaders at two zoos formed an ad hoc committee of collaborative colleagues who had the same vision of returning wolves to the wild, sharing standards of professional husbandry, and implementing strategies to minimize inbreeding (Bernal-Stoopen 1999). Through personal contact, enthusiasm for the captive breeding effort grew within the zoo community, enticing more institutions to join and to support exchanges with Mexican colleagues. Support of key leaders in environmental groups was galvanized when government agencies could not agree on a site for re-introduction and announced that the federal recovery effort would be terminated (Bowden 1992; Burbank 1990; Savage 1995). Today the captive population is distributed among institutions participating in the Species Survival Plan (Dinon 1998).

The controversial issues extended beyond potential impacts on agricultural production and sport hunting. The Mexican wolf was only one of several endangered species that sparked debate over local versus state and federal rights (Savage 1995). Private citizens expressed concern that their needs were not being considered in the decisions of government agencies. Previous experience had eroded their confidence in being able to work with diverse federal and state agencies in resolving the problems that inevitably arise during recovery efforts. Extremely polarized opinions were presented by both those who loved and those who hated wolves (Savage 1995).

Several decisive actions broke the deadlock: (1) a legal suit, (2) innovative state programs in Arizona, and (3) a task force to facilitate communication among citizens groups and government agencies. First, a coalition of environmental groups sued the FWS in 1990 on the grounds of noncompliance with the legal mandate to recover endangered species (Savage 1995). In the settlement, the issue changed from "if" reintroduction would occur to "when," "how," and "where" it would occur. Two full-time federal employees were hired to coordinate and implement the reintroduction process (Bowden 1992).

Second, a leader within the non-game program of the Arizona Game and Fish Department pledged a moratorium on further decisions regarding reintroduction until he was able to complete an informal process of dispute resolution (Johnson personal communication). He met with representatives of each stakeholder group in order to (1) understand more fully the concerns of people who felt their lives would be impacted by re-introduction of wolves, (2) enhance two-way communication, and (3) explore elements of common ground.

For stakeholders in Arizona, working with their state agency was the lesser of two evils. A public survey in Arizona indicated that urban residents were in favor of wolf recovery, although rural residents were not. After systematically evaluating potential release sites within Arizona, the state agency took the lead in writing a recovery plan that was acceptable to the FWS (Groebner et al. 1995).

In 1991 the Arizona Cattle Growers Association (ACGA) met to review its 1986 resolution opposing wolf re-introduction (Dale 1992). Some members argued that as long as their organization opposed the re-introduction, they would remain outside the decisionmaking loop. Other ACGA members decided it was better to be key players in the program and approved a statement specifying seven conditions under which wolf re-introduction would be acceptable to them.

Biologically, the logical site for re-introduction was a large forested fragment (4.6 million acres) crossing from the Blue Primitive Area in the Apache

National Forest of Arizona to the Gila National Forest in New Mexico (Savage 1995; Stolzenberg 1998). Politically, the Arizona side of this forest fragment was more acceptable as a re-introduction site because attitudes toward federal agencies were less polarized there than in New Mexico (Savage 1995). Because wolves disperse hundreds of miles, there was a high probability that the forest in New Mexico would be recolonized by wolves from Arizona.

A leader of an environmental group, Preserve Arizona's Wolves, organized the Blue Range Task Force (Holoday personal communication), an innovative experiment in private initiative. Her vision was to provide a monthly forum for concerned citizens to communicate directly with federal and state representatives in an atmosphere of collaborative problem-solving. Initially, discussions were led by a professional facilitator, making it "safe" for Task Force members to honestly express their own emotions and viewpoints while listening to the perspectives of others. Together, the group explored what it meant to them to live in the Blue Range during the past, the present, and the future. In addition to the mandated public hearings, the Task Force provided input to the preparation of the Draft Environmental Impact Statement (U.S. Fish and Wildlife Service 1997) and design of a livestock-damage compensation fund (Savage 1995).

In April 1998, 11 Mexican wolves in three family groups were re-introduced to the Blue Range (Dinon 1998). By December, none remained in the wild: five were dead (or presumed dead), and six had been captured and returned to captivity because they had wandered outside the re-introduction area or their mates had died (Benke 1998). Many lessons were learned during this experimental phase, which was scheduled to continue for 3 to 5 years. One of the most important lessons was the value of collaboration between the public and private sectors (Bass 1998).

Such collaboration also was a key element in the recovery effort for Mexican wolves in Mexico (Bernal-Stoopen 1999). Primarily through the personal efforts of interested professionals, participation in the binational Species Survival Program expanded to the stage at which 40 institutions (public and private) managed 185 captive Mexican wolves in 1998); 11 of these sites were in Mexico (Dinon 1998). The prognosis for the captive population has improved since genetic studies verified that both lineages— one maintained in American zoos, and one in Mexican zoos—were suitable to include in the certified captive breeding program (Hedrick et al. 1997). If re-introduction is possible in Mexico, it is likely to result primarily from private initiative coordinated with federal and state programs.

FUTURE AND PROGNOSIS

The future recovery of Mexican wolves will both shape, and be shaped by, public debate over the rights of local versus state and federal interests.

The more polarized the debate becomes, the longer the time frame for recovery. However, the more sincere, two-way communication among stakeholders there is, the more positive the prognosis.

Decisive action by leaders in the public and private sectors has sent a clear and consistent message that Mexican wolves will have a role in the landscapes of the future (Bass 1998). However, just what role that will be remains to be determined. Increases in predator populations will require adjustment on the part of many people, including hunters (Barsness 1998), livestock producers (Dale 1992), rural residents (Savage 1995), land managers (Bass 1998), and government agencies responsible for services addressing the risks of animal damage to human property and human damage to the functioning of healthy ecosystems (U.S. Fish and Wildlife Service 1997).

Biologically, the prognosis for Mexican wolf recovery is positive. Wolves reproduce readily, adapt to wild conditions within a few generations, and are effectively managed under both captive and wild conditions. The goal of the captive breeding program—long-term maintenance of a population of 240 Mexican wolves in captivity—will be achieved with the addition of 10 to 20 facilities.

Socioeconomically, the "Old West" is moving from a colonial to a postcolonial phase. More ranchlands in the former range of Mexican wolves are being managed with goals and resources that extend beyond livestock production (Bass 1998; Page 1992). The key to Mexican wolf recovery lies in the hands of those who shape the "New West." Ranchers, hunters, environmentalists, rural residents, and urban nature-enthusiasts share common ground: concern over fragmentation of habitat and increasing property taxes.

In a precolonial Mexican fable, a man was terrified by a wolf in the woods. He was reassured by the wolf that no harm would come to him and they could co-exist. Overextended, the man fell off a precipice. Encountering the scattered bones of the man, the wolf howled and the fragments reassembled, bringing renewed life to the man. During the Mexican colonial era, the wolf was a symbol of liberty and wilderness frontiers, and its near extinction is a reflection of loss of liberty to exploit resources. In this postcolonial era, the recovery of Mexican wolf populations in fragmented forest habitats will depend on the abilities of stakeholder groups to discover common ground and to work together in restoration of forest systems that protect many other species in addition to wolves.

Mountain Plover

Stephen J. Dinsmore

Common Name: mountain plover
Scientific Name: *Charadrius montanus*
Order: Charadriiformes
Family: Charadriidae
Status: Currently being considered for Threatened or Endangered status under the U.S. Endangered Species Act; listed as an Endangered Species in Canada.
Threats: Loss of prairie dogs (*Cynomes* spp.), which create important habitat for plovers; conversion of shortgrass prairie to agriculture; possibly the alteration of grazing regimes on native shortgrass prairie.
Habitat: Occupies a wide variety of habitats including native, grazed shortgrass prairie, prairie dog colonies, fallow agricultural fields, and semidesert areas. This species is endemic to the Great Plains and is often considered an associate of the shortgrass prairie.
Distribution: Western Great Plains of the United States from Montana southward to the Oklahoma Panhandle and northeastern New Mexico. Isolated breeding populations occur in northeastern Utah and the Davis Mountains of Texas. The mountain plover may breed locally in Mexico (Knopf & Rupert 1999a). Its historic range extended eastward into the central Great Plains and into southern Canada. It winters from the central valleys of California eastward across northern Mexico, in southeastern Arizona, and in southern Texas.

DESCRIPTION

The mountain plover is a fairly large (90–130 g), drab plover that was described in detail by Knopf (1996). Mountain plovers are monomorphic (i.e., the sexes look similar), although the sexes may often be distinguished by the brighter plumage of the male when both are present. In breeding plumage, the head pattern is distinct with a black crown patch, white forehead and throat, and black eyeline. The upperparts are sandy brown with occasional buff or rufous color on the nape. The underparts are white and lack the characteristic dark breast bands of other plovers. In flight, the bird shows a faint white stripe along the bases of the upper surface of the flight feathers. The underwings are pure white. Vocalizations were described by Graul (1974) and include two primary calls; the *"Wee-wee"* song given during courtship and a *"Kip"* flight call.

NATURAL HISTORY

Mountain plovers formerly occupied a much larger range on the western Great Plains. Fossil records (Quaternary period), which were summarized by Knopf (1996), document a broad historical range throughout the Great Plains. The current breeding range is greatly reduced, especially along eastern edge in South Dakota, Nebraska, and Kansas. Within their breeding range, mountain plovers are very locally distributed in suitable habitat. Birds along the eastern edge of the breeding range tend to occupy fallow agricultural fields, possibly because native, heavily grazed landscapes there are lacking.

Mountain plovers occupy habitats of bare ground with or without some coverage by low vegetation, including prairie dog colonies, fallow agricultural fields, and native grasslands. They prefer relatively flat areas with a minimum of 30% bare ground for nesting. The bare ground component may be the most important habitat characteristic (Knopf & Miller 1994). In addition, plovers tend to nest near a conspicuous object such as a rock or pile of cow manure (Graul 1975; Knopf & Miller 1994).

They are thought to be monogamous (Knopf 1996), although some polyandry may occur (Graul 1975). At least some birds breed during the first year after hatching. The breeding season extends from mid-April through July with some longitudinal variation. At least some females lay a clutch for themselves and a second clutch for the male, an unusual mating system among birds. A single adult incubates each nest. The nest is a shallow scrape lined with pebbles, prairie dog droppings, and bits of dried vegetation (Knopf 1996). Clutch size is typically three eggs, and incubation lasts about 29 days (Graul 1975). Hatching success is variable but probably averages 40 to 50% throughout their range. The young are precocial, leave the nest site immediately, and fledge in about 35 days (Graul 1975; Miller & Knopf 1993). The diet is almost strictly insectivorous. Preliminary estimates of annual survival are 25% for juveniles and 80% for adults. (Dinsmore unpublished data). Plovers are site-faithful, with many birds returning to the same breeding site in subsequent years.

CONFLICTING ISSUES

Historically, mountain plovers were widely distributed throughout the western Great Plains of North America. There they were apparently adapted to a disturbed prairie ecosystem. Bison, prairie dogs, and other grazers altered the native grasslands, creating a mosaic of habitat that was suitable for plovers. However, this region has experienced dramatic changes in the last 100 years. It is estimated that prairie dogs have declined by more than 98% since settlement, and the bison that once numbered 30 million or more on the Great Plains have disappeared entirely from this region (see Knopf

1994). In addition, many of the native grasslands have been converted to agriculture or altered through changes in grazing practices.

Because of these changes in the Great Plains, mountain plovers have declined in the 20th century. Breeding Bird Survey data have documented a 66% decline in the period 1966–1993 (Knopf 1996). The present continental population is now thought to number 8,000–10,000 birds (Knopf 1996). The major breeding strongholds appear to be Weld County and South Park in Colorado and Phillips and Blaine Counties in Montana.

There are two major threats to the continued survival of mountain plovers: habitat loss or degradation, and the loss of prairie dogs. The loss and degradation of breeding habitat appears to be the greatest threat. As much as 30% of the native grasslands in the Great Plains have been lost since settlement (see Knopf 1994). Much of this region is still threatened by future development. An estimated 4 million hectares of the remaining native grasslands in Montana, Wyoming, and Colorado might be converted to cropland by the year 2002 (Laycock 1988).

The remaining, often highly fragmented, native grasslands have also been altered. Throughout the southern portion of their range, mountain plovers nest on native, grazed grasslands, especially in eastern Colorado. Mountain plovers require specific habitat characteristics for their continued survival. Low vegetation and bare ground are the two most important features. On public grazing allotments, grazing pressure is managed to maintain a uniform grass cover favorable to cattle. In addition, these grasslands are often seeded with non-native grasses to enhance forage for cattle. This large-scale, homogeneous environment is unfavorable to plovers because it lacks short grass and bare ground components. On the Pawnee National Grassland in northeastern Colorado, public pressure to reduce cattle grazing has left much of the grassland habitat unsuitable for plovers. Indeed, plovers were much more numerous there in the 1970s and appear to have declined as lighter grazing pressure on public lands has lowered plover habitat quality. Consequently many plovers have shifted to more heavily grazed private lands surrounding the National Grassland. Management of these public grasslands for cattle and non-game species, such as the mountain plover, may require changes in grazing practices. Grazing pressure may need to be rotated to create a spatially heterogeneous environment favorable to multiple species, including the plover. Such a change would require cooperation among ranchers, land managers, and biologists. Land managers are willing to alter grazing regimes, but they fear a negative reaction from the ranching community.

Habitat loss is also occurring on the principal wintering grounds in the Central Valley of California. In the post-settlement period, an estimated 97% of the native habitat in this region has been lost to agriculture and other forms of development. The few remaining patches of native grassland are highly preferred by wintering plovers (Knopf & Rupert 1995). However, despite these large habitat losses, Knopf and Rupert (1995) documented extremely high over-winter survival of mountain plovers and concluded that continental declines should probably be attributed to factors on the breeding grounds.

Recently, significant numbers of mountain plovers have been found nesting on fallow agricultural fields, especially in the southern part of their range (Knopf & Rupert 1999b). Changes in agricultural practices on the western Great Plains in the last quarter-century have probably impacted mountain plover productivity (Knopf 1996). Preference for agricultural fields may further complicate management of the species. Potential conflicts arise when fields are worked during the nesting season, resulting in a direct loss of nests and possibly even young. Knopf (1994) speculated that the direct take of nests on agricultural lands might explain the annual decline in plover numbers. These farming practices are not likely to change in the near future. Within the region, emphasis should be placed on enhancing remaining native habitat for breeding mountain plovers in an attempt to draw birds away from agricultural fields that may act as population sinks. (i.e., areas where death rates exceed birth rates). It is unrealistic to attempt to convince farmers to delay planting until after the plover nesting season.

The loss of prairie dogs, especially black-tailed prairie dogs (*C. ludovicianus*), represents a major threat to mountain plovers and is highly contentious. In the northern part of their range in Wyoming and Montana, much of the native grassland is a taller, mid- to short-grass community. Except in areas that are heavily grazed or occupied by prairie dogs, this region contains little plover habitat. The majority of mountain plovers in Montana now occur on prairie dog colonies in Phillips and Blaine Counties. Because this area contains possibly the second largest mountain plover population in existence (Knopf & Miller 1994), it is important to conserve these colonies. Approximately 70% of these prairie dog towns are on public lands administered by the Bureau of Land Management (BLM) and the U.S. Fish and Wildlife Service. In the early 1990s, the surface acreage of colonies in the area declined by more than 75% owing to an apparent outbreak of sylvatic plague. Prairie dogs now appear to be recovering but are still threatened by losses from recreational shooting. Complicating matters further, the BLM was recently petitioned to stop all recreational shooting of prairie dogs. Polarized attitudes toward recreational shooting include strong support from ranchers and varmint hunters and strong opposition from the general public, animal rights activists, and biologists concerned about the prairie dog ecosystem. The outcome of this legal petition will ultimately affect mountain plovers because of their association with black-tailed prairie dogs.

Management of prairie dogs in Phillips County is complicated by agreements between the Bureau of Land Management and ranchers (Bureau of Land Management 1992). Prairie dogs are classified as agricultural pests in

Montana and receive no formal protection. The ranching community supports an active control program aimed at capping the prairie dog population at 22,000 acres on BLM lands. The agreement calls for poisoning and a recreational shooting program to assist in prairie dog control (Bureau of Land Management 1992). Ranchers' attitudes toward mountain plovers are similar to their attitudes toward the Endangered black-footed ferret. However, because mountain plovers are not as imperiled as the ferret, ranchers do not perceive plovers as a threat to their livelihood.

At present, mountain plovers are not the focus of intense recovery efforts. Instead, state and federal agencies have made some effort to enhance breeding habitat. Habitat factors can be addressed by active management of existing grasslands through burning, alteration of grazing regimes, and possibly other forms of management. Several National Forest Service grasslands in Colorado and Kansas have successfully used early season burns to provide plover nesting habitat. However, these small-scale burns may act as a population sink by concentrating large numbers of breeding plovers in areas susceptible to predators, including the swift fox. Other manipulations such as mowing have shown little promise as management tools. The conservation of prairie dogs is more complicated. Existing management of prairie dogs for black-footed ferrets will continue to benefit plovers. In 1998 the National Forest Service restricted recreational shooting of prairie dogs on the lands it manages. This action may ultimately benefit plovers if prairie dog numbers are allowed to rise.

FUTURE AND PROGNOSIS

Concern for declining grassland bird species in the 1990s (Knopf 1994) has helped focus interest on the mountain plover. Long-term declines in plover populations resulted in the U.S. Fish and Wildlife Service being petitioned to list the species under the U.S. Endangered Species Act. This action may be a necessary measure to ensure the long-term persistence of the species. The process will focus on current knowledge of the species and what can be done to aid recovery. Management of federally listed species on private lands will continue to be debated prior to any listing action.

Several management strategies are likely to emerge from the listing process. High quality breeding habitat will need to be protected and actively managed for plovers. Issues surrounding prairie dog management in the Great Plains will need to be solved through the interaction of federal and state agencies and special interest groups. The groups seeking to protect prairie dogs, exploit them for recreational shooting, or eradicate them in the interest of agriculture and ranching will have to agree on how to best manage them. A recent proposal to list the black-tailed prairie dog under the U.S. Endangered Species Act may impact mountain plovers, especially in the northern part of their range where they occur almost exclusively on prairie dog colonies.

Future research would be useful on demographic parameters such as annual survival and recruitment, population genetics, and the use of various management tools such as burns and prairie dog enhancement. The interaction of land managers and biologists, and the implementation of sound management techniques, will be necessary if mountain plovers are to persist.

Northern Rocky Mountain Wolf

Douglas W. Smith and Michael K. Phillips

Common Name: gray wolf
Scientific Name: *Canis lupus nubilus*
Order: Carnivora
Family: Canidae
Status: Endangered under the Endangered Species Act of the United States—except in Minnesota, Michigan, and Wisconsin where it is listed as Threatened, and in Wyoming and portions of Montana and Idaho where reintroduced populations are listed as Experimental-Nonessential; the species is not listed in Alaska.
Threats: Persecution by humans owing to livestock depredation; competition for ungulates; a misunderstanding of wolf and its ecological role.
Habitat: Wherever there are large ungulates.
Distribution: Reintroduction sites in Idaho and the Greater Yellowstone Area have approximately 220 wolves; a naturally recolonizing population in northwestern Montana numbers between 50 and 70; Minnesota has about 2,000 wolves; and Michigan and Wisconsin support roughly 100 wolves each. There are reports of wolves in North Dakota and Maine, but no verified populations. Wolves are common throughout Canada and Alaska.

DESCRIPTION

Gray wolves, described by Mech (1970), are the largest member of the dog family (Canidae). Adult males weigh from 30 to 60 kg and females from 25 to 50 kg. Long-legged with oversized feet, the wolf is difficult to confuse with the much smaller but similar looking coyote. Wolves are typically gray but can be black or white. The dental formula is 3/3 incisors, 1/1 canines, 4/4 premolars, and 2/3 molars.

NATURAL HISTORY

Wolves are obligate carnivores (i.e., they can only eat meat), that rely almost exclusively on ungulates (Mech 1970). Once considered a wilderness species (Theberge 1975), wolves are now known to exist wherever there are adequate prey and human tolerance.

Wolves breed in February through March, depending on latitude (Mech 1970). Gestation is approximately 63 days (Mech 1970). Females remain receptive for 65 days if not bred. Young are born in a den that can be a hole in the ground, an excavation under a rock, a log, an old beaver lodge, or a cave. Wolf pups have also been born above ground (Phillips & Smith 1996). Litters average five pups, which are helpless at birth. Pups emerge from the den at 10 to 15 days of age and are weaned in 6 to 8 weeks (Mech 1970).

Wolves live in family groups known as packs that form because of delayed dispersal. Young may disperse as early as 10 months or delay dispersal for years, but most leave their natal pack by 3 years of age (Gese & Mech 1991). Variable dispersal ages make most packs multigenerational.

Wolves are territorial, marking and defending their areas of use from other wolves. Territories are large, ranging from 10,000 hectares to over 250,000 hectares depending on prey availability. Changes in prey density are related to expansion or contraction of territory—and therefore population density of wolves. Prey availability may be directly linked to wolf mortality, either through starvation or through intraspecific strife (i.e., fights between individuals of the same species).

Wolves are usually monogamous but will re-pair if a mate dies. It was once believed that all breeding was done by the dominant male and female in a pack (i.e., alphas). Although this tendency is the norm, recent research has shown more than one female breeding in a pack is not as uncommon as once thought (Mech et al. 1998). The number of females in a pack that breed is probably related to pack size, social relations within the pack, and availability of food.

CONFLICTING ISSUES

Young and Goldman (1944), Lopez (1978), and McIntyre (1995) best described the decline of wolves across North America and Eurasia. Wolf eradication was conducted with an almost religious fervor, and sometimes horrific methods (e.g., setting wolves on fire, cutting off their lower jaws and releasing them, etc.), were used (Lopez 1978; McIntyre 1995).

A change in values in the 1960s brought a greater awareness of predators and of ecosystems (Mowat 1963). Watershed studies revealed important aspects of wolf ecology (Murie 1944), which prompted biologists to question long-held beliefs about predators and the need for wolf control (Errington 1946). During the last three decades wolf ecology has been exhaustively researched, and there is now a better understanding of the topic (Allen 1979; Carbyn et al. 1995; Peterson 1995; Mech et al. 1998).

This understanding, however, has not simplified wolf recovery, which is an important issue because the species' current distribution is much reduced. During the last two decades significant effort has been expended to recover wolf populations to appropriate habitats under the authority of the Endangered Species Act (U.S. Fish and Wildlife Service 1994; Bangs & Fritts

1996; Fritts et al. 1997). Indeed, for many conservationists wolf recovery has become an important benchmark for measuring the U.S. commitment to conserving imperiled species.

Two issues are largely responsible for the contentious nature of wolf recovery: livestock depredations, and competition with humans for wild ungulates. Wolf control (i.e., the purposeful reduction of wolf populations) has been the most frequently applied solution for both issues. However, because of a sympathetic public that often opposed control, strong-minded farmers and ranchers who often supported control, and intense debate among scientists about the efficacy and need (because other factors may contribute to prey declines) for wolf control, the practice has been hounded by controversy (Gasaway et al. 1983; Mech 1995; Haber 1996; National Research Council 1997).

Depredation of livestock may be the single most significant issue related to wolf recovery, because it was the impetus behind the original eradication of wolves. Historically in North America, wolf-induced livestock losses were much greater than current losses (Lopez 1978). As North America was settled by Europeans, a significant reduction in most populations of native ungulates occurred, leaving wolves with an increased need to kill livestock (Lopez 1978). In recent times, active wildlife conservation programs (U.S. Fish and Wildlife Service 1987) facilitated the recovery of native ungulate populations, leading to a reduction in depredations. Wolves in Minnesota and Alberta live near farming and ranching operations, yet few wolves there actually kill livestock (Gunson 1983). Wolves are now settling areas in Minnesota that support numerous people and domestic animals, yet depredations have not increased proportionally.

Although several methods have been developed to minimize or prevent depredations, few have proven successful. Guard dogs have been used widely, but with marginal results (Coppinger & Coppinger 1995). Generally one guard dog is not sufficient, as several dogs seem necessary to deter a wolf attack (Coppinger & Coppinger 1995). Another approach requires farmers and ranchers to intensify husbandry of livestock (e.g., confine sheep to structures overnight, develop calving areas near ranch headquarters, or monitor open range stock daily). Ultimately, killing the wolf or wolves responsible for the depredation is often the only long-term solution (Mech 1995).

Depredation of livestock was a major concern for ranchers in the northern Rocky Mountains, where the U.S. Fish and Wildlife Service (USFWS) is actively promoting recovery of gray wolves. In response, the USFWS developed management protocols that rely on lethal control after a wolf has been involved in two depredation events (U.S. Fish and Wildlife Service 1994). This "two strike" rule will likely be in effect only while the wolf population is small. Once recovery has been achieved (i.e., when ten breeding packs have each produced offspring for three successive years in each of three areas: northwestern Montana, central Idaho, and the Greater Yellowstone Area), then the gray wolf in the northern Rocky Mountains will be removed from the list of endangered species and managed as resident wildlife by the states of Montana, Wyoming, and Idaho (Bangs & Fritts 1996). It is likely that these states will adopt liberal protocols for managing wolves, including recreational harvest and lethal control whenever wolves are near livestock.

Wolf conservation is also contentious because the species' reliance on native ungulates often conflicts with human use (National Research Council 1997). In vast areas of Canada and Alaska, thousands of wolves have been purposefully killed because of concern over the ability of local ungulate populations to support both wolf predation and human harvest (Gasaway et al. 1983; Haber 1996). Many individuals believe that every wild ungulate a wolf kills is one less for the human hunter (Lopez 1978).

There is much disagreement among wildlife conservationists over the need to control wolves to promote the growth of ungulate populations (Bergerud et al. 1988; Thompson & Peterson 1988). It is unclear if wolf predation adds to the overall mortality burden placed on ungulate populations or is simply compensatory (i.e., that wolves prey mostly on animals that would have died at about the same time owing to some other cause) (Gasaway et al. 1983). Additionally, even if wolf predation restricts ungulate population growth, there is reason to question the efficacy of wolf control as a cost-effective wildlife management tool. A recent review of wolf management in Alaska concluded that there was no long-term evidence to support wolf control to increase prey populations, and it stated that of 11 studies examined, limited data and experimental design flaws made conclusions tenuous (National Research Council 1997). The report did acknowledge short-term increases in moose and caribou numbers for hunters when a large percentage of the wolves in an area were killed. Accordingly, the debate over killing wolves for purposes of enhancing hunting opportunities for humans rages as strong as ever throughout Alaska and Canada (Bergerud et al. 1988; Thompson & Peterson 1988; Haber 1996).

FUTURE AND PROGNOSIS

Despite intense controversy, wolf conservation has been successful. Widespread lethal control of wolves is no longer practiced without biological justification and social input. And nonlethal methods of controlling the size and distribution of wolf populations are being sought. As a result, there are more wolves in North America today than there were 30 years ago. In Montana, Minnesota, Michigan, and Wisconsin wolves have naturally reclaimed significant portions of their historic ranges. Re-introductions to central Idaho and the Greater Yellowstone Area have been successful beyond expectations and have prompted discussions about initiating such projects in

Adirondack State Park in New York, in northern Maine, and in Olympic National Park in Washington. Controversy will be lessened and success maximized if wolf recovery focuses on large wildland areas with low human population density (Smith et al. 1999). Also, wolf recovery will be most successful if public education about management issues is emphasized so that a significant proportion of the public supports recovery while tolerating some form of control. It is important that public education programs include the message that widespread recovery of wolves will ultimately result in a need to control them (Fritts et al. 1995; Mech 1995).

Orinoco River Dolphin

Esmeralda Mujica-Jorquera and
Ernesto O. Boede

Common Name: Orinoco River dolphin, pink dolphin
Scientific Name: *Inia geoffrensis* (de Blainville 1817)
Order: Cetacea
Family: Iniidae
Status: Considered Vulnerable by the IUCN (IUCN 1996) and listed on Appendix II of CITES; in Venezuela it is listed as Near Threatened (Rodríguez & Rojas-Suárez 1995).
Threats: Major threats include habitat destruction; alteration of habitat by commercial navigation; destruction of gallery forests for agricultural purposes; use of pesticides and herbicides; pollution produced by urban factories; and mercury used in mining activities. In Brazil, population fragmentation caused by hydroelectric dams prevents genetic flow between subpopulations (Best & Da Silva 1989, 1993; Rodríguez & Rojas-Suárez 1995). Fishing becomes a threat when *Inia* are accidentally caught in nets. Minor threats include isolation or stranding in channels and small lagoons during critically dry seasons (Best & Da Silva 1989; Boede & Mujica-Jorquera 1992; Kendall et al. 1995; Trebbau & van Bree 1974).
Habitat: The slow, sandy, muddy waters of the Amazon and Orinoco Rivers and their tributaries; and the clear and amber-colored rivers such as Rio Negro and Ventuary (Trebbau & Van Bree 1974; Trebbau 1975; Best & Da Silva 1993; Boede et al. 1998). In the rainy season these dolphins can be seen in flooded savannahs and gallery forests.
Distribution: These dolphins are found in the continental fresh waters of the Orinoco and Amazon River basins, where they are widely distributed and endemic. They are limited by rapids, waterfalls, and cold waters of the rivers on the eastern watershed of the Andes Mountains.

DESCRIPTION

The Orinoco River dolphin is the biggest freshwater river dolphin, growing to 2.7 m in length and 170 kg in weight. It has a strong and flexible body and is capable of turning its head and neck up to 90°. The back of *Inia* is blue-gray; the ventral part of the body from the snout to the fluke is light pink in adults and dark gray in juveniles. The head has a prominent melon (i.e., rounded protrusion on their foreheads), an elongated and cylindrical snout, and very small eyes. The dorsal fin is elongated and poorly developed, the pectoral fins are long and wide at the base, and the body ends in a wide caudal fluke (Best & Da Silva 1993). Three subspecies of Orinoco River dolphin are recognized: *I.g. humboldtiana*, *I. g. geoffrensis*, and *I. g. boliviensis* (Best & Da Silva 1993).

NATURAL HISTORY

Some authors state that *Inia* is predominantly solitary (Best & Da Silva 1989), but we have observed groups of up to eight animals in all riverine habitat types in the Orinoco and Amazon basins of Venezuela and Colombia; these are primarily females with their calves and juveniles (Kendall et al. 1995; Mujica & Boede unpublished data). Adult males are solitary except during the breeding season, when they interact with females in heat.

During the dry season Orinoco River dolphins are restricted to major river channels and deep lagoons or seasonal lakes (*prestamos*). Also in the dry season, small groups of dolphins inhabit main channels at the intersection of two rivers where it is easy to catch prey. Movements and seasonal migrations are correlated with the annual flooding cycle of the rivers and subsequent fish migrations (Da Silva 1995). In the rainy season, river dolphins are found in flooded forest and flooded savannahs, which they seasonally invade when sufficiently flooded. The invasion of smaller lagoons is typical of *Inia* in the Venezuelan Llanos grasslands. When the waters recede during the dry season, dolphins sometimes become trapped in isolated floodplain pools and die if they are not rescued (Trebbau & van Bree 1974; Mujica & Boede unpublished data).

Based on our research and experience, dolphin reproduction in the Orinoco River basin takes place from October to December and probably extends to January and February as well. In other words, reproduction occurs at the end of the rainy season and beginning of the dry season. At this time, food is most plentiful and easiest to catch because there are high concentrations of fish and water levels of the rivers of the region are relatively low. This restriction of fish movements facilitates the feeding of young, inexperienced dolphins.

CONFLICTING ISSUES

Of all of the world's freshwater river dolphins, *Inia* is the one with the most stable population. The distribution and abundance of Orinoco River dolphins has not changed significantly over the last 20 years. Nevertheless, threats from human activities exist.

In Venezuela, Peru, and Colombia indigenous people and fishermen consider river dolphins sacred and mystic animals (Kendall et al. 1995). People from Orinoco communities affirm that *Inia* save drowning children. When a dolphin is accidentally drowned in a fishing net, many people remove some of the fat from the carcass and bury the animal out of respect (the fat is viewed as a cure for asthma; see below). Among the different tribes that inhabit the region, there are different beliefs about *Inia*; for example, that they help shamans cure illnesses or that they sing and enchant men, drawing them into the water and transforming them into river dolphins (Boede 1991). In Venezuela, people generally know that these river dolphins are not dangerous and therefore do not harm them. Indeed, for many local people, the Orinoco River dolphin represents their historic legacy. They perceive the dolphin as a wild animal that should be preserved for future generations and whose continued survival contributes to the preservation of their own cultural identity (Correa-Viana & O'Shea 1992).

Most countries of Amazonia and Orinoquia suffer severe social and economical crises that are characterized by high levels of unemployment, illiteracy, human misery, insecurity, and illegal traffic of drugs, wild animals, minerals, weapons, and so on across country borders. This situation facilitates poaching of animals such as the Orinoco River dolphin.

For years, the indigenous people of Venezuela have used the fat of the *Inia* as a popular medicine to cure diseases such as asthma. River dolphin fat is collected primarily from animals that accidentally drown in fishing nets. It is possible to find river dolphin fat, as well as products from other wild animals, in the markets of many villages. The local demand for dolphin fat is not very great, as it is used only in traditional medicine; therefore its price is a relatively modest US $1.80 per 100 ml. However, there is a strong black market for the organs (eyes, penis, and teeth) of river dolphins in Brazil, where the parts have strong superstitious values (Kramer personal communication). In fact, poaching of Orinoco River dolphins in Venezuela has proliferated greatly over the last 8 years to supply the black markets of Brazil and the demands of increasing numbers of illegal immigrant miners from Brazil, Guyana, and Colombia. This problem has grown to the point that river dolphin products are actually being commercialized in Puerto Ayacucho, the capital of Amazons State, Venezuela.

Fishermen are killing increasing numbers of river dolphins. The use of illegal nets in the principal channels of the Orinoco and its tributaries has caused the number of river dolphin deaths in recent years to rise (Kendall et al. 1995). In addition, some commercial fishermen use dried dolphin meat as bait for big catfish (*Pseudoplatystoma faciatum* and *Callophisus macropteus*).

Free commercial shipping access to the Amazon and Orinoco Rivers creates additional risks for the fish fauna and other species, such as the river dolphin. Little scientific research has examined the environmental impacts of this threat. The Orinoco-Apure (Venezuela) and Orinoco-Meta (Colombia) River commercial shipping waterways already exist, and efforts are under way to open additional commercial waterways on the Orinoco-Amazon-La Plata Rivers. Yet, there have been no environmental impact assessments of the possible effects to *Inia* and other wild flora and fauna of those rivers and flooded savannah ecosystems.

Deforestation of gallery forests for agricultural purposes leads to several problems. First, forest loss destabilizes the soil, causing riverbank erosion. In addition, forest loss negatively affects many species of fish that depend on fruits and seeds that fall from trees during the annual floodplain inundation. This threat is critical to several of the most important commercial species of fish of the Amazon and to many fish in the river dolphin's diet. Conversion also results in pollution emanating from a variety of sources, such as pesticides and insecticides used in agriculture, with harmful impacts to the entire aquatic ecosystem. Other types of contaminants include trash (plastic, glass, cans, etc.) from rural river villagers, who dispose of garbage directly into the rivers. Gold mining activities produce heavy metal pollutants, such as mercury, that end up in the river and become concentrated the further they progress up the food chain (Best & Da Silva 1989). Finally, oil exploration threatens the upper regions of the Amazon in Peru, Ecuador, and Colombia, as well as the central and eastern savannah oil field region and Orinoco delta of Venezuela.

FUTURE AND PROGNOSIS

Conservation education is improving throughout South America. Each day more information about the importance of conservation becomes available to a wider array of people, including farmers, ranchers, and other people inhabiting the countryside far from big cities. In addition, conservation efforts are increasingly conducted on private lands, including private landholdings of the Venezuelan plains.

At the local level, freeing stranded river dolphins during the dry season has become an increasingly common practice in many areas of Venezuela. The vast Venezuelan plains are crossed by tributary rivers of the Orinoco River. During the rainy season (May–November) these low plains are inundated by flooding rivers. At this time of year the fish fauna disperses across the flooded plains, and predators, such as the Orinoco River dolphin, migrate into the flooded savannahs in search of prey. During the dry season (December–April) the water withdraws, leaving lagoons and estuaries with a high fish density. River dolphins sometimes remain in these lagoons and estuaries, with survival to the next rainy season dependent on the water levels and food supply. Occasionally farmers and other local people find such dolphins, which they capture and release into the nearest river. For example, in April 1992 six river dolphins were trapped in a lagoon, causing great concern to people living in the nearest community, San Juan de Payara, located in Apure State, Venezuela. The mayor of the village contacted the federal wildlife management agency (PROFAUNA) in an effort to save the animals from sunburn and dehydration. PROFAUNA personnel, San Juan de Payara's mayor, local fishermen, researchers, and a wildlife veterinarian rescued the animals (three males and two females, one of which had a nursing calf 1.43 m in length). Morphometric data (i.e., body and body part measurements and weights) and blood and skin samples were taken for scientific studies.

Other conservation initiatives include passage of protective legislation, creation of protected areas, and scientific research. For example, the Venezuelan Wildlife Protection Law of 1970 prohibits *Inia* hunting, and populations are protected in national parks that include important Orinoco River tributaries. Since the 1960s important research has been conducted on Orinoco River dolphins, greatly increasing knowledge of the species (Hershkovitz 1963; Trebbau & van Bree 1974; Trebbau 1975; Gewalt 1978; Best & Da Silva 1989, 1993; Da Silva 1994; Kendall et al. 1995; Boede et al. 1998). Recently, regular censuses are being done in some protected areas (Gonzalez and Carantoña personal communication).

Despite recent progress toward the conservation of the Orinoco River dolphin, illegal activities (e.g., poaching, polluting dolphin habitat, and using illegal fishing methods), continue to negatively impact the dolphins. The continued occurrence of illegal activities results from several factors, including the difficulty of adequately controlling activities across a vast region, lack of political will, insufficient economic resources and equipment, and poor regulatory skills on the part of guards and managers. Scientists have proposed a variety of actions to further the conservation of *Inia* (see Best & Da Silva 1989; Rodríguez & Rojas-Suárez 1995). Political discussion, scientific research, and the education of people in rural communities will be critical and should be the priorities for the next 10 years. Important and necessary actions include the following:

1. Formulation and adoption of political agreements to conserve the Orinoco River dolphin among the South American nations (members of The Amazon Cooperation Trait) that have *Inia* populations.

2. Development of environmental education programs within the schools of the rural river communities that share riverine habitats with *Inia*.

3. Research into river dolphin population ecology, migratory movements, behavior, and threats.

   a. Determination of the genetic variability and differences (if any) in natural *Inia* populations of the Orinoco and Amazon basins.

   b. Compilation of biological data and experience on the reproductive management of captive river dolphins being held in aquariums and zoological parks throughout the world.

   c. Assessment of the impacts of hydroelectric dams, mining activity, and other human developments on the biology and ecology of river dolphins and their habitat.

d. Studies of the impacts of pesticides and heavy metal pollution on river dolphins and their principal prey species. Useful research would examine the relationship of the impacts of pollution to demographic traits of dolphins, such as age, sex, and reproductive condition. The results of these studies may cause mining activity to be halted in appropriate Orinoco tributaries.

4. Support of conservation staff and national guards by increasing salaries, provisioning with necessary equipment, and providing training in vigilance, conservation education, and professional ethics.

ACKNOWLEDGMENT

Nancy de Boede translated the manuscript of this entry.

Piping Plover

Lauren C. Wemmer and Francesca J. Cuthbert

Common Name: piping plover
Scientific Name: *Charadrius melodus*
Order: Charadriiformes
Family: Charadriidae
Status: Endangered in the Great Lakes region; Threatened on the Atlantic Coast and the northern Great Plains under the U.S. Endangered Species Act; Endangered under the Canadian Committee on the Status of Endangered Wildlife; categorized as Vulnerable by the IUCN.
Threats: Destruction of habitat; disturbance by pedestrians, pets, and off-road vehicles; increased predation owing to garbage on beaches; incompatible water level management practices; contamination by pollutants.
Habitat: Sandy, sparsely vegetated beaches and lakeshores; alkaline wetlands; sandbars of major rivers within its distribution.
Distribution: Breeds on the Atlantic Coast from Newfoundland southward to North Carolina, in isolated sites on Lakes Michigan and Superior in the Great Lakes region, and in the northern Great Plains from Alberta and Saskatchewan southward to Nebraska and Iowa. In fall the piping plover migrates to the southern Atlantic Coast from the Carolinas to southern Florida, the Gulf Coast of Florida westward to Texas and southward to Mexico, and the Caribbean Islands (Haig & Plissner 1993).

DESCRIPTION

A small, stocky shorebird, the piping plover is 17 to 18 cm in length, weighs 40 to 65 g, and has a wingspan of 38 cm (Palmer 1967). Light sand-colored upper plumage and white undersides blend in well with the bird's preferred beach habitat. During the breeding season, the legs and bill are bright orange and the bill has a black tip. A single black band extends across the upper breast and a smaller black band across the forehead. Females often have duller plumage and a thin or incomplete breast band (Wilcox 1959; Haig 1992). The species was named for the characteristic flute-like alarm call uttered by both sexes. During winter the legs pale, the bill turns black, and darker markings are lost. Chicks have speckled gray, buff, brown, and white down. The coloration of fledged young resembles that of adults in winter.

NATURAL HISTORY

The piping plover's natural history places it in conflict with coastal development and recreation. Arriving on the breeding grounds in early spring, plovers often begin nesting on quiet beaches that become busy tourist destinations later in the summer. The nest, a shallow pebble-lined scrape in the sand, holds three or four speckled, tan eggs that are well camouflaged from predators but may be crushed by unaware humans. At times, eggs or chicks are washed away by storms or artificial water level changes. If disturbed repeatedly, plovers may abandon their nests. Human disturbance also alters the behavior of chicks and may reduce their survival. Although they usually replace clutches of eggs that are destroyed, piping plovers produce only a single brood per year.

Piping plovers are monogamous, and males and females share incubation of the eggs for 28 to 30 days. Within 24 hours of hatching, the downy chicks permanently leave the nest to run about and feed. Both chicks and adults are visual predators attracted to movements of insects and other small invertebrates, which they chase, capture, and consume. The first week after hatching, chicks are most vulnerable to the elements and to predators, including crows, gulls, domestic dogs and cats, and wild mammalian predators. The parents guard and periodically brood the young for 3 to 4 weeks until the chicks can fly. In late July, plovers begin to migrate to the wintering grounds. Plovers spend 8 to 9 months feeding and roosting on sand beaches and mudflats and thus accumulate enough fat reserves for spring migration and subsequent reproduction.

CONFLICTING ISSUES

Historical records of piping plovers are few, but they suggest these birds were once fairly common. In the early 1900s their numbers declined owing to hunting and egg collecting but then they began to rebound after protection by the Migratory Bird Treaty Act (Bent 1929). Populations began declining again in the 1940s as shoreline development accelerated after World War II. Russell (1983) estimated the Great Lakes population to have originally numbered as many as 800 pairs. By 1986, when the piping plover was listed as Endangered under the ESA, only 17 pairs were known to nest in the region; all occurred in northern Michigan (USFWS 1988). The Atlantic Coast and Great Plains breeding populations, although much larger, had also declined significantly and were designated Threatened.

Conservation efforts began in most states after listing; efforts vary by geographic region and the agency conducting them. They have included annual surveys, protection of nests with fencing to exclude predators, limiting human activity on beaches (USFWS 1996), predator control, captive rearing of eggs that are abandoned (Powell et al. 1997) or in danger of being flooded, enhancement of breeding habitat (Currier & Lingle 1993), nest guardian programs, and public education (Cuthbert & Wemmer 1999). Additionally, a range-wide census of breeding and wintering piping plovers was first undertaken in 1991 and is conducted every 5 years (Haig & Plissner 1993). As a result of an emergency, large-scale captive rearing initiative on the Missouri River in 1995, several individuals are now housed in a few American zoos, which are in the process of developing public education programs about piping plovers (McPhillips et al. 1996). The stringent use of predator enclosures and intensive protection efforts on the Atlantic Coast have led to recent increases in the number of breeding pairs (Plissner & Haig 1997). Currently, however, the Great Lakes population remains very small and the Great Plains population continues to decline.

For the most imperiled breeding population, human disturbance and small population size are the two greatest threats to continued survival. Numbering fewer than 100 individuals, the Great Lakes population of piping plovers is extremely vulnerable to chance events. An extended period of stormy weather that destroys nests or young, an exploding predator population, genetic defects, or individuals simply failing to encounter one another at widely scattered breeding sites could rapidly lead to the bird's disappearance from the Great Lakes. At a minimum, breeding plovers, their nests, and their offspring must be protected from human disturbance to boost the population from its current dangerously low level.

In Michigan, about two-thirds of piping plovers nest and raise their young on public land, and one-third nest on privately owned land. Protection of breeding plovers therefore involves many different parties, including federal and state agencies, researchers, private organizations, and private citizens. Extent of protection to plovers has been determined by the attitudes of individual landowners rather than by a standard protocol. Recently, negative sentiment toward the plover has increased among some Michigan citizens as a result of several thwarted coastal development projects. Fearing public outcry, some managers have been slow to limit human activity on popular swimming beaches. Contradictory opinions among individuals about how to achieve recovery goals have sometimes translated into conflict among field personnel. Government restructuring, inadequate funding, and reorganization of the recovery planning process have slowed progress on recovery plan revision. All these factors have hindered progress toward recovery during the 12 years since the plover was listed.

In 1996, Defenders of Wildlife brought a lawsuit against the U.S. Fish and Wildlife Service (USFWS) for failing to implement the recovery plan and declare habitat critical to the plover's survival (Defenders of Wildlife 1996). The Service's failure to designate critical habitat for the majority of listed species (Bean et al. 1991) is understandable given that it often has little effect on the actions of federal agencies beyond that already provided by major sections of the ESA (Rohlf 1989). In addition, it is time consuming and may be perceived negatively by the public. The litigation has resulted in USFWS taking steps toward developing a critical habitat designation for the piping plover; however, many political hurdles may need to be overcome before the designation becomes a reality.

Even though the Great Plains breeding population is better off than the Great Lakes population in terms of numbers, major threats to it may be politically more difficult to address. Water management policies significantly impact the presence and productivity of breeding populations. Ill-timed releases of water from reservoirs have resulted in the loss of almost all nests at Lake Diefenbaker, Saskatchewan (Haig & Plissner 1993). Heavily channelized and impounded by dams, the Missouri River offers much less river sandbar habitat than it did under natural conditions. The limited remaining habitat is deceptively attractive to the birds; after plovers choose to nest on the sandbars, their nests may be washed away when spring runoff is released from reservoirs upriver.

In 1989, the USFWS determined that the U.S. Army Corps of Engineers' river management was jeopardizing the plover's survival and required the Corps to increase reproductive success on the river (USFWS 1990). The Corps has tried various approaches including habitat improvement, predator management, and large-scale captive rearing. To protect plovers nesting on the river effectively, the Corps must address the root of the problem—its water management policy. Not only is modification of water regulation logistically difficult and costly, it is also unpopular with the public who reside and make a living near the river. Unless this problematic situation is tackled, however, it is unlikely that the decline in the Great Plains population will be reversed.

On the Atlantic Coast, threats to piping plovers are similar to those in the Great Lakes. Construction of seawalls, piers, and homes; government-subsidized beach stabilization; and dredging continue to destroy or degrade breeding habitat (USFWS 1996). Harassment or even death owing to pedestrians, pets, beach-raking machines, and other motorized vehicles on beaches is another peril. In some states, off-road vehicle (ORV) user groups have been vocal opponents of plover conservation. The Atlantic recovery team and state and federal agencies have been successful at working with ORV user groups to allow ORV access while protecting areas of beach for plover reproduction (Canzanelli & Reynolds 1996). Yet, not all uses of beaches are compatible with plover conservation, and legal regulation has been the only recourse, albeit generally an effective one, in those cases.

Conservation successes on the breeding grounds have little meaning if the plover is not adequately protected during winter. Plovers spend three-quarters of their lives on the wintering grounds, a critical place for resting and refueling for subsequent breeding. In Texas, where the majority of all piping plovers winter along the Laguna Madre, development projects and a tradition of vehicles on beaches have defeated plover conservation efforts.

Without the benefit of a critical habitat designation, it is difficult for the USFWS to check the Army Corps' validation of development permits or to curb vehicle use by the public. In addition to these more widespread threats, the possibility of an oil spill that could wipe out hundreds of birds looms ever-present.

FUTURE AND PROGNOSIS

Although diverse and problematic issues surround piping plover conservation, recent increases in the Atlantic breeding population give hope that major hurdles can be overcome. Even though reorganization of recovery team roles somewhat slowed progress toward recovery, creation of a new International Piping Plover Working Group holds promise for more effective, range-wide coordination of conservation efforts. The lawsuit against the USFWS has had positive aspects, highlighting conservation needs and potentially catalyzing concerted efforts to address habitat issues, especially those of winter habitat. In Michigan the recovery program has gradually become more coordinated and assertive. Institution of a statewide protocol for protection of nesting areas and increased coordination of protection will improve the program even further. Still, self-sustaining populations of piping plovers will never exist anywhere if humans continue to develop and populate coastal areas at the current pace and fail to protect adequate habitat that meets the plover's needs throughout all stages of its life cycle.

Po'o-uli (Black-faced Honeycreeper)

Paul Conry, Peter Schuyler, and Mark Collins

Common Name: po'o-uli or black-faced honeycreeper
Scientific Name: *Melamprosops phaeosoma*
Order: Passeriformes
Family: Fringillidae (perching birds)
Status: Listed as Endangered under the U.S. Fish and Wildlife Service; Endangered in the state of Hawaii; Rare under the IUCN.
Threats: The only documented loss of po'o-uli was a nestling abandoned owing to foul weather, but suspected primary threats include nest predation by rats (*Rattus rattus*); competition for food (invertebrates) by the Polynesian rat (Rattus exulans); avian disease; habitat modification by feral pigs (*Sus scrofa*); small population demographic factors; and natural catastrophes (e.g., storms). Secondary threats may include predation by feral cats, mongoose, and Hawaiian owl; inadequate specialized food resources; and competition from introduced birds.
Habitats: Upper-elevation wet 'Ohi'a-lehua (Metrosideros polymorpha) forest with a well-developed understory of mixed native tree species, shrubs, and tree ferns. Po'o-uli formerly inhabited lower elevation dry to mesic forests (James et al. 1987).
Distribution: On the island of Maui, the northeastern slope of Haleakala Volcano from 1,650 to 2,100 m elevation. The current known population of three individuals is limited to Hanaw'i Natural Area Reserve (NAR) and Haleakala National Park (Figure 7). Fossil records indicate previous distribution on Haleakala's drier southwestern slopes from 300 to 1,500 m elevation.

DESCRIPTION

The po'o-uli is a medium-sized, stocky Hawaiian honeycreeper, distinguished by a black triangular mask, whitish cheek-patch and throat, brown dorsal plumage, and a stubby tail. Adult birds weigh about 25 g and have similar plumage. Juveniles have smaller masks than adults, with fewer defined plumage characteristics (Pratt et al. 1997).

NATURAL HISTORY

Extremely rare, cryptic, and secretive, and inhabiting a remote and harsh environment that receives 300 to 400 inches of rain a year, the po'o-uli

Figure 7. Distribution of the Po'o-uli

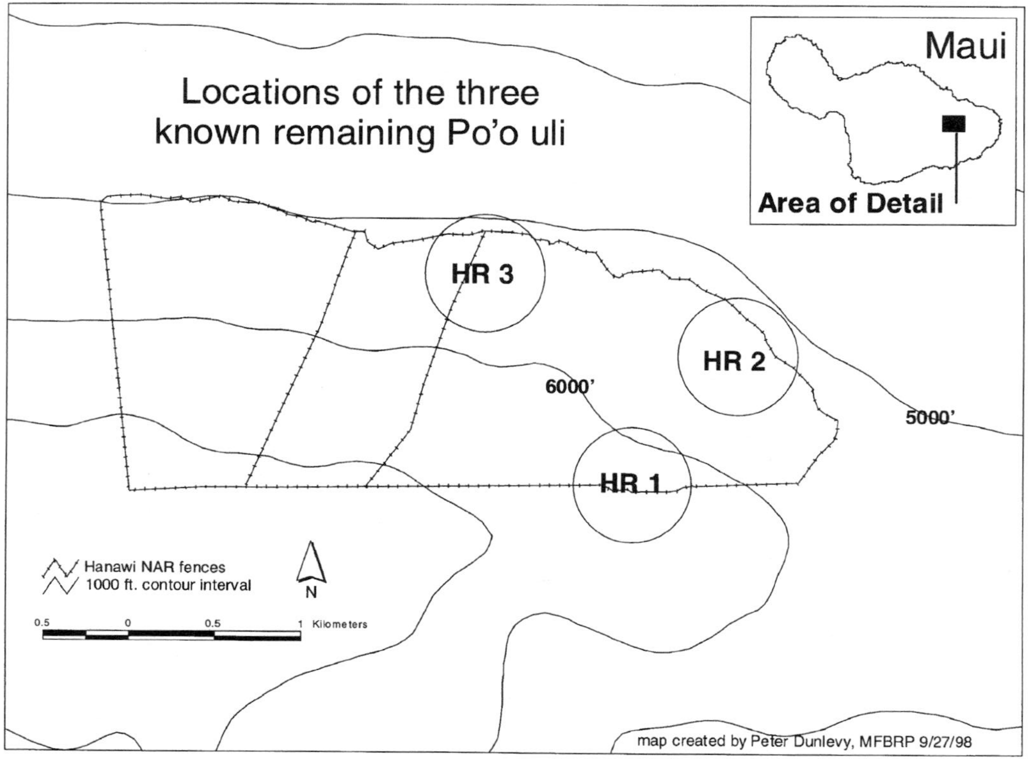

Locations of the three
known remaining Po'o uli

Maui

Area of Detail

HR 2

HR 1

HR 3

6000'

5000'

Hanawi NAR fences
1000 ft. contour interval

N

0.5   0   0.5   1 Kilometers

map created by Peter Dunlevy, MFBRP 9/27/98

HR–Home Range

remains one of the least studied and most endangered birds in the world. Although the discovery of this monotypic genus (i.e., genus with only 1 species) in 1973 by a group of undergraduate students made worldwide news, only minimal descriptive work and incidental observations were made in the following 10 years (Casey 1978; Baldwin & Casey 1983). Initially, nine individuals were located in a 60-hectare area of which two were taken as voucher specimens (Casey & Jacobi 1974). In 1980, following the first systematic east Maui forest bird surveys, the population was estimated at 140–280 birds, although only three individuals were actually detected on transects (Scott et al. 1986). This imprecise figure established the baseline for discussions of the status and trend of po'o-uli for the next 15 years. Since that time the population estimate has plummeted to three known individuals.

Mostly insectivorous, the po'o-uli methodically forages on branches and twigs as it gleans insect larvae and adult arthropod prey, primarily snout beetles, moth and butterfly larvae, parasitic wasps, and spiders. The po'o-uli is unique among native Hawaiian song birds in that it preys on tree snails (Baldwin & Casey 1983), although they may no longer represent a primary food source in the remnant population.

Active throughout the day, po'o-uli commonly move through the forest in mixed foraging flocks of alauahio (*Paroreomyza montana*), and endangered Maui parrotbill (*Pseudonestor xanthophrys*), at times investigating the foraging sites of other flock members. Foraging in mixed flocks may provide protection from avian predators (Mountainspring et al. 1990).

Knowledge of the breeding behavior of po'o-uli is limited to observations of one breeding pair that made two nesting attemps in one nesting season (Kepler et al. 1996). Both parents contributed nesting material. Only the female incubated the egg(s). The male fed the female while on the nest. Both parents attended to the fledgling.

CONFLICTING ISSUES

Even at the time of its discovery, the po'o-uli was so rare that it was considered threatened with extinction and officially listed as Endangered in 1975 (USFWS 1975). In-depth field studies only began in 1985 when USFWS researchers investigated population status, behavior, and threats facing forest birds in Hanaw'i Natural Area Reserve (NAR). Po'o-uli were only incidentally encountered and conspicuously rarer than on previous surveys. Although discovery of the first po'o-uli nest was significant, funding for the project was minimal and the po'o-uli slipped back into obscurity until research resumed in 1994.

Even though early research efforts were inadequate, habitat protection, conservation efforts, and the development of recovery tools progressed. By the 1980s the Nature Conservancy (TNC) established the 1,200-acre Waikamoi Preserve and the Hawaii Department of Land and Natural Resources (DLNR) established the 7,500-acre Hanaw'i NAR. Active programs of fencing, ungulate removal, and weed control soon followed (USFWS 1984; DLNR 1988). Captive breeding and surrogate programs for endangered birds initiated in the 1980s continue to this day. Techniques to collect and incubate eggs, rear and release young, and breed and maintain captive stock were developed. Research directed toward understanding and controlling limiting factors also progressed. Van Riper et al. (1986) and Atkinson et al. (1995) documented the prevalence and virulence of avian malaria and pox in native forest birds and the relationship between mosquito abundance and malaria. By 1994 the rodenticide diphacinone was registered for forest use and provided a tool to reduce rat predation in forest bird habitat (Mosher et al. 1996).

Comprehensive bird surveys for east Maui birds resumed in 1992 (DLNR 1992b). The lack of sightings of po'o-uli caused some to conclude that it was already too late for this species and that conservation efforts would be better focused on the endangered, but more numerous, akohekohe (*Palmeria dolei*) and Maui parrotbill (DLNR 1992a). However, in 1994 searches located a few po'o-uli, prompting DLNR, USFWS, TNC, and Biological Resources Division (BRD) staff to initiate new field studies. Over the next 2 years extensive searches located seven birds, including an adult pair with one fledgling, although only four could be found with any regularity (BRD 1997). In 1997 state, federal, and private sector stakeholders formed a working group to develop a po'o-uli action plan (USFWS 1997).

This action plan emphasized locating, capturing, banding, and regularly monitoring all remaining po'o-uli. By 1998 three po'o-uli were successfully captured and banded in separate, disjunct home ranges. Feathers from each were submitted for DNA sexing analysis, but results were conflicting. Continued search efforts have failed to detect any additional birds. The USFWS and DLNR now face the difficult decision of determining the best management course to save this species.

The conflicting or inadequate biological knowledge of the po'o-uli has made it hard to clearly define the problems and even harder to outline recovery efforts. This lack of information stems mainly from the small population size, the habits of the bird, and the nature of the terrain it inhabits. There is general agreement in the scientific community that the po'o-uli is facing extinction, but no consensus on what to do next. Opinions include captive propagation, translocations, management of wild nests, expanded habitat management and predator control, and just allowing nature to "take its course."

Another missing critical component is public support and a sense of ownership in the po'o-uli's ultimate fate. Dealing with the possible extinction of a species is highly emotional and ideally should pull a community together. Too often, as with the po'o-uli, it can be a divisive and contentious issue, particularly when recovery actions affect an entire community or involve a difference of scientific opinion. For example, the potential use of toxicants in the large East Maui Watershed has the possibility to affect a much larger area than just the po'o-uli's home range. The community is not in agreement as to the appropriateness of this strategy, and a number of concerns relating to unintended consequences have been raised. In addition, ecosystem level protection efforts, such as fencing large areas of the mountain, have elicited concerns from different community members. In general, agencies are typically conservative and seek public support, particularly if needed actions may be controversial and ineffective in spite of all efforts (Clark 1997). An environmental assessment was initiated in May 1998 to gain broader public review and provide feedback on proposed management actions. The lengthy review process (6–10 months if not challenged in court) places the po'o-uli at risk of going functionally extinct during this time. Some proposed recovery actions, such as aerial broadcasting of rodenticides, require review by regulatory agencies and experts. Striking a balance between (1) the need to act quickly to avert extinction, and (2) the need to fully evaluate the fiscal, social, and environmental ramifications and gain community support remains unresolved for the po'o-uli.

The larger questions posed in the single species versus ecosystem management debate are well illustrated by the po'o-uli situation. Efforts to protect the po'o-uli include potentially negative impacts to pristine rain forests. What level of disturbance is acceptable in the name of single species management? Should the most critically endangered species come first, even if it is admittedly on the brink of extinction and may be unrecoverable despite all efforts? How many species are people willing to let go extinct while pursuing an ecosystem solution? Is the decision to "pull the plug" on a species biologically or morally defensible?

Natural resource agencies do not have adequate mechanisms to handle these tough questions or to determine how best to allocate limited funding and staff among multiple endangered species. The current approach of working on the most critical species first is a defensive one. However, this approach may place other rare species at greater risk by not addressing their needs while they are readily achievable. Is it better to focus on fencing and other protective measures for many east Maui endangered species instead of expending heroic efforts on three po'o-uli with questionable success? Answers to these questions remain unknown.

FUTURE AND PROGNOSIS

When a species such as the po'o-uli is threatened with imminent extinction, a high performance field-based recovery team should be formed to frame recovery strategies that will be adopted by policy-makers. The team will then implement the approved field actions and make tough, on-the-spot decisions based on their best judgment. As Clark et al. (1994) have pointed out, the team should be independent and be able to take high-risk actions without the fear of continual interference, delay, or reprisals by peripheral stakeholders. The po'o-uli's critical status requires that recovery decisions be made with less than perfect information and with a high risk of failure. A po'o-uli recovery team will need to (1) provide recommendations on the advisability of captive propagation versus wild management versus habitat management, and all the various permutations of each scenario, (2) initiate surrogate programs to develop and test controversial or risky actions (e.g., habituating insectivorous birds to captivity), and (3) speed up the process of ecosystem safety and regulatory review to allow use of an aerially broadcast rodenticide to control predators over large, remote areas.

At the same time that the recovery team is empowered to take emergency action, all stakeholders and the general public should be kept informed of the efforts. This will help build public ownership and support of the project so that once the current crisis is resolved, the critical long-term support needed for the full recovery of the species will be in place.

Although they are probably too late for the po'o-uli, recovery actions to aid other very rare species should be initiated well before the species is on the brink of extinction and any action or decision that affects individual birds affects the fate of the entire species. This would allow managers the time and flexibility to try different approaches and obtain needed information to make decisions outside of the highly emotional and high-risk crisis management scenarios.

In conclusion, society needs to discuss the dilemmas posed by species on the brink versus the realities of limited funding and staff. Without diluting the Endangered Species Act we must find a way to realistically address the needs of multiple endangered species. With the po'o-uli, we are facing a high-probability extinction event in which a monotypic genus will likely be lost just 25 years after it was discovered. Only time will tell if the po'o-uli recovers.

Red-cockaded Woodpecker

Jerome A. Jackson

Common Name: red-cockaded woodpecker
Scientific Name: *Picoides borealis*
Order: Piciformes
Family: Picidae
Status: Endangered under the Endangered Species Act of the United States, not included on CITES Appendices.
Threats: Destruction and fragmentation of old-growth southern pine forests, either for nonforest uses or short-rotation commercial forestry; simplification of southern pine forest ecosystems by even-aged management; elimination/restriction of natural fire patterns; concentration of species on federal lands as a result of translocating them from private lands; rigid management plans that do not adjust across different regions and pine species types.
Habitat: Extensive, open, old-growth, fire climax, southern pine forests.
Distribution: Southeastern United States from eastern Texas and southeastern Oklahoma, across the Gulf states to southern Florida, northward to western and southern Arkansas, south-central Kentucky, southeastern Virginia, the coastal plain and sandhills of North Carolina, South Carolina, and Georgia; formerly extended to southern Missouri, central Tennessee, Maryland, and southern New Jersey.

DESCRIPTION

The red-cockaded woodpecker is a medium-sized (20–23 cm in length) black-and-white woodpecker that is closely related and intermediate in size to the more common downy (*Picoides pubsecens*) and hairy (*P. villosus*) woodpeckers of North America. It is readily distinguished from other North American woodpeckers by its large white cheek patches. The red-cockaded woodpecker has a ladder-back, white breast, black-streaked sides and flanks, and white-spotted wings. It is notable for a lack of readily visible red on the head of both sexes. Males have 12 to 16 tiny red feathers that form a streak behind the eye. This streak separates the black cap and white cheek patch, but the red feathers are kept concealed beneath black crown feathers except when the bird is excited (e.g., during courtship or agonistic encounters). In profile on a tree, the red-cockaded woodpecker seems to hold its body farther from the trunk than other woodpecker species, an adaptation likely to minimize contact with the sticky gum flowing down the surfaces of cavity trees.

NATURAL HISTORY

The evolution and history of the red-cockaded woodpecker is intimately tied to several species of southern pines. Whereas other woodpeckers in eastern North America are solitary, the red-cockaded woodpecker lives in family groups and has a cooperative breeding system. This social system is considered part of a suite of adaptations for living in the fire climax pine ecosystems (i.e., pine dominated ecosystems that rely on fire). Groups include a breeding pair, recently fledged young, and often one or more male (rarely female) offspring from previous years. Extra adults in the group serve as helpers at the nest, assisting with incubation and feeding of nestlings and fledglings. Groups with helpers tend to fledge more young than those without (Lennartz et al. 1987; Ligon 1970). Helpers also assist with cavity excavation, and the birds forage as a group (Jackson 1994). Because each individual in a group roosts in a separate cavity each night, the availability of more usable cavities increases group size (Carrie et al. 1998).

Characteristically red-cockaded woodpeckers use fire-resistant living pines for nest and roost cavities, excavating into the softer heartwood of older (ca 100+ years) trees that are infected with the red heart fungus (*Phellinus pini*). This dependence on diseased, old-growth trees is a primary reason that red-cockaded woodpeckers are endangered. Because old-growth southern pine trees are less valuable economically, they are typically harvested before they become diseased.

A significant characteristic of red-cockaded woodpeckers is the differential use of their foraging niche by the sexes (Ligon 1968; Ramey 1980). Males characteristically hunt for arthropod prey on the trunk and branches near the top of pines, whereas females most often hunt for food on the trunk below the lowest branch. This difference may reduce competition between the sexes and enhance the pair bond. From a conservation perspective, however, it means the foraging needs of each sex need to be considered. Males may find adequate food among the branches of younger pines, but young pines lack the large surface area and flakey bark that are characteristic of the trunks of older pines where females forage. Thus females are negatively influenced by the loss of older trees (Jackson & Jackson 1986; Jackson & Parris 1995).

CONFLICTING ISSUES

The decline of the red-cockaded woodpecker has only been chronicled during the last half of the 20th century. One can only speculate on populations before that time. Prior to the arrival of aboriginal humans, the southern pine forest habitat used by red-cockaded woodpeckers was probably limited to Florida, southern Georgia, and the south Atlantic coastal region (Jackson 1989). Palynological evidence (from spores and pollen) suggests that early humans used fire and that tactic spread the southern pines to the north and west. Thus more habitat was available for red-cockaded woodpeckers. Since the arrival of Europeans, however, red-cockaded woodpecker populations have likely declined as virgin pine forests were cleared for settlements and as roads began serving as unnatural firebreaks (the latter favored growth of hardwood understories).

Precipitous local declines in populations of red-cockaded woodpeckers likely occurred between the 1880s and World War I when virgin pine forests were logged (Jackson 1989). This may have forced the woodpeckers through a genetic bottleneck. The decline was not as drastic as it could have been because chainsaws were not yet commercially available, trees damaged by woodpecker cavity excavations (cull trees) were left standing, and there were frequent fires owing to accumulated slash (i.e., broken limbs and other tree parts) and a belief that fire could improve grazing.

By the 1950s red-cockaded woodpeckers again declined rapidly. Human population grew, and chainsaws became commercially available. In addition, public campaigns such as "Smokey the Bear" controlled fire. As a result, red-cockaded woodpecker habitat was fragmented. In the 1960s "preferred" management for both industrial and public forests centered on clear-cutting and even-aged, short-rotation pine monoculture. As a result, habitat quality and quantity for red-cockaded woodpeckers continued downward.

In 1968 the red-cockaded woodpecker was recognized as a species in trouble (U.S. Department of the Interior 1968). It was then listed as Endangered under the Endangered Species Act of 1973 and has maintained that status in spite of lobbying efforts on the part of the forest industry to remove it (Jackson 1995).

Despite federal protection, four areas of misunderstanding have fostered an optimistically false picture of species viability and population numbers. First, red-cockaded woodpeckers were referred to as "colonial" with the cavity trees being a "colony." Actually, each site referred to as a "colony" is composed of only a breeding pair and their offspring. Second, male red-cockaded woodpeckers are strongly philopatric; that is, they remain at a site for years in the absence of a female, thus precluding reproduction. The effects of this behavior are exacerbated by forest fragmentation that isolates woodpecker groups, leaving little potential for females to disperse. Third, red-cockaded woodpecker cavities in living pines are conspicuous and persist long after the woodpeckers are gone, creating census errors. Fourth, yellow-bellied sapsuckers (*Sphyrapicus varius*) also peck holes in living pines, causing them to ooze wet gum. Their work has often been mistakenly identified as that of red-cockaded woodpeckers.

Perhaps because these factors contribute to a false sense of security about numbers, misguided management practices offer a current threat to woodpecker viability. For example, managers fail to fully consider, and appropriately respond to, the geographic diversity of habitats used by the species, the normal timing of natural events (e.g., fire) within those habitats, and the differential niche use by the sexes.

Federal managers also cling to clear-cutting as preferred forest management for the species (e.g., Rudolph & Conner 1996), whereas uneven-aged management provides better natural habitat stability (Engstrom et al. 1996; Jackson 1997). Current federal management for the red-cockaded woodpecker seems to rely more on using artificial cavities and translocations than on restoring old-growth ecosystems. Inadequate records of translocations have been kept, and some translocations have moved birds between grossly different ecotypes.

Although a new recovery team was appointed in 1995 to revise the Fish and Wildlife Service Recovery Plan for the species, it suffers from large size, diverse interest group representation, and tightly controlled functioning. As of January 1999 it has met several times but has not produced a revised plan. Federal management of the species continues to be dominated by internal guidelines that have not been subjected to external scientific review (e.g., Costa 1992; Richardson & Costa 1998). Minimum habitat quality guides management on federal lands, whereas less than minimum quality habitat is accepted for private lands (Jackson 1997). Optimum habitat, rather than minimums, should be the target of management efforts (Ligon et al. 1986, 1991).

Another controversy is the use of Habitat Conservation Plans (HCPs) to reduce conflict between conservation and development or forestry interests. Commonly, HCP allows habitat to be destroyed in exchange for moving birds to federal lands. Thus real woodpecker habitat is being lost, populations are becoming even more fragmented, and most of the "eggs" are being dumped into a few federal "baskets."

A related controversy, Safe Harbor agreements, involves agreements between the U.S. Fish and Wildlife Service and private landowners that give private landowners "safe harbor" (i.e., exemption) from the Endangered Species Act for periods of up to 99 years. These agreements are being highly touted in some circles (e.g., Bonnie 1996), but questioned in others (e.g., Jackson 1997). "Safe Harbor" can present problems for species that are limited to old-growth habitat, such as the red-cockaded woodpecker. Although landowners agree to manage the species on their land, they are free to destroy old growth. Because old-growth trees are economically less valuable, they will be cut before red-cockaded woodpecker habitat reaches its optimum state (one of the primary causes of decline listed earlier).

Another suggested approach for private lands is the trading of exemptions like trading stocks (Bonnie 1996). This alternative has not been fully explored but seems to rely on the ability to move birds around and create artificial cavities.

FUTURE AND PROGNOSIS

Throughout the 1970s and 1980s red-cockaded woodpeckers continued to decline (Costa & Escano 1989; McFarlane 1992). Conservation advocates and forest industry and pro-development forces became highly polarized, and the species became known as the "Spotted Owl of the Southeast" (Jackson 1995). In the latter half of the 1990s, however, the controversy has calmed. Does this mean the situation has improved? It is true that increases in some populations are being seen. It is also true that forest industry and developers have been entering unprecedented, and highly publicized, agreements with the U.S. Fish and Wildlife Service in the name of "conserving the woodpecker." But are red-cockaded woodpeckers in better condition?

First, a disaster led to some very positive developments. On 22 September 1989, Hurricane Hugo hit the Francis Marion National Forest of South Carolina. This devastated what had been the largest existing population of red-cockaded woodpeckers. An estimated 87% of the active red-cockaded woodpecker cavity trees were destroyed and 63% of the red-cockaded woodpeckers were killed (Watson et al. 1995). Technology came to the rescue when Copeyon (1990) and Allen (1991) developed techniques for providing artificial cavities to the remaining birds. A massive, and expensive, effort resulted in the provision of 537 artificial cavities by the beginning of the 1990 nesting season. That year, 45% of the red-cockaded woodpeckers living on the Francis Marion National Forest nested in artificial cavities (Watson et al. 1995). Artificial cavities immediately became a powerful tool for assisting woodpeckers (where trees were old enough). The ultimate lesson from Hurricane Hugo, however, was that red-cockaded woodpecker populations should not be concentrated in a few areas, as is being done by Habitat Conservation Plans (see previous discussion).

A second tool that was developed at about the same time was translocation, or the moving of birds from one site to another. During the 1970s and 1980s biologists noted that a high percentage of isolated red-cockaded woodpecker sites held only males. As mentioned earlier, the social system is composed of a single female per group, and males are strongly philopatric to a cluster of cavity trees. When the breeding female dies, males wait for a new female to disperse into the area, but fragmentation has limited female movement (see Jackson & Parris 1995; Engstrom & Mikusinski 1998).

Given that females normally disperse from their natal group during their first winter, I suggested that we might be able to capture those females prior to their dispersal and introduce them to males-only clusters. It worked (DeFazio et al. 1987). Efforts to move older females and young and old males have met with some, but much less, success (Allen et al. 1993; Jackson 1997).

To date, all population increases can be linked to very intensive and expensive management, including the translocation of birds from other sites and the construction of artificial cavities. Some, such as that on Noxubee National Wildlife Refuge (Richardson & Stockie 1995) in Mississippi, are real, resulting primarily from productivity of local populations; others, such as that on the Savannah River Site in South Carolina, are highly suspect (Gaines et al. 1995). In the latter case, population growth reflects the number of birds introduced from other populations and the construction of artificial cavities. Because forests of this area are too young, it is unlikely that this population will be self-sustaining for many years.

While biologists and land managers struggle with innovations and searches for easy answers, they should always consider the full range of impacts their efforts have on the biology of the species, both in the short and long term. The red-cockaded woodpecker is a social bird whose cavity clusters are traditional sites, passed on from generation to generation. Although the tools exist to create cavities and move birds, one must never forget that they are tools, not answers. For example, trees with artificial cavities may be more vulnerable to disease and insects than are trees with cavities that birds have excavated (Conner et al. 1998). And how is translocation affecting the complicated social system? The true answer to red-cockaded woodpecker conservation is not to apply technological fixes. It is to restore the ecosystems in which it evolved.

The immediate future of the red-cockaded woodpecker seems to have been purchased with innovative tools. The real future of the species may be very much in doubt as a result of increasing reliance on these tools as answers, and increasing willingness to forget populations on private lands. Despite the successes wrought by these tools and by federal land populations growing with birds translocated from private lands, the species' populations are becoming even more fragmented and vulnerable on federal "islands" whose management can shift with the political winds.

Red Wolf

Brian T. Kelly and Michael K. Phillips

Common Name: red wolf
Scientific Name: *Canis rufus*
Order: Carnivora
Family: Canidae
Status: Endangered under the Endangered Species Act of the United States; Critically Endangered (CR) on the 1996 IUCN Red List.
Threats: Hybridization with other members of the Canidae family; lack of public acceptance of large carnivores.
Habitat: Highly variable throughout southeastern and eastern North America where sufficient prey and minimal human development occur.
Distribution: Currently free-ranging in a reintroduced population in northeastern North Carolina and on (2) islands in South Carolina and Florida. The red wolf's historic range was recently redefined to extend from eastern Texas northward to Missouri and eastward and northward to the northeastern United States.

DESCRIPTION

The red wolf is intermediate in size between the coyote (*Canis latrans*) and gray wolf (*C. lupus*). Male red wolves range from 23 to 38 kg, and females range from 19 to 34 kg. Coloration is typically brownish with black shading on the back and tail.

NATURAL HISTORY

The red wolf was first described during the 18th century. However, its natural history remains poorly understood. Knowledge of red wolves prior to restoration efforts is based on relatively small samples from remnant and probably atypical red wolf populations. Historical data and restoration data indicate that the red wolf is a monestrous species (i.e., goes into estrous once each year) that typically becomes sexually mature by its second year. Red wolves live in extended family groups similar to gray wolves, and litters average three to five pups (Riley & McBride 1972; Shaw 1975). Data from the restored population indicate that offspring from a breeding pair are tolerated in their natal home range until they disperse, with dispersal apparently related to social factors most typically associated with the onset of sexual maturity. Principal prey items prior to the red wolf's extinction in the wild included nutria (*Myocastor coypus*), rabbits (*Sylvilagus* spp.), and rodents (*Sigmadon hispidus*, *Oryzomys palustris*, *Ondatra zibethicus*) (Riley & McBride 1972; Shaw 1975), whereas in the re-introduced population in North Carolina, white-tailed deer (*Odocoileus virginianus*), raccoon (*Procyon lotor*), rabbits, and small rodents (*Mus musculus*, *Sigmadon hispidus*, *Peromyscus* spp.) are the primary prey, with resource partitioning evident within packs. Data from the restoration program indicate that dens can be located both above and below ground and that mortality is owing to a variety of factors, including vehicles, parasitism, and intraspecific aggression (i.e., aggression among wolves).

CONFLICTING ISSUES

In 1973 a decision was made to place wild red wolves in captivity for managed breeding and eventual restoration (U.S. Fish and Wildlife Service 1989). This action led to the extinction of the red wolf in the wild by 1980. In 1987 captive-born descendants of the animals removed from the wild 14 years earlier were released onto Alligator River National Wildlife Refuge (NWR) in northeastern North Carolina (NENC). This was the first attempt to restore a carnivore declared extinct in the wild to a portion of its former range. Currently a free-ranging population of red wolves, estimated at 80 individuals, inhabits approximately 1 million acres of federal, private, and state lands in northeastern North Carolina.

Despite this success, red wolf recovery has had setbacks and challenges. Land Between the Lakes (LBL) in Kentucky and Tennessee was the initial choice for the re-introduced of red wolves. However, in 1984 the proposal to release wolves at LBL was abandoned owing to lack of public support. This failure to address private landowner concerns helped precipitate an amendment to the Endangered Species Act (Public Law No. 93–205) that allows for the designation of re-introduced populations as experimental and nonessential (section 10(j) of the Act). Duly designated populations are managed as threatened species with specific management rules drafted to address public concern. The ability to designate re-introduced populations in this manner made red wolf re-introduction possible, as it removed many of the potential social and bio-political conflicts associated with re-introduction (Parker & Phillips 1991).

By the mid-1960s habitat alteration and loss, as well as predator extermination campaigns, effectively reduced the free-ranging red wolf population to a small remnant population in southeastern Texas and southwestern Louisiana (U.S. Fish and Wildlife Service 1989). Extensive hybridization between the red wolf and coyote threatened those wolves that remained (U.S. Fish and Wildlife Service 1989). This hybridization was the ultimate factor that caused red wolves to be removed from the wild. The hybridization also fueled a debate regarding whether the red wolf is of hybrid origin or a unique taxon that hybridized with coyotes owing to a dwindling population and a concomitant expansion of the coyote population (Nowak et al. 1995).

The ability of wolves to colonize large areas quickly is a fundamental biological reason for the success of the restoration program in NENC. However, the lack of federal land relative to a conglomeration of many small private land holdings, many of which are utilized as farms and hunt clubs, has presented a challenge to red wolf recovery. Although there is strong public support for the red wolf program, landowners can and do request that wolves be removed from their land, often simply because hunters who pay for access to the land believe the wolves have an adverse effect on their hunting success. Public attitude surveys have indicated widespread support for the red wolf program (Mangun et al. 1996; Quintal 1995; Rosen 1997), and the projected economic impact of the program to the re-introduction area may be to bring as much as $184 million into the local economy annually (Rosen 1997).

Two primary conflicting issues challenge red wolf recovery: (1) the interface with private landowners, and (2) hybridization with other members of the *Canis* genus.

First, section 10(j) of the Endangered Species Act has been and will continue to be a critical tool to affect restoration of wolves. A publicly reviewed rule package that addresses management of wolves on private lands accompanies a re-introduction under section 10(j) of the Act. It is important to remember that the release of red wolves on Alligator River NWR in 1987, and the eventual expansion of the population throughout the designated five-county restoration area in NENC, represents the first successful restoration of a carnivore extinct in the wild. However, there was no model regarding the management of wolves on private lands when wolves were released in NENC. The public had no firsthand experience with wolves and was apprehensive regarding the possible threats to personal safety, the potential for depredations of livestock and pets, and the likelihood of land use restrictions.

This, combined with a general lack of biological knowledge of the red wolf, resulted in a set of 10(j) rules that are biologically and politically problematic. To address local landowners' concerns about wolves leaving federal land, the rules associated with the restoration allow for landowners to request the capture and removal of a wolf from their land when there is not an associated, wolf-caused problem. Whereas during the initial years of the re-introduction this may have been feasible, the current demographics of the re-introduced population (80 animals over 1 million acres) make it impossible to resolve such requests, and such removals may represent a threat to recovery. Reasons for this include the following: (1) wolves disperse widely and are a fluid resource, (2) wolves thrive in a variety of habitats, (3) the current rule requires such wolves be released back to the wild,

(4) the unknown effect such removals may have on hybridization rates, and (5) the diversion of manpower away from monitoring hybridization and achieving recovery goals.

Second, issues regarding hybridization fall into two categories. First, is the red wolf of hybrid origin? Two petitions to delist the red wolf have been filed on the basis of its being of hybrid origin. In 1991 the American Sheep Industry Association filed a petition based on mitochondrial DNA analysis, and in 1995 the National Wilderness Institute filed a petition based on nuclear DNA results. Both petitions were found to be untenable based on current data (Henry 1992, 1998). Recent genetic and morphological evidence supports these findings (Theberge 1998). Second, what is the potential for and effect of hybridization with coyotes, hybrids, or feral dogs? Such hybridization is not unique to the red wolf (Wayne et al. 1995; Theberge 1998). A better understanding of the cause and significance of hybridization among canid species is needed.

FUTURE AND PROGNOSIS

The prognosis for landowners and red wolves to co-exist is good. Most landowners have come to understand that wolves do not represent a significant threat to person or property. The implementation and use of the experimental nonessential designation has illustrated how endangered species so designated can represent little if any threat to loss of private property rights. There is currently nothing landowners cannot do on their land that they could not do prior to the presence of red wolves. Furthermore, landowners may, and do, request that wolves be captured and removed from their land. It must be recognized that with the current population demographics, removal of nonproblem wolves is typically no longer possible. Additionally, such efforts cost the taxpayer money and divert recovery personnel away from the program's ultimate goal—delisting the red wolf. The removal of wolves that are established and not affecting personal safety or personal property (e.g., livestock) contradicts the goals of the program, and diverts manpower away from monitoring and managing hybridization. Furthermore, when wolves are removed, social bonds may be disrupted and/or a vacant territory may result. These factors alone threaten recovery, but they also potentially facilitate the establishment of a resident coyote population. The degree to which hybridization occurs in a wolf population may depend on having enough wolves established to exclude coyotes or maintain them at relatively low levels.

Specific recommendations on removing red wolves are problematic. Until the current rule is changed, Service personnel will continue to attempt to trap and relocate red wolves that inhabit private land on which the landowner requests their removal. Rules written for additional red wolf reintroductions should reflect that wolves will be removed only to resolve a depredation or related problem. Rules written for gray wolf re-introduction to Yellowstone National Park include this provision and have worked well (Phillips & Smith 1998). However, such a change in NENC may be viewed unfavorably by some landowners, instead of as a natural evolution of the program. In contrast to removing nonproblem wolves, the current rules allow landowners, or their agents, to take a problem wolf when Service efforts have not been successful. Written permission from the Service is required before a landowner can take such wolves. However, this option has, to date, not been used. Part of living with wolves is realizing that some wolves will need to be taken (Mech 1996). It remains, however, sociologically and biologically problematic that wolves that have not caused a problem, and are critically endangered, may be included in this realization—especially given the implications of such removals to hybridization rates between coyotes and red wolves. Recently a North Carolina law was passed that would allow the taking of nonproblem red wolves. Although a federal court recently upheld the Service's authority to regulate such taking of red wolves, the decision is currently being appealed.

The prognosis for addressing hybridization in red wolf recovery is uncertain. Too little is currently understood. The advent of DNA analysis has raised questions about how species are defined and how such data are applied to taxonomic classification (Dowling et al. 1992; Nowak et al. 1995).

Interbreeding between wolves and coyotes may be the result of a small remnant or expanding population of wolves. This paradigm was the basis for choosing re-introduction sites for red wolves without coyotes present (Parker 1987). Such restoration sites would give red wolves the opportunity to establish a population without the potential for hybridization. Thereafter, the potential for hybridization should be minimized. However, it is doubtful there are any potential re-introduction sites within the historic range of the red wolf that are free of coyotes.

To date, hybridization between coyotes and red wolves has occurred in the NENC red wolf population; however, the circumstances under which mixed pairs occur require better understanding. Coyotes were not present in NENC in 1987 when red wolves were first released, but they are now being seen frequently. Space use studies of sympatric coyotes and red wolves (i.e., where both species live in the same location) are being undertaken. Such studies are part of a monitoring program designed to help understand (1) the degree to which hybridization occurs between red wolves and coyotes, (2) the circumstances under which hybridization occurs, and (3) the contribution hybrids make to a population. It is not known with certainty whether red wolf/coyote hybrids are reproductively viable. The red wolf program has an opportunity to study these issues that it did not have in the 1970s when the red wolf was recognized as being endangered.

The threat that hybridization represents to red wolf recovery is not unique to red wolves. Indeed, the traditional definition of species should be revised and/or the role hybridization plays in canid populations and evolution should be re-examined by the scientific community. Is some level of hybridization in a population "natural" or "acceptable"? By definition, wolves and coyotes are different species, yet they interbreed.

With respect to the red wolf, if hybridization occurs at levels that are unacceptable, can a population be managed such that acceptable levels are maintained? If hybridization occurs at acceptable levels or under circumstances that are manageable in a red wolf population that established itself essentially in the absence of coyotes, can future populations of red wolves be established in areas with established coyote populations with the same result? Clearly, more information is needed on sympatric interactions between red wolves and coyotes, and with respect to acceptable levels of hybridization in canid populations in general. The red wolf recovery program is in a unique position to provide data to help clarify these issues for the red wolf and other canid species.

Scarlet Macaw

Katherine Renton

Common Name: scarlet macaw
Scientific Name: *Ara macao cyanoptera* (Central American subspecies)
Order: Psittaciformes
Family: Psittacidae
Status: Endangered in Mexico (NOM-059-ECOL-1994), Guatemala (Decree 4–89), Belize (Wildlife Protection Act 4–81), Honduras (Law #004-78, 206–82), Nicaragua (Wildlife Law 1983), and Costa Rica (Wildlife Law #7317); Extinct in El Salvador; CITES Appendix I; IUCN Category 1 (Vulnerable).
Threats: Habitat destruction and capture for the wildlife trade were principal threats to the species during the 1970s and 1980s. Extensive illegal poaching of wild nestlings continues to be a severe problem for remaining populations in the wild.
Habitat: Tropical humid and dry forests below 1,000 m in elevation.
Distribution: Currently restricted to small remnant populations in southern Chiapas, Mexico; western Peten, Guatemala; southwestern Belize; northeastern Honduras; eastern Nicaragua; and the Pacific coast of Costa Rica. Formerly distributed throughout Central America from northeastern Mexico along the Pacific and Atlantic coasts to Panama.

DESCRIPTION

The scarlet macaw is the third largest of the 16 species of macaw in the New World tropics or Neotropics, measuring 85 cm from head to tail. Average adult weight is 1,200 g, with a wing length of 41 cm and tail length of 53 cm (Forshaw 1989; Wiedenfeld 1994). The general plumage is bright red with a distinctive yellow band on the wing, which is tipped with blue (Forshaw 1989). The scarlet macaw in Central America is considered a separate subspecies to that in South America, as there is no green band separating the yellow and blue on the wing and birds from Central America are larger than those from South America (Wiedenfeld 1994).

NATURAL HISTORY

Scarlet macaws are generally seen in bonded pairs or small family groups with one or two fledged young, although large groups of 20 to 30 individuals may flock together at feeding trees. The breeding season extends from late November through the end of May, and nest sites are located in cavities in tall live or dead trees (Marineros & Vaughan 1995; Iñigo-Elias 1996).

Scarlet macaws are highly territorial around nest sites and may experience pressures from nest site limitation (Iñigo-Elias 1996).

Nesting macaws usually lay one to three eggs, which the female incubates for approximately 28 to 34 days (Iñigo-Elias 1996). The chicks hatch asynchronously (i.e., at different times), with eyes fused shut and only a light covering of feather down. The nestlings develop in the nest cavity over two to three months and attain adult size and plumage prior to fledging (Iñigo-Elias 1996). Only 20% of the wild population may breed in a given year, with 60 to 70% of nests producing one or two young (Munn 1992; Marineros & Vaughan 1995; Iñigo-Elias 1996). The main cause of nest failure is predation on eggs or young chicks by reptiles, small raptors, and medium-sized mammals, as well as human poachers (Marineros & Vaughan 1995; Iñigo-Elias 1996).

The diet of scarlet macaws consists principally of immature seeds from a variety of plant species, although they also consume fruits, flowers, and leaf stems (Munn 1988; Marineros & Vaughan 1995). Scarlet macaws may be highly adaptable in diet and range widely in search of food resources, but nothing is known of their movements and area requirements.

CONFLICTING ISSUES

Scarlet macaws were once widespread in the tropical humid and dry forests along the Pacific and Atlantic coasts of Central America from northeastern Mexico through Panama (Forshaw 1989; Wiedenfeld 1994). However, in the 1970s government-sponsored development and recolonization policies, such as the National Deforestation Program (Programa Nacional de Desmontes) in Mexico, and the Peten Promotion and Development Association (FYDEP) in Guatemala, resulted in extensive deforestation. In Mexico, government policy aimed to promote agricultural expansion by donating land and financial subsidies to community groups known as *ejidos*. However, in order to maintain ownership, members of the *ejido* had to demonstrate that they were developing the land. The government definition of development was to deforest land for agriculture, and financial subsidies were provided to encourage this. Moreover, border disputes with Guatemala led to government-sponsored recolonization programs to populate and develop tropical forest areas along the frontier.

By the late 1970s agricultural development, hardwood extraction, and recolonization had resulted in the destruction of tropical forests, and the elimination of the scarlet macaw from eastern Mexico and the Pacific coasts of Guatemala, Honduras, and Nicaragua. In Costa Rica only 20% of original macaw habitat still exists, with the remaining scarlet macaw populations located in three main areas on the Pacific slope (Marineros & Vaughan 1995). Civil wars in El Salvador and Nicaragua also decimated forests and wildlife of those countries. Natural disasters such as Hurricane Hattie in 1961 and

Hurricane Joan in 1988 further impacted scarlet macaw populations in Belize and Nicaragua. At present, the impact of the 1998 Hurricane Mitch on scarlet macaw populations in Honduras is unknown.

The principal threat to wild populations, however, has been the commercial exploitation of scarlet macaws for the international wildlife trade. Many of the problems of international trade derive from the disparity between local and international market values, with major profits from trade going to a few middlemen. In Mexico a local trapper may receive $19 for a scarlet macaw, whereas a trader in the country of origin may receive $450. However, once it arrives in the United States, a scarlet macaw may be sold for as much as $4,000 (Iñigo-Elias & Ramos 1991). Hence, it is at the level of the U.S. importer that the trade becomes most highly organized and profitable, with only four distinct companies controlling 74% of U.S. live bird imports in 1988 (Swanson 1992). The enormous profits that may be made by a few highly organized importers constitute a driving force behind the international trade.

The scarlet macaw was placed on Appendix I of CITES in 1986. This prohibited international trade but did not control internal trade. CITES is in essence a trade agreement rather than a conservation treaty. Hence although CITES may limit legal international trade, it is ineffective against illegal or local trade. Classification of a species under CITES Appendix I may also raise the market value and demand for that species, which is then considered rare. At the local level, enforcement of wildlife laws is hampered by a lack of resources in wildlife departments and a lack of importance attributed to wildlife laws. Thus enforcement is frequently nonexistent in the remote, rural, or border regions where most macaw populations occur. There remains an extensive and overt illegal commerce in scarlet macaws in Guatemala, Honduras, Nicaragua, and Costa Rica, even though the species is protected by national laws in all these countries. In Costa Rica, traditional scarlet macaw nest trees located along roadsides have ladders built into the trunks to aid poachers (Marineros & Vaughan 1995).

The Central American scarlet macaw—estimated at a total of 4,000 individuals—has now been reduced to small, discontinuous populations in Mexico, Guatemala, Belize, Honduras, Nicaragua, and Costa Rica (Wiedenfeld 1994). Conservation of the scarlet macaw in Central America therefore involves a number of countries with potentially differing conservation structures, national agendas, and socioeconomic profiles, many of which have only recently resolved longstanding disputes over sovereignty. Role players in scarlet macaw conservation include various wildlife departments, researchers from different institutions, national and international conservation organizations, and economically poor rural communities situated around national parks and macaw areas.

Captive breeding and re-introduction of scarlet macaws may be an option for conservation. However, captive breeding requires enormous logistical and financial investment, and it has potentially detrimental impacts on wild populations through the spread of disease or genetic inbreeding (Snyder et al. 1996). In addition, captive-reared individuals frequently lack the behavioral skills required to locate food and evade predators in the wild (Snyder et. al. 1994), hence release programs have been most successful where captive-reared birds are able to join wild populations (Sanz & Grajal 1998).

Above all, captive breeding and re-introduction do not address the socio-economic problems in scarlet macaw conservation. There is an inherent conflict in attitudes between (1) those organizations, institutions, and individuals desiring to protect scarlet macaw populations, and (2) residents of poor rural communities that view the scarlet macaw as an economic resource to be exploited. The most effective conservation measures are likely to be those that attack the underlying problem of poaching through education and work with local communities.

FUTURE AND PROGNOSIS

Effective conservation of the scarlet macaw must be based on ecological data of population trends and habitat requirements of macaws, and it must address socioeconomic aspects through educational outreach programs and community-based development. Field studies demonstrate that scarlet macaw populations have low reproductive rates, making them vulnerable to decline. However, little is known of the dynamics, resource, and area requirements of wild populations, which is necessary to determine appropriate conservation strategies.

Providing economic benefits to local communities may encourage them to conserve scarlet macaw populations. Sustainable harvesting of wild macaws has been proposed as a means of providing an economic incentive for conservation. However, this requires a comprehensive database on population density, dynamics, and limiting factors (Beissinger & Bucher 1992), which is not available for any scarlet macaw population. Moreover, scarlet macaw populations in Central America are recorded to be in severe decline (Forshaw 1989; Wiedenfeld 1994; Marineros & Vaughan 1995; Iñigo-Elias 1996), rather than stable or increasing, and they exhibit conservative breeding strategies, making them unsuitable for sustainable harvesting. In addition, harvesting of macaws destined for international trade is not comparable with harvesting for local markets, such as iguana farming. High commercial values in international markets drive local harvesting rates. An increasing scarcity in the wild further raises market values and creates powerful incentives for increased harvesting of declining populations. This type of runaway positive feedback in international trade makes harvesting of threatened species, such as the scarlet macaw, inherently unsustainable.

Ecotourism may be the most appropriate nonconsumptive use of scarlet macaws (Munn 1992; Marineros & Vaughan 1995). Tourism is one of the largest industries in the economies of Mexico, Guatemala, Belize, and Costa Rica; however, the majority of that income frequently returns to foreign-owned companies (Boo 1990). In Costa Rica, the Carara Biological Reserve receives only 1% of the income generated by tourists, and rural communities located around the reserve receive no benefit (Marineros & Vaughan 1995).

The development of community-based nature tourism may provide an opportunity for local people to extract economic benefit from the tourist appeal of scarlet macaws. Proposals for community-based nature tourism in Costa Rica involve (1) tourists paying local communities to see macaw nests, and (2) training of local guides (Marineros & Vaughan 1995). In Belize a community project is also being developed that provides facilities for tourists to view large feeding groups of scarlet macaws. However, the success of community-based nature tourism depends on effective organization, training, infrastructure, services, and promotion, and it should involve all members of the community (Norris et al. 1998). Outreach programs as developed in the Caribbean (Butler 1992) and Belize (Coc et al. 1998) should be implemented to educate schools, local communities, and the visiting public about scarlet macaw ecology and conservation.

The dispersed, increasingly isolated nature of remaining scarlet macaw populations, many of which are located close to national frontiers, raises the need for cooperation among governments, institutions, conservation organizations, and individuals. An initiative is currently under way to develop a regional strategy for conservation of the scarlet macaw in the Selva Maya of Mexico, Guatemala, and Belize. However, this will require collaboration among government departments and agencies in each country to establish effective policies and procedures for conservation. The challenge will be to develop an integrated approach to conservation of the scarlet macaw in Central America that addresses the socioeconomic problems of poaching and habitat destruction.

Snow Leopard

Rodney M. Jackson

Common Name: snow leopard
Scientific Name: *Uncia uncia* (Schreber 1778)
Order: Carnivora
Family: Felidae
Status: Listed as Endangered under the U.S. Fish and Wildlife Service and on the 1996 IUCN Red Data List; CITES Appendix I.
Threats: Illegal hunting and sale of pelts, bones, and body parts for the fur trade and traditional Chinese medicine trade; depletion of natural prey populations; killing by herders for depredation of livestock (i.e., retribution); habitat degradation and fragmentation.
Habitat: Relatively steep, broken terrain; shrub-grassland; open forest habitats of alpine and subalpine zones.
Distribution: Mountain ranges in China, Bhutan, Nepal, India, Pakistan, Afghanistan, Tajikistan, Uzbekistan, Kyrgyzstan, Kazakhstan, Russia, and Mongolia. Patchy and fragmented distribution across nearly 3 million km² of central Asia.

DESCRIPTION

The snow leopard is a large cat (adult shoulder height 60 cm, body–tail length 1.8–2.3 m) with a smoky-grey pelage tinged with yellow and patterned with dark grey, open rosettes, and black spots (Hemmer 1972). Besides its superb camouflage for life among bare rocks and patchy snow, it has a well-developed chest, short fore-limbs with sizable paws, long hind-limbs, and a noticeably long tail (75–90% of its head–body length) that gives the species its renowned agility for negotiating steep terrain and narrow rocky ledges (Nowell & Jackson 1996). Adaptations for cold include an enlarged nasal cavity, long body hair with a dense, woolly underfur (belly fur up to 12 cm in length), and a thick tail that can be wrapped around the body (Hemmer 1972). Males average 45–55 kg compared with 35–40 kg for females.

NATURAL HISTORY

The wild population of snow leopards is crudely estimated at 4,500 to 7,500 individuals. Snow leopards can be found at elevations of 3,000 to 4,500 m and occasionally as high as 5,500 m in the Himalaya or as low as

600 m in northerly latitudes. Densities range from 0.1 to 10 or more per 100 km² (Jackson & Ahlborn 1989; Nowell & Jackson 1996). The species favors steep terrain broken by cliffs, ridges, gullies, and rocky outcrops, although in Mongolia and on the Tibetan Plateau it occupies relatively flat or rolling terrain as long as there is sufficient cover. Mountain ridges, cliff edges, and well-defined drainages serve as common travel routes and sites for social marking including the deposition of scrapes, scats, and scent (Ahlborn & Jackson 1988).

An opportunistic predator, the snow leopard is capable of killing prey up to three times its own weight. This excludes only adult camel, kiang, and yak as potential food items. Snow leopards also prey on marmot, pika, hare, small rodents, and game birds (Chundawat & Rawat 1994; Oli et al. 1993). The snow leopard's distribution coincides closely with that of its principal large prey: blue sheep (*Pseudois nayaur*) and ibex (*Capra sibirica*). Annual prey requirements are estimated at 20 to 30 adult blue sheep, with radio-tracking data indicating such a kill every 10 to 15 days. A solitary leopard may remain at its kill for up to a week.

The species' life history is not well documented. A typical solitary felid (cat), the snow leopard is crepuscular (i.e., active at dusk and dawn) if minimally disturbed. However, in prime habitat in Nepal, home ranges overlapped widely between and within sexes, averaging 10 to 40 km² (Jackson & Ahlborn 1989). Home ranges are considerably larger in Mongolia, where the terrain is less broken and prey density is significantly lower (McCarthy personal communication). Animals may remain within a small area for an extended period before shifting to another part of their home range. In Nepal, leopards moved up to 7 km (straight-line distance) in a single day, averaging around 1 km (Jackson & Ahlborn 1989). In contrast, animals in Mongolia covered a much greater daily distance.

Mating occurs between January and mid-March, a period of intensified marking and vocalization (Ahlborn & Jackson 1988). In captivity, estrus lasts 2 to 12 days, with a cycle of 15 to 39 days (Nowell & Jackson 1996). Age at sexual maturity is 2 to 3 years, but there is no information on longevity in the wild. Litter size is usually two to three and exceptionally as many as seven. Dispersal is said to occur at 18 to 22 months of age, and sibling groups may remain together briefly on independence. This explains sightings of as many as five snow leopards in a group.

CONFLICTING ISSUES

Historically, habitat remoteness served to insulate the species from humans, but snow leopards are no longer present in many areas that they formerly occupied. The species continues to decline as a result of hunting, depletion of natural prey, habitat destruction and fragmentation, and low population density, especially across more disturbed or isolated mountain ranges (Fox 1994; Nowell & Jackson 1996). Available records are vague but indicate that large numbers of snow leopards were trapped or hunted in the former USSR, Mongolia, and China prior to the early 1970s for the world's fur market and zoos. In the mid-1970s, 70 to 80 snow leopard pelts passed through Afghanistan's Kabul bazaar annually en route to European and North American fashion-seekers. Local hunters earned from U.S. $10 to $50 in a remote district of Nepal to U.S. $350 in tourist centers like Srinagar. In contrast, a high quality fur coat, requiring as many as a dozen animal pelts, still sells for U.S. $50,000 or more on the black market.

Increased complaints about domestic stock damage from carnivores usually accompany a decline in prey numbers, especially outside of national parks and other protected areas that are viewed by some herders as a refugium (place of refuge) from which problem predators disperse. However, conflict between humans and large cats dates back over 9,000 years when animals were first domesticated (Nowell & Jackson 1996). Before possessing modern firearms and traps, herders employed simple but effective methods for minimizing depredation losses, including maintaining close watch over their livestock, avoiding predator-rich habitat, using guard dogs, keeping livestock in predator-proof night-time corrals, and favoring breeds with well-developed anti-predator traits (Jackson et al. 1996). Erosion of traditional knowledge, reduced herder vigilance, increased livestock numbers, and other animal husbandry changes have aggravated the situation.

Although the enactment of CITES (Convention of Trade in Endangered Species) in 1973 led to reduced international trade in Appendix I species such as the snow leopard, it has not been eliminated. Barnes (1989) found 12 snow leopard pelts for sale in Kathmandu, with one five-star hotel offering a fur coat for $3,000. Heinen and Leisure (1993) found increased numbers of fur shops over those reported by Barnes, but an apparent decline in demand despite higher tourist visitation. Increased public awareness and anti-fur sentiment among consumers are probably responsible for this trend.

Snow leopard bone is rapidly replacing fur as a marketable item. Like tiger (*Panthera tigris*), bone, it is used in traditional Chinese medicine to alleviate maladies ranging from epilepsy and rheumatism to impotence. Easily dried, preserved, and transported, bone can be taken to lucrative city markets and traded with varying degrees of official ignorance, disregard, or even open collusion. Snow leopard bones reach mainland China from India and Nepal, as well as economically hard-pressed states like Kazakhstan and Kyrgyzstan, often with the assistance of development workers. The opportunity to earn U.S. $50 to $300 from the skeleton of a snow leopard that preyed on their livestock is an obvious incentive for rural people whose per capita annual income totals under U.S. $100 to $200. Herders living in Nepal have been known to exchange bones for domestic sheep breeding stock (Shah personal communication).

Some governments have sanctioned snow leopard trophy hunting in an effort to generate revenue and provide people with incentive to protect the targeted species. Thus Mongolia listed the hunting fee for a snow leopard at U.S. $11,200. O'Gara (1988: 224) argued that "a managed harvest of snow leopards in a country with a stable prey base will increase the value of the cats to the government and to the local people, providing incentive to wisely manage all wildlife." Clearly there are strong moral, biological, and political arguments against hunting endangered species, notwithstanding the fact that such revenue rarely reaches the local people or funds wildlife management activities (PEER 1996).

Domestic animals commonly far outnumber natural mainstay food items such as blue sheep or ibex, and they are also easier to kill. Livestock constitutes between 15 and 25% of the food items found in feces, even in national parks with dense prey populations (Oli et al. 1993; Chundawat & Rawat 1994). Livestock forms an even higher proportion of the snow leopard's winter diet when marmots are hibernating. Stock losses average between 2 and 9.5% in some "depredation hotspots" (Jackson et al. 1996). Schaller et al. (1987) found that 7.6% of sheep and goats were predated in one valley of Xinjiang's Taxkorgan Nature Reserve. Depredation is a serious hindrance to conservation initiatives, largely because of its economic impact. The worst-case scenario involves "surplus killing"—a catastrophic incident resulting when a predator enters a poorly made livestock pen during the night, becomes confused, and then kills many animals. A single family may lose as many as 50 to 100 sheep or goats in a single event, which could be avoided if corrals were better constructed.

Jackson et al. (1996) found that depredation was not evenly distributed but rather associated with the presence of cliffs, rocky areas, and good cover. Near protected areas, the most likely stock raiders are dispersing subadults seeking to establish their own home range. Snow leopards that bring their cubs to a kill may be reinforcing the taking of livestock as prey, although the tendency of snow leopards to remain at a kill and consume all available meat increases their vulnerability to human retribution.

Oli et al. (1994) found herders to have strongly negative feelings toward snow leopard and wolf owing to loss of livestock. Of 102 households interviewed, 52% considered total eradication of the snow leopard as the "only remedy worth considering." An additional 35% believed that eradication should be attempted first and compensation schemes implemented only if this failed. Other possible options, such as the selective removal of problem individuals or changes in animal husbandry, were universally viewed as unacceptable. Herders typically held the government responsible for their loss because of its prohibition on hunting. Historically, herders suffering from excessive depredation solicited help from *shikaris*, or professional hunters, who were rewarded with gifts of food, alcohol, and livestock for trapping habitual stock predators.

FUTURE AND PROGNOSIS

To date, conservation efforts have been centered on the creation of protected areas, localized law enforcement, and public education (Fox 1994; Jackson & Fox 1997). Increasing attention is being given to resolving depredation conflicts while also improving reserve management. However, status and distribution surveys remain inadequate and illegal fur trading continues. The latter appears to be rapidly increasing in the central Asian republics, driven by rampant local currency devaluations and economic hardship, including the failure to pay reserve staff salaries (Koshkarev 1994).

The accelerated consumption of tiger bone will lead to increased pressure on other large cats, including snow leopard. Although such use may constitute a status symbol, it is also a way of retaining ancient customs in the face of rapid change. As Asia's economy booms, there has been a resurgence in traditional Asian cures among a user population of more than a billion people. Actions to protect snow leopards from unsustainable harvesting include (1) encouraging better enforcement of protective measures, (2) offering rural people alternative sources of income, and (3) conducting research to better understand consumer market forces and underlying motivations so that acceptable synthetic or herbal substitutes can be promoted.

National recovery plans have not been developed, with the exception of India's Snow Leopard Conservation Scheme that listed 14 core snow leopard reserves totaling 15,000 km^2 or 16% of potential snow leopard habitat. Although modeled on Project Tiger, it lacks political clout, for snow leopards are not as threatened as Bengal tigers (*P.t. bengalensis*), which inhabit densely populated lowlands rather than the remote and sparsely peopled mountains. Along with lack of trained staff, rugged terrain hampers status and distribution surveys. Mountain parks tend to have limited personnel who must contend with harsh winters, so it is not surprising that park management and law enforcement are weak.

These and other factors provide a clear argument for a new approach to conservation that places increased responsibility with local residents. It may be possible to use religious sentiment as a powerful conservation tool. As Buddhists, Tibetans are loath to kill wildlife, especially if that wildlife is protected by monasteries or inhabits areas deemed as sacred. Because hunting is usually undertaken by the poor, management is largely a socioeconomic issue. However, underlying biological factors must also be addressed, such as poor animal husbandry practices, depletion of prey species, home-range overlap between predators and livestock, and habitat degradation or fragmentation. Although governments establish protected areas, it is the local people who must live with the consequences, co-exist with the predators, and preserve high-mountain biodiversity, often without realizing tangible benefits. Whether it affects many or a few families, livestock damage undermines local willingness to protect wildlife and tolerate the presence of a nature reserve. This highlights the importance of implementing procedures and policies that effectively address pastoralists' concerns.

Ultimately the continued presence of snow leopards and other carnivores depends on the willingness of local people to tolerate some loss of their livestock. On one level, better animal husbandry practices are needed, perhaps reinstating some of the traditional livestock practices mentioned previously. In addition, conservationists must work to diversify local people's economic opportunities and standards of living. Solutions are best packaged in multifaceted programs offering compensation for losses; tax incentives; shared tourist revenues; improved health, education, and basic services; and other actions to enhance quality of life for herders while also conserving snow leopards and associated wildlife of the region. The International Snow Leopard Trust and the Mongolian Association for the Conservation of Nature and Environment are providing tea, noodles, clothing, and other items to south Gobi herders with the understanding they will protect wildlife. Similar efforts in Pakistan and Tibet focus on improving night-time corrals to make them predator-proof. There is an urgent need to extend such pioneering efforts to other parts of the snow leopard's vast range.

Southern Sea Otter

Katherine Ralls

Common Name: southern or California sea otter
Scientific Name: *Enhydra lutris nereis*
Order: Carnivora
Family: Mustelidae
Status: Threatened under the Endangered Species Act of the United States; Depleted under the Marine Mammal Protection Act of the United States; Completely Protection under the California Fish and Game Code; CITES Category I.
Threats: Historically, the commercial exploitation for fur. Currently, oil spills; entanglement in fishing gear; possibly infectious diseases and pollutants.
Habitat: Rocky and sandy-sediment marine near-shore communities.
Distribution: Central California coast from approximately Capitola in the north to Point Conception in the south.

DESCRIPTION

The southern sea otter is one of three subspecies of sea otters. Sea otters resemble other otters except for their flattened tail and flipperlike hind feet, which help them swim on their back at the surface of the water. The smaller forelegs are used primarily for grooming and locating and handling invertebrate prey. The blunt, rounded canines and broad, flat molars are used to crush hard-shelled prey such as clams. Adult males average about 2,900 g and females 1,950 g. Sea otters have dark to reddish-brown fur, although some individuals have lighter, almost white fur on the head, neck, chest, and forearms. Their water-resistant coat consists of flattened outer guard hairs protecting a dense undercoat.

NATURAL HISTORY

The natural history of sea otters is reviewed in detail by Riedman and Estes (1990). Sea otters occur in near-shore habitats up to depths of about 100 m. Sea otters eat large invertebrates such as sea urchins, abalones, crabs, and clams, and they often use rocks to break the shells of their prey. They are a keystone species because they promote the growth of kelp (by preying on herbivorous invertebrates). This in turn has a variety of community- and ecosystem-level consequences (Estes 1996).

Because of their small size and lack of blubber, sea otters depend on their high metabolic rate and the insulative properties of their fur to maintain normal body temperature in the cold marine environment. They can consume an amount of food equal to as much as one-third of their body weight daily. They are extremely vulnerable to oil spills because oil destroys the insulative properties of their fur, leading to hypothermia and death.

Sea otters are polygynous; that is, one male mates with several females. Some adult males maintain territories in areas inhabited by breeding females, whereas nonterritorial males gather in groups away from the breeding areas. Females give birth to a single pup each year after a gestation period of about 6 months. Pups are born throughout the year and are dependent on their mother for 5 to 6 months.

The California sea otter population has never increased at a rate of more than 5% per year, whereas some populations farther north have increased at 17 to 20% per year (Estes et al. 1996). As all sea otter populations have similar reproductive rates, the lower growth rate in California must be owing to increased mortality.

CONFLICTING ISSUES

Sea otters once ranged along the rim of the Pacific Ocean from northern Japan to central Baja California, Mexico. The historical, worldwide population was probably about 150,000 to 300,000 otters and the California population about 16,000 to 20,000 otters. Commercial hunting of sea otters for their fur began in 1741, and by 1911 the worldwide population may have consisted of only 1,000 to 2,000 individuals. Sea otters were thought to be extinct in California until a group of 32 was seen near Big Sur in 1914. This population steadily increased in size and geographical range at about 5% per year until the early 1970s. It declined from the mid-1970s to the early 1980s as a result of entanglement and drowning in fishing gear. The state imposed restrictions and closures on gill and trammel net fishing, and the population resumed growth until the early 1990s, when it numbered approximately 2,400.

California legally protected sea otters soon after their rediscovery and established a Sea Otter Game Refuge in 1941, which it later expanded to include the entire range of the sea otter population as of 1959. However, the California Department of Fish and Game (CDFG) is charged with both protecting the sea otter population and maintaining shellfisheries. These goals began to conflict once the sea otter population expanded southward into the range of the fisheries. It soon became clear that sea otters and humans could not simultaneously exploit shellfish populations, which had increased to unusually high levels during the time that sea otters had been absent. In fact, some scientists predict that these shellfisheries will be unsustainable even in the absence of sea otters unless harvest levels are reduced (VanBlaricom 1996). In an effort to achieve both goals, the CDFG suggested zonal management in which geographically separate zones would be maintained for sea otters and shellfisheries.

In 1972 management authority for sea otters and other marine mammals was transferred to the federal government by the Marine Mammal Protection Act (MMPA). This law created the Marine Mammal Commission (MMC) to advise federal agencies on marine mammal management. The MMPA also specified that marine mammal populations be maintained or recovered to their "optimum sustainable population" level, which was later interpreted as a population between 60% to 100% of carrying capacity. A marine mammal population is considered "depleted" under the MMPA if it is below the optimum level. In 1977 the southern sea otter was listed as Threatened under the Endangered Species Act (ESA). Marine mammals listed under the ESA are automatically depleted under the MMPA. Thus the U.S. Fish and Wildlife Service (FWS) was charged not only with recovering the sea otter population until it could be delisted under the ESA, but also with increasing the population size to the optimum level.

The transfer of management to the federal government and the opposing mandates of the state and federal agencies led to conflicts. As one former participant put it, "Federal-state interactions regarding California sea otter conservation often have shown all the sophistication of a rugby match on the mud flats of Elkhorn slough" (VanBlaricom 1996:86). The current draft recovery plan proposes to delist otters when the annual spring count reaches 2,650 (based on the probability that sea otters could survive a major oil spill), whereas the best estimate of the number of otters required to reach the optimum population level is 8,100 (estimated via modeling; see Demaster et al. 1996). Neither the state nor shellfishers are pleased with the prospect of this many otters.

The main reasons that the sea otter population was listed as Threatened were its small size and its vulnerability to oil spill. Tanker traffic along the coast poses the greatest oil spill threat. However, unless there is a permanent moratorium on oil development off the coast of California, commercial interests wishing to explore and develop offshore oil deposits will be concerned with sea otter issues. Oil developers were skeptical that otters were highly vulnerable to oil spills until 1989, when the Exxon *Valdez* spill in Alaska caused thousands of sea otters to die.

In response to the Exxon *Valdez* spill, the California state legislature imposed a tax on oil transported or processed in California, and it specified that part of the proceeds be used to rescue and rehabilitate birds, sea otters, and other marine mammals affected by oil spills. The CDFG built and staffed a large facility for this purpose. The state maintains that this effort is vital for protecting the sea otter population from the effects of oil spills. However, based on the results of attempting to rehabilitate oiled otters following the Exxon *Valdez* spill, the recovery team believes that such efforts will be of little benefit should another spill occur (Estes 1991).

The 1982 recovery plan specified that a new sea otter colony be established to minimize the risk that an oil spill would affect the entire population. Conservation organizations endorsed this plan, but fishermen, oil interests, and the state opposed it. In 1986 Congress attempted to resolve the conflicts by passing Public Law 99–625, which authorized the FWS to establish new colonies of sea otters but also called for zonal management, now endorsed by the state, FWS, and MMC.

From 1987 to 1991 the FWS translocated 139 sea otters to San Nicolas Island off the coast of Southern California and established an otter zone around the Island and a no-otter zone including the waters surrounding the remainder of the Channel Islands and all the mainland coast south of Point Conception. Otters in the otter zone were fully protected, but any otters entering the no-otter zone were to be removed by nonlethal means. Thus the recovery of sea otters would be promoted, but existing shellfisheries and the possibility of additional oil development south of Point Conception would be protected.

Because otters are a near-shore species and it was not yet known that they are capable of long-distance movements over deep water (Ralls, Eagle, & Siniff 1996) and have strong homing tendencies (Benz 1996), the FWS assumed that the otters would remain near San Nicolas. However, most of them soon left the Island and some traveled back to the mainland or other islands (Benz 1996). Many of the otters moving into the no-otter zone were captured and released in the mainland sea otter range, but it soon became apparent that it was impossible to maintain a no-otter zone by nonlethal means. The failure of the FWS to keep otters out of the no-otter zone aroused resentment, particularly among fishermen and the state. Furthermore, the 1998 population at San Nicolas was only about 20 independent otters, despite the birth of 47 pups since the translocation. Many suspect that the reason for the population's lack of growth is entanglement of sea otters in some of the hundreds of lobster pots set around the Island each season (Benz 1996).

In 1989 the FWS appointed a recovery team (consisting entirely of scientists) to revise the recovery plan, and a group of technical consultants (representing other interested parties) to advise it. This division of labor resulted in a scientifically credible recovery plan but did little to solve conflicts because the technical consultants failed to produce consensus advice to the team (Ralls, Demaster, & Estes 1996). In revising the recovery plan, the team noted that the Exxon *Valdez* spill spread over 670 km—a distance much greater than the present length of coastline occupied by both the mainland sea otter population and the San Nicolas colony. The team concluded that the San Nicolas colony did not provide a reasonable safeguard against a major oil spill and recommended a passive recovery strategy of letting the mainland population grow and expand its range rather than attempting to establish new colonies.

This strategy was based on the assumption that the sea otter population would continue growing at a rate of 5% per year. However, by 1998 it became clear that the population had stopped growing for unknown reasons and was likely declining. Factors that may be contributing to the lack of growth are a new and rapidly expanding live-fish fishery that sets traps near shore, pollutants (Jarman et al. 1996), and infectious diseases and parasites (Thomas & Cole 1996).

In 1998 over 100 otters moved southward into the no-otter zone. Fishermen protested vigorously. FWS legal counsel advised capturing and returning the errant otters back into the mainland otter zone. The recovery team strongly advised against this because it would have a detrimental effect on an already declining population by disrupting the existing social structure and increasing competition for food resources. Furthermore, it would not solve the fishermen's problem because many of the otters would return to the no-otter zone.

FUTURE AND PROGNOSIS

The prospects for recovery of the southern sea otter population are brighter than those for many endangered species. The population is capable of growing at 5% per year, and unoccupied habitat extends north and south of its current range. There are many advocates for its recovery, including the Friends of the Sea Otter, the MMC, the recovery team, and a large public constituency. Although conflicts with fisheries have existed for decades, they have not seriously hampered recovery.

The most pressing need is to identify and correct the factors responsible for the current lack of population growth. The recovery plan should be revised to reflect current circumstances. Furthermore, the FWS should declare the translocation a failure and thus eliminate the no-otter zone. All the stakeholders need to admit that zonal management by nonlethal means is impossible.

If the southern sea otter population resumes growth, it should reach the proposed ESA delisting criterion of 2,650 individuals within a relatively short time. The major societal issues regarding sea otters, such as conflicts with shellfisheries, will probably not be resolved under the ESA, but rather in the process of determining that the population is no longer depleted under MMPA.

Sturgeons and Paddlefishes (Acipenseriformes)

Vadim J. Birstein

This case deals with a group of threatened animals: 25 species of sturgeons and 2 species of paddlefishes (Pisces, *Class*: Actinopterygii, *Order*: Acipenseriformes). Their common English and scientific names, as well as their distribution and threatened status, are given in Table 2.

DESCRIPTION

Sturgeons and their close relatives, paddlefishes, are the most numerous group of ancient fishes named "living fossils" (Gardiner 1984). Sturgeons are usually large (on average, 1.5–2.0 m), spindle-shaped fishes with a developed snout. They have five longitudinal rows of large bony scales (scutes). Caudal fins (tails) have a well-developed upper lobe. Skeleton and skull are cartilaginous. The mouth faces downward. Paddlefishes do not have scutes and have long, paddle-shaped snouts.

NATURAL HISTORY

Many sturgeon species belonging to the genera *Acipenser* and *Huso* are anadromous fishes; that is, they live mainly in the sea and migrate into rivers for reproduction. Others are amphidromous fishes; that is, they live mostly in the rivers and migrate to semi–salt water estuaries. Species of *Scaphirhynchus, Pseudoscaphirhynchus*, and two paddlefishes are potamodromous; that is, they migrate within a river system to breed and forage.

All sturgeons and paddlefishes have long life cycles and mature late (at 7 to 18 years). Eggs are usually deposited on gravel. Embryonic development lasts several days. Larvae of anadromous species migrate downstream to the sea for feeding. Most sturgeon species feed on bottom organisms, but the beluga (*Huso huso*) is predatory. Sturgeons hybridize easily, and some hybrids, especially between beluga and sterlet (*A. ruthenus*), are fertile.

Sturgeons and paddlefishes live in the Northern Hemisphere only (Bemis et al. 1997). Currently all species are endangered (Birstein 1993). Several species are on the brink of extinction in the wild (Table 2). The phylogenetic or evolutionary relationships between different sturgeon species, based

Table 2. Sturgeons and Paddlefishes of the World: Their Current Distribution and IUCN Status (1996)

Species (Latin Name)	English Name	Distribution	IUCN Status	CITES[†]
Family: Acipenseridae	**Sturgeons**			
1. *Acipenser baerii*	Siberian Sturgeon	Siberia (Russia), Kazakhstan, China	Vulnerable	II
A. b. baerii	Siberian Sturgeon	Ob River system (Siberia)	Endangered	
A. b. baikalensis	Baikal Sturgeon	Lake Baikal (Siberia)	Endangered	
A. b. stenorrhynchus	Lena River Sturgeon (Siberia)	Yenisei, Lena, Indigirka, and Kolyma Rivers	Vulnerable	
2. *Acipenser brevirostrum*	Shortnose Sturgeon	Rivers, estuaries, and ocean along the eastern coast of North America from Indian River (Florida) to Saint John's River (New Brunswick)	Vulnerable	I
3. *Acipenser dabryanus*	Yangtze/Dabry Sturgeon	Yangtze River (China)	Critically Endangered	II
4. *Acipenser fulvescens*	Lake Sturgeon	Great Lakes, lakes of southern Canada, and river systems connected with them	Vulnerable	II
5. *Acipenser guldenstaedtii*	Russian Sturgeon	Caspian, Black, Azov Seas, and rivers entering into them	Endangered	II
		Caspian Sea stock	Endangered	
		Sea of Azov stock	Endangered	
		Black Sea stock	Endangered	

Species (Latin Name)	English Name	Distribution	IUCN Status	CITES†
6. *Acipenser medirostris*	Green Sturgeon	Pacific coast of North America from the Aleutian Islands and Gulf of Alaska to the Ensenanda (Mexico), Rogue (Oregon), Sacramento, and Klamath (Calif.) Rivers	Vulnerable	II
7. *Acipenser mikadoi*	Sakhalin Sturgeon	Pacific Ocean from the Amur River to northern Japan, Bering Sea, Tumnin (Datta) River	Endangered	II
8. *Acipenser naccarii*	Adriatic Sturgeon	Adriatic Sea; Po and Adige Rivers	Vulnerable	II
9. *Acipenser nudiventris*	Ship Sturgeon	Caspian, Black, and Aral Seas, and rivers entering into them	Endangered	II
		Black Sea stock	Endangered	
		Caspian Sea stock	Endangered	
		Aral Sea stock	Extinct	
		Danube River population	Critically Endangered	
10. *Acipenser oxyrinchus*	Atlantic Sturgeon	Eastern coast of North America	Lower Risk	II
A. o. desotoi	Gulf Sturgeon	Gulf of Mexico and northern coast of South America	Lower Risk	
A. o. oxyrinchus	Atlantic Sturgeon	Rivers, estuaries, and ocean along the east coast of North America from the St. John's River (Florida) to Hamilton Inlet (Labrador)	Lower Risk	

Table 2. (*continued*)

Species (Latin Name)	English Name	Distribution	IUCN Status	CITES[†]
11. *Acipenser persicus*	Persian Sturgeon	Black and Caspian Seas and rivers entering into them	Endangered	II
		Black Sea stock	Endangered	
		Caspian Sea stock	Vulnerable	
12. *Acipenser ruthenus*	Sterlet	Most of central and eastern European and western Siberian rivers	Vulnerable	II
		Caspian and Black Sea drainage stock (Volga, Danube)	Vulnerable	
		Ob, Irtysh, and Yenisei Rivers (Siberia, Russia)	Vulnerable	
13. *Acipenser schrenckii*	Amur River Sturgeon	Amur River system (Siberia, Russia, and China)	Endangered	II
14. *Acipenser sinensis*	Chinese Sturgeon	Yangtze River (China)	Endangered	II
15. *Acipenser stellatus*	Stellate Sturgeon or Sevruga	Caspian, Black, and Azov Seas, and rivers entering into them	Endangered	II
		Caspian Sea stock	Vulnerable	
		Sea of Azov stock	Endangered	
		Black Sea stock	Endangered	
16. *Acipenser sturio*	European Atlantic or Baltic Sturgeon	Baltic, Eastern North Atlantic, Mediterranean, and Black Seas, and rivers entering into them	Critically Endangered	I

Species (Latin Name)	English Name	Distribution	IUCN Status	CITES[†]
17. *Acipenser transmontanus*	White Sturgeon	Rivers and Pacific coast of North America from the Gulf of Alaska to Baja California	Lower Risk	II
		Kootenai River population (USA)	Endangered	
18. *Huso dauricus*	Kaluga Sturgeon	Amur River system (Siberia: Russia, and China)	Endangered	II
19. *Huso huso*	Giant or Beluga Sturgeon	Caspian, Black, and Azov Seas, and rivers entering into them	Endangered	II
		Caspian Sea stock	Endangered	
		Sea of Azov stock	Critically Endangered	
		Black Sea stock	Endangered	
		Adriatic Sea stock	Extinct	
20. *Pseudoscaphirhynchus fedtschenkoi*	Syr-Dar Shovelnose Sturgeon	Syr-Darya River (Kazakhstan, Central Asia)	Critically Endangered	II
21. *Pseudoscaphirhynchus hermanni*	Small Amu-Dar Shovelnose Sturgeon	Amu-Darya River (Uzbekistan and Turkmenistan, Central Asia)	Critically Endangered	II
22. *Pseudoscaphirhynchus kaufmanni*	Large Amu-Dar Shovelnose Sturgeon	Amu-Darya River (Uzbekistan and Turkmenistan, Central Asia)	Endangered	II
23. *Scaphirhynchus albus*	Pallid Sturgeon	Missouri and Mississippi River basins	Endangered	II

Table 2. (*continued*)

Species (Latin Name)	English Name	Distribution	IUCN Status	CITES[†]
24. *Scaphirhynchus platorynchus*	Shovelnose Sturgeon	Missouri and Mississippi River basins	Vulnerable	II
25. *Scaphirhynchus suttkusi*	Alabama Sturgeon	Mobile River basin (Alabama and Mississippi)	Critically Endangered	II
Family: Polyodontidae	**Paddlefishes**			
1. *Polyodon spathula*	American Paddlefish	Mississippi River system	Vulnerable	II
2. *Psephurus gladius*	Chinese Paddlefish	Yangtze River (China)	Critically Endangered	II

[†] The listing of all sturgeon and paddlefish species on the Convention on the International Trade of Endangered Species of Flora and Fauna (CITES) appendices was adopted at the CITES COP10 meeting in Harare, Zimbabwe in June 1997. It came into force on April 1, 1998. Species listed on Appendix I are threatened with extinction. Species on Appendix II may become threatened with extinction if trade is not regulated.

on DNA sequence data, were determined only recently (Birstein & DeSalle 1998).

CONFLICTING ISSUES

Historically, three main factors caused sturgeon and paddlefish depletion: dam and channel construction in rivers, overfishing, and pollution (Birstein 1993). Construction of dams cut sturgeons off from their spawning sites. Thus the beluga sturgeon has not been able to reproduce in the Volga River for 30 years. Pollution in rivers and seas has had a serious impact on sturgeons and their reproduction (Birstein, Bemis, & Waldman 1997; Khodorevskaya et al. 1997). During the last few years the level of pollution has decreased in the major rivers of sturgeon spawning, although for varying reasons. The Volga River has seen a drastic drop in industrial activity since the breakup of the former Soviet Union. Pollution in the Danube River and the Missouri-Mississippi river system has decreased owing to local environmental action.

The recent rising of the Caspian Sea water level resulted in an increase in oil pollution throughout the whole basin, but especially in the western part that is used by young sturgeons (Dumont 1995). Rising water covered oil pools and other wastes close to shore. In the past year the sea has stopped rising, so the pollution level may stabilize. However, development of oil production in the northern (Kazakh and especially Russian) part of the Caspian Sea is planned by 2001, and this could have a highly destructive impact on sturgeons because the feeding grounds of sturgeons during their first year are located in the area of oilfields. The drying of the Aral Sea in Central Asia has caused the disappearance of the ship sturgeon (*A. nudiventris*) from the region, the possible extinction of the endemic Syr-Dar shovelnose sturgeon (*Pseudoscaphirhynchus fedtschenkoi*), and a sharp depletion in the population size of two Amu-Dar shovelenose sturgeons, *P. hermanni* and *P. kaufmanni* (Zholdasova 1997).

Today, legal and illegal overfishing is the most devastating factor. Overfishing at the turn of the 20th century caused the almost complete disappearance of the European Atlantic sturgeon (*A. sturio*) from its entire historic range (Birstein, Betts, & DeSalle 1998). At the same time, overfishing on the U.S. Pacific coast resulted in a sharp decline in the commercial sturgeon catch and caviar production (Ryder 1888/1890). Caviar export from the United States to Europe stopped and has not recovered since then. At present the international caviar market stimulates poaching in the Black, Azov, and Caspian Seas and in the Amur River (Birstein 1997). The decrease in natural stocks of the beluga, the Russian sturgeon (*A. gueldenstaedtii*), and the stellate sturgeon (*A. stellatus*) has already led to the use of eggs of other sturgeon species as replacements for expensive types of caviar (DeSalle & Birstein 1996; Birstein, Doukakis, et al. 1998).

Overfishing is aggravated by two factors. First, the number of countries catching sturgeons in the Caspian, Black, and Azov Seas rose after the breakup of the former Soviet Union. Second, endemic corruption in the former Soviet Union has led to the involvement of governmental agencies and officials in the illegal sturgeon catch and caviar trade (Anonymous 1997c). Beginning in 1990, the number of sturgeons killed for caviar increased dramatically, and markets for sturgeon products have expanded into new areas (De Meulenaer & Raymakers 1996; Taylor 1997). However, according to experts, the quality of caviar has dropped partly because it is produced from the roe of immature fish.

The current sturgeon crisis has become a challenge for the international conservation community, including the IUCN (World Conservation Union) and TRAFFIC International. In 1994 the International Conference on Sturgeon Biodiversity and Conservation (New York) attracted the attention of conservationists to the crisis. In 1996, at the 13th meeting of the CITES Animals Committee (Pruhonice, Czech Republic), the sturgeon crisis was discussed, as was the possibility of using a molecular method of caviar species identification developed at the American Museum of Natural History (New York) as a possible tool for the CITES implementation (DeSalle and Birstein 1996). At the 10th meeting of the Conference of the Parties in June 1997 (Harare, Zimbabwe), a proposal for inclusion of sturgeons in CITES Appendix II (controlled trade) was presented by Germany and the United States. However, the scientific community was not given an opportunity to review the final version of the proposal, and it was adopted with mistakes (Mrosovsky 1997). The CITES sturgeon listing came into force on 1 April 1998.

Most parties to the CITES convention were not prepared to implement the sturgeon listing, and coordination between countries on implementation efforts is not presently achievable. The U.S. Fish and Wildlife Service announced that it had its own method of caviar species identification in November 1997 (Anonymous 1997a), although this method was developed without tissue samples from the three commercial sturgeon species and/or the necessary DNA sequence data (which were not publicly available at the time of the announcement). Additionally, the method has not been published or scientifically peer reviewed. Other countries, such as Germany and France, have decided to develop their own methods of identification. The opportunity to use an available, precise (only a single caviar egg is required), and inexpensive method (DeSalle & Birstein 1996; Birstein, Doukakis, et al. 1998) as a standard international tool for the CITES implementation has been lost.

Another problem for implementation is the lack of scientifically based data on the population size of sturgeons in the Caspian and Black Seas. The CITES Secretariat needs these data for international, independent regulation of the catch and caviar trade quotas. In the former Soviet Union all sturgeon issues, from estimation of stocks and catch quotas to caviar production, were "top secret" and controlled by the Ministry of Fisheries. In a decree issued in September 1997, President Yeltsin ordered that responsibility for overseeing sturgeon catch be moved from the successor of this ministry, the Department of Fisheries, to the Border Guard Troops (Anonymous 1997b). At that time he characterized the Department of Fisheries as one of the most corrupt structures in Russia. The previous history of secrecy surrounding sturgeons resulted in the use of the outdated, inaccurate population size estimation methods, which have never been peer reviewed by international experts. Therefore there are no reliable data on sturgeon populations for the CITES Secretariat. Despite this history, the World Bank plans to put Russia in charge of controlling fisheries and legal regulations for the whole Caspian Sea basin (Wilczynski 1998).

Continuation of restocking programs in the Caspian, Azov, and Black Seas is one more controversial issue. In the early 1960s the Soviet Union began a massive program to release sturgeon juveniles from 28 specially built hatcheries (Khodorevskaya et al. 1997). These juveniles were obtained from sturgeon spawners caught in the wild and artificially bred at the hatcheries. In the 1980s approximately 100 million hatchery-raised juveniles were released annually into the Caspian Sea. Since 1990 the number of released juveniles has dropped by 50% because of economic problems in Russia. Only half of the hatcheries are operational now, and all the hatcheries desperately need modernization.

The efficiency of the Soviet sturgeon restocking program has never been evaluated by means of modern experimental methods. The only evidence that it worked at all is the presence of beluga sturgeon in the Caspian Sea despite the end of natural reproduction in the Volga River 30 years ago. However, the size of beluga sturgeon from stocked generations is considerably smaller than that of naturally reproduced individuals (Khodorevskaya & Novikova 1995). It is not known which factor caused this reduction. American conservationists are very cautious about sturgeon restocking because of genetic concerns (Waldman & Wirgin 1998).

Despite the questions it raises, restocking may be the only opportunity to maintain and/or restore the main wild populations of Eurasian and American sturgeons and paddlefishes. For the Caspian, Azov, and Black Seas, and the Amur and Yangtze Rivers (China), restocking would only be possible through international projects and funding. A genetic study of sturgeons from these basins has begun recently (Birstein, Hanner, & DeSalle 1997; Birstein & DeSalle 1998; Birstein, Doukakis et al. 1998). Of all countries in these areas, only China and Romania are working together with American and European experts to estimate sturgeon stocks and find spawning sites by using modern technology.

Despite the urgent need for international professional geneticists and sturgeon biology experts to work together in the Caspian and Black Seas, the

IUCN continues to fund conferences and project plans full of generalities. Examples of that can be taken from the list of Pilot Projects suggested by the IUCN Task Force on the Caspian Sea Problems at its meeting in February 1998, a meeting that no professional sturgeon biologist attended.

FUTURE AND PROGNOSIS

There is little hope that CITES listing will improve the sturgeon crisis soon, because no country is ready to implement the listing. Illegal overfishing, at least in the Caspian Sea, Siberian rivers, and the Amur River, will probably continue until sturgeons are eliminated from these basins. The lack of sturgeons for caviar production in Eurasia might stimulate sturgeon poaching in the United States.

Ironically, it seems that business interests, not international conservation effort, are providing the only ray of hope for sturgeon survival. Sturgeon aquaculture is a rapidly growing industry in Europe, the United States, and Latin America. Caviar from aquacultured white sturgeon (*A. transmontanus*) is produced in California and Italy. In 1997 the French company Gie l'Esturgeon d'Aquitaine produced 1 ton of the aquacultured Siberian sturgeon (*A. baerii*) caviar, and it plans to increase caviar production to 10 to 15 tons (Sabeau 1997). The availability of, and lower prices for, aquacultured caviar may turn caviar consumers from the wild sturgeon caviar to aquacultured caviar. Perhaps this will lead to a decrease in the level of sturgeon overfishing in Eurasia. However, by the time the aquacultured caviar can be produced in sufficient quantities to take a significant part of the caviar market share, it may be too late for wild sturgeon populations.

Yuma Clapper Rail

Courtney J. Conway and William R. Eddleman

Common Name: Yuma clapper rail
Scientific Name: *Rallus longirostris yumanensis*
Order: Gruiformes
Family: Rallidae
Status: Endangered under the Endangered Species Act of the United States; Endangered in Mexico; Endangered in Arizona and California.
Threats: Habitat loss owing to water diversion and channelization for urban and agricultural uses and conversion of wetlands to crop land and urban development. Habitat quality is reduced when flood control by dams and fire suppression allows vegetative succession to different community types, contaminant accumulation from agricultural run-off, and invasion of marshlands by exotic salt cedar (*Tamarisk chinensis*).
Habitat: Emergent freshwater and brackish-water riverine wetlands exposed to periodic flooding (Todd 1986). Most commonly found in shallow, early and mid-successional cattail (*Typha* spp.) and bulrush (*Scirpus* spp.) marshes with high interspersion of both dense vegetation and open water/mudflat habitats (Conway 1990).
Distribution: Southwestern United States and northwestern Mexico. Historically, rails were found in extensive marshlands of the Colorado River delta and in backwater marshes and old oxbows of the lower Colorado and Gila Rivers. Currently, they reside in isolated Colorado River marshes from Needles, California, southward to the delta in Mexico, along the lower Gila River in Arizona, and in marshes associated with the Salton Sea in California (Todd 1986).

DESCRIPTION

The Yuma clapper rail is the size of a crow, with long, gray-brown legs and toes. The orange bill is long, thin, and slightly down-curved. The head, neck, and breast are gray-brown, and the back feathers are darker brown with black centers. Both the flanks and undertail covert feathers are distinctly marked with alternate black and white bars. Males and females are similar in plumage coloration. Adult mass varies seasonally, but males average 269 g and females 210 g. The body is laterally compressed, allowing rails to run quickly through dense emergent vegetation. The tail and wings are noticeably short and the legs are large and strong, evolutionary adaptations that allow birds to run through dense reeds or swim underwater to avoid danger.

NATURAL HISTORY

Unlike most races of clapper rail, *yumanensis* inhabits freshwater marshes. Yuma clapper rails feed on a variety of wetland invertebrates and small vertebrates, primarily crayfish (*Procambarus clarki* and *Orconectes virilis*), clams (*Corbicula* sp.), isopods, water beetles (Hydrophilidae), and small fish (Ohmart & Tomlinson 1977; Eddleman & Conway 1998). Birds forage by picking prey off vegetation and the surface of the ground, probing soft mud and sand with their long bill, and spearing prey beneath the water surface.

Male and female birds give loud, distinctive mate-attraction calls: "*Kek*" for males, "*Kek-burr*" for females. Pairs vocalize simultaneously, forming a duet that resembles the rapid clapping of hands and gives the species its name. Pairs are at least annually monogamous. Nesting occurs in February through May. Males may build multiple nests, and the female chooses one for egg-laying. Alternate or "dummy" nests are often used for preening, loafing, and as brood platforms, but they are used for incubation if predators or high water disturb the primary nest. Adults have the extraordinary ability to carry eggs in their bills to a new nest. Females lay large clutches (mean = 7.8 eggs; Bennett & Ohmart 1978). Male and female birds alternate incubating the clutch, with males typically incubating eggs throughout the night. Incubation begins before the last egg is laid and requires 21 to 29 days. Chicks can leave the nest almost immediately after hatching, and juveniles become independent at 5 to 6 weeks of age. Estimates of annual adult survival are rare but probably are 49 to 67%, with most mortality attributed to predation during fall and winter (Eddleman & Conway 1994).

Yuma clapper rails are mostly nonmigratory, although some winter movement may occur (Eddleman 1989). Birds only defend a small area around their nest site (Conway 1990). Home range size averages 7.6 hectares for males and 10.0 hectares for females (Conway et al. 1993). Densities may be as high as 0.15 birds per hectare (Smith 1975).

CONFLICTING ISSUES

Wetlands used by clapper rails also serve as settling basins for a wide variety of contaminants (Eddleman et al. 1988; Eddleman & Conway 1998). Selenium levels in Yuma clapper rail tissues were at levels that caused hatching defects in many other birds (Rusk 1991). Other clapper rail subspecies have relatively high residues of DDE (dichlorodiphenylethane), PCBs (polychlorinated biphenyl), and selenium in eggs, and moderate to high levels of mercury, DDD (dichlorodiphenyldichloroethane), DDT (dichlorodiphenyltrichloroethane), DDE, dieldrin, heptachlor epoxide, and PCBs in body tissues (Roth 1972; Odom 1975; Klaas et al. 1980; Jarman 1991; Lonzarich et al. 1992; Eddleman & Conway 1998).

Little is known about either the historic or current distribution in Mexico. The historic range of Yuma clapper rails in the United States was probably not significantly larger than it is today (Tomlinson & Todd 1973). However, one thing is certain: suitable habitat, and hence population size, has declined with the rapid and aggressive control of water on the Colorado River (Eddleman & Conway 1998). Historically, marsh habitat along the Colorado River shifted annually with existing marshes being wiped out by flood or fire and new marshes growing elsewhere (Sykes 1937; Ohmart et al. 1975). As a result, high quality, early-successional marshlands rich in productivity were probably always available, and Yuma clapper rails evolved behavioral adaptations and life history strategies to deal well with the dynamic nature of riverine wetlands.

In 1902 the U.S. Congress created what is now the U.S. Bureau of Reclamation (BOR) to develop water projects for farmers in the West. With water diversion, Congress hoped the arid West could support large cities and millions of small farms. The first Colorado River dam (Laguna) was completed in 1909. Early dams profoundly altered the dynamics of marshland creation and succession along the Colorado River (Grinnell 1914). Moreover, conversion of the Colorado River floodplain to agriculture reduced the number of oxbow river marshes that rails traditionally used for nesting and foraging. As croplands expanded in the arid desert regions, clapper rail habitat declined.

Upstream dams have altered hydrology and have profoundly impacted expansive rail habitat in the Colorado River delta. Approximately two-thirds of the formerly extensive marshlands of the delta quickly disappeared following completion of the first major Colorado River dam (Hoover Dam in 1935; Sykes 1937). As time passed more dams were built, and ever more water was taken from the river, leaving less to cross the Mexican border into the delta. Although small marshland habitats eventually became established behind the large U.S. dams, the once massive river delta marshes were gone. In recent years not a drop of Colorado River water typically reaches the delta. The water that is delivered to remnant marshes in the delta is mainly discharge from first-generation uses in the United States, such as irrigation runoff and municipal sewage effluent (Glenn et al. 1996).

Prior to 1922 federal water appropriation laws were based on "beneficial use." Once a state diverted water and used it to some "benefit," the state owned future rights to that water as long as it continued to divert. California's rapidly increasing diversion of Colorado River water left other western states worried that there would soon be none left. Thus in 1922 the U.S. Department of Commerce brought the seven Colorado River basin states together and established the Colorado River Compact. Fifteen million-acre feet (MAF) were appropriated among the seven states, presumably leaving 2 to 3 MAF excess both as a buffer for low flow years and for Mexico.

Those appropriations, still in effect today, were based on a 1922 estimated annual flow of 17 to 18 MAF. However, average annual flow since 1922 has only been 13.9 MAF. The overestimate was initially not a problem, but as states diverted more of their appropriations, water rights issues led to contentious legal battles. Not only is the struggle over Colorado River water a battle among states, but rapidly growing urban cities in the Southwest are fighting against agricultural interests within states. Hundreds of legal cases over water rights have been fought, with dozens reaching the Supreme Court.

It is in this intense legal battle over water that the Yuma clapper rail must persist. By the time the ESA was passed, the lower Colorado River was already a series of consecutive dams separated by water storage reservoirs and narrow delivery channels. Yuma clapper rails were declared Endangered in 1967, soon after the ESA was passed. Yet fish and wildlife have no legal right to water under current appropriation laws and no legitimate status as beneficiaries. Indeed, in efforts to store and divert more river water, the BOR considered removing all existing vegetation along the lower Colorado River and replacing it with vegetation that requires less water. Moreover, the agency responsible for endangered species recovery (the U.S. Fish and Wildlife Service or FWS) has no legal footing by which to stop the BOR. Few specific efforts aimed at recovery have been proposed, much less implemented.

In addition to issues of quantity of water, quality of the marshes that remain in the lower Colorado River basin is threatened. Irrigation runoff that re-enters the Colorado River contains high concentrations of salts, mineral deposits, and pesticides. At the same time, additional rail habitat has been lost in subsidiary river basins. Faced with dwindling flows of Colorado River water, farmers began to pump groundwater for irrigation. Extensive pumping has lowered groundwater aquifers, causing entire river basins in the Southwest (e.g., the San Pedro River) and their marshland habitats to dry up completely.

A recovery team was formed in 1972 (Gould 1975), and the team completed a recovery plan for the Yuma clapper rail in 1983 (U.S. Fish and Wildlife Service 1983). However, the recovery team was unable to address the large political issues. Membership on the recovery team was based on agency representation rather than on expertise. Members represented the BOR, FWS, Bureau of Land Management, Arizona Game and Fish Department, and California Department of Fish and Game. Some of these agency representatives were local rail experts, but others were not. If a recovery team member took a new job or switched positions within the same agency, his or her position on the team was usually filled by a replacement within the agency. Consequently, the recovery team had high turnover. Often, new members had no knowledge of or experiences with Yuma clapper rails. Such turnover and inexperienced membership limited the team's ability to promote recovery efforts.

Veteran recovery team members who had the knowledge and experience to push for active recovery efforts did not; however, the ability of these individuals to change river water flow and release schedules was probably nil. Although the FWS is legally mandated to establish and attempt to meet recovery objectives under the ESA, wildlife biologists within state natural resource agencies in California and Arizona assumed long-term leadership roles on the recovery team. These state agencies are administered by legislatures that have been immersed in legal battles to divert *more* water from the Colorado River for their expanding urban and agricultural needs. It is unlikely that state legislators in the West would urge their employees to fight to keep more water in-stream for rails while they sued other states to take more out.

When faced with difficult decisions, managers frequently stall by asking for additional information (Nichols 1999), and this is what the Yuma clapper rail recovery team did. After the recovery plan was completed, the only purpose of annual recovery team meetings was to coordinate annual population surveys. Methods of increasing population size, improving current habitat conditions, restoring historic habitat, or changing water flows to benefit rails were seldom, if ever, discussed. By 1990 the recovery team had dissolved. The water needs of agricultural and urban growth in the West were politically, socially, and economically too big to fight. Besides, rail populations in the United States appeared relatively stable (700–1,000 birds), and annual surveys could be coordinated among the agencies involved without holding meetings.

Consequently the primary efforts of the recovery team over the past 25 years have been to coordinate population surveys. Monitoring alone, however, is not likely to lead to better understanding or ensure long-term viability (Nichols 1991) unless it is coupled with experimental work to understand the mechanisms driving population change (Krebs 1991). Fortunately the U.S. Yuma clapper rail population has been relatively stable and appears healthy for the short term, with substantial numbers breeding in several state and federal wildlife management areas.

FUTURE AND PROGNOSIS

Despite the lack of active recovery efforts, the future for the Yuma clapper rail looks relatively good. The total U.S. population is estimated at 1,500 birds, and counts in the United States have remained relatively stable since the late 1980s (Anderson & Ohmart 1985; Shuford 1993; Small 1994; Piest & Campoy 1998). Numbers of birds in Mexico were until recently unknown, but surveys from 1998 suggest as many as 5,300 birds in the largest remaining delta wetland, Ciénega de Santa Clara (Piest & Campoy 1998). If these estimates are correct, the Yuma clapper rails' future appears promising as long as Mexican marshes are not developed or degraded. One requirement for delisting in the recovery plan is a signed agreement between Mexico and the United States for cooperative management of Yuma clapper rails and their habitat; such an agreement has not been reached. Habitats in Mexico require protection, and U.S. and Mexican officials will have to work together to prevent development or degradation of Ciénega de Santa Clara. Cooperation is essential because the quantity and quality of rail habitat in the delta are dependent on water from the United States.

Any future management or recovery efforts in the United States will be hampered by separate federal agencies working under different mandates. Federal agencies managing first-generation water uses (irrigation, flood control, hydroelectric power generation) are different from those managing second-generation uses (water pollution prevention, management of endangered species dependent on in-stream flow). Hence the agencies responsible for water flow (BOR, Army Corps of Engineers) have expended little effort in conserving endangered species, and the agency responsible for species recovery (FWS) has no authority over agencies controlling water flow. The separate missions of these agencies should be integrated into a comprehensive plan to manage the Colorado River ecosystem. Efforts to do this have begun recently.

Federal and state wildlife refuges on the lower Colorado River have traditionally managed marshlands exclusively to benefit migrating waterfowl. Refuges frequently drained, plowed, and mechanically treated marshes to increase waterfowl foraging opportunities. However, the state and federal wildlife refuges that support the largest number of rails have only recently included habitat requirements for Yuma clapper rails into their master management plans for the future. Additional efforts might include prescribed fire and occasional flooding to scour older marshlands, remove invading salt cedar, and stimulate new emergent growth for the benefit of rails. Many of the remaining habitats have contaminants in water inflows, sediments, or both (Eddleman & Conway 1998), and efforts to reduce contaminant inflows from agricultural runoff must be discussed. The recent discovery of the large rail population in Mexico should be taken as an opportunity to address the organizational and ecological problems preventing real progress on recovery and restoration on the lower Colorado River.

Interdisciplinary Problem Solving in Endangered Species Conservation: The Yellowstone Grizzly Bear Case

Tim W. Clark

Endangered species recovery and conservation require problem-solving strategies. Demands for natural resources professionals to be more interdisciplinary, policy relevant, and effective are being made on many fronts (e.g., Pool 1990). If one thinks of a policy process (or management process) as the development and implementation of strategic aims, then professionals have a lot to offer society to improve conservation. For example, impacts on Yellowstone National Park and the surrounding national forests and wildlife refuges are receiving increasing attention and eliciting demands for improved management and policy (Clark & Minta 1994). Grizzly bear (*Ursus arctos*) management in the Greater Yellowstone Ecosystem (GYE) is one particularly high profile case. Threats to grizzly bears may be defined in terms of habitat, population fragmentation, and the biological measures needed to maintain or restore populations (e.g., Knight et al. 1999). However, they may also be considered an interdisciplinary management problem in that grizzly bear conservation is only partly a technical problem and largely an outcome of complex human social dynamics—a policy process. Understanding this policy process and making it more effective are key to achieving effective grizzly bear conservation.

This chapter offers a brief overview of the policy process and then examines three interdisciplinary problem-solving concepts that can be applied to endangered species conservation. These concepts can be stated as questions about a given policy or management effort (e.g., Is it rational? Is it politically practical? Is it morally justified?), and they can be answered by using a set of logically comprehensive, conceptual tools. The toolkit to problem solving includes rational problem orientation, social and decision process mapping, and basic belief analysis. The chapter also discusses the roles or standpoints of professionals in the social and decision process. Finally, the text focuses on integrating knowledge about rationality, politics, morality, and standpoint. Even a little knowledge of these concepts and how to use them practically can enhance professional effectiveness dramatically. Grizzly bear conservation in the GYE is illustrative (Primm 1996).

BACKGROUND AND OVERVIEW

Natural resource professionals are deeply concerned about degradation of the biological world and its consequences. Professionals possess discipline-based knowledge that is important to conservation and management issues, but they need another kind of knowledge as well if they are to be successful. The second kind of knowledge is skill at interdisciplinary problem solving within the entire policy process. The more a professional knows about both kinds of knowledge—disciplinary and interdisciplinary—and how they are interrelated, the more successful he or she will likely be. Just as several disciplines offer theory about natural resources management, there are theory and experience about interdisciplinary problem solving. The latter kind of knowledge is seldom taught in the biological curricula of universities, and an interdisciplinary approach has not yet been applied in grizzly bear conservation.

During the 20th century in the GYE, the policy process led to many bears being killed and eventually to the species being listed as Threatened under the 1973 Endangered Species Act. After all, people and organizations (and nations) seek to maximize power, wealth, or some other human value, and in this case the bears got in the way. They were exploited well beyond sustainable levels, and policy ignored their needs until recently. Over the last few decades, however, efforts began at restoring the GYE population. Today some people regard the GYE bear population as recovered, whereas others think it is in need of further, intense conservation. The difference of opinion is important, because the motivation for a rational grizzly bear policy is the perception that a problem exists and needs a solution (see Weiss 1989). A *problem* is a discrepancy between what one prefers to have happen and what is most likely to happen—a difference between goals and trends. A *policy* is a commitment to a program (an alternative to procedures causing a problem) that is intended to reach an aim or preferred outcome. The current grizzly bear conservation effort is a complex policy process, and it appears to be mired in conflict.

What Is Meant by Managing the Policy Process?

Regardless of how grizzly bear conservation is interpreted, it can best be resolved if its entire context is fully understood. For example, an exclusive focus on biological causes will do little to resolve the underlying human causes of the problem (Brewer & Clark 1994). Only by understanding the human process of management, and learning how to analyze it practically, can policy be improved (Clark et al. in press). This is true not only for professionals working in offices on policy formulation, but also for those working in the field on policy implementation.

In general terms, the *policy process* is a human social dynamic that determines how the "good and bad things in life" are partitioned out, and who gets what, how, and why. Many people misunderstand the policy process because it is "often treated as an abstraction, associated with the dry prose and dusty volumes of government documents" (Culhane 1981:30). Such a view is highly misleading. Policy is not the same as legislation or government action. Instead, it is what government and private bodies do for or to citizens and the environment. What professionals and other people *do* matters far more in the long run than what is *said* in formal government documents. Real policy is made in the field through the collective actions of many people. The grizzly bear policy process has spanned decades and is ongoing. However, different people have different conceptions of what it is and what to do about it.

A more complete and realistic definition of the policy process is needed. Policymaking is a sequence of many actions by many actors, each with potentially different perspectives, values, and strategies (Ascher & Healy 1990). No one can guarantee that any policy process will be optimal. Each phase is populated with different people, organizations, and interest groups. Most texts that describe policy leave out the analytic and political challenges facing the people involved, as well as the difficulty of coordination and communication. Yet all these interactions constitute the policy process, and how they unfold spells the difference between success and failure. Some processes work better than others, and evidence suggests that the grizzly bear process is not working very well.

What Is Interdisciplinary Problem Solving?

One obstacle to finding effective solutions is the fact that a lot of knowledge about problems and the policy (management) process is highly fragmented and dispersed. Knowledge tends to be partitioned according to disciplines, organizations, and other interests. Consequently a wildlife biologist may interpret the grizzly bear conservation problem quite differently from a sociologist, and a federal official may view bear management from a much different perspective from that of a state administrator. Traditionally the policy problem is subordinated to the perspective of one scientific discipline or managing organization. That limits the way the problem is defined and the options for solution.

The conceptual framework for interdisciplinary problem solving that follows minimizes distortions and helps professionals see the entire picture. This kind of skill is essential for integrating diverse knowledge. In Lasswell's (1971:181) words, these tools enable their users to "study the process of deciding or choosing and evaluate the relevance of available knowledge for

Figure 8. Interdisciplinary Guidelines for Addressing Conservation Policy and Management Issues

#1 Is it reasonable? Problem Orientation	#2 Is it possible? Social & Decision Process[1]	#3 Is it justified? Social Process[1]
Value task (goals) Historic task (trends) Scientific task (conditions) Futuring task (projecting) Practical task (alternatives)	Participants Perspectives Situation Base values Strategies Outcomes Initiation Estimation Selection Implementation Evaluation Termination Effects	Participants[2] Identification Expectations Demands Participants' Myths[3] Doctrine Formula Symbol

#4 What is my standpoint?
Standpoint Clarification

In terms of epistemological, disciplinary, organizational, and parochial biases.

#5 How will I integrate what I know?
Knowledge Integration

Synthesize knowledge from #1-4 to improve understanding and judgment for action.

1. A social process involves people pursuing values (i.e., power, wealth, knowledge, skill, respect, well-being, affection, rectitude) through institutions by using resources. Human social process includes participants, their perspectives, situations, base values, strategies, outcomes, and effects.
2. People are likely to act in their self-interest, to complete acts that they perceive leave themselves better off than if they had completed them differently (maximization postulate).
3. Myth comprised doctrine (philosophy, basic beliefs), formula (constitution, laws), and symbols (lore heroes, flags, grizzly bears). Myths are constantly being readjusted through social and decision-process.

the solution of particular problems." The tools are directly applicable to the grizzly bear case but have not been used to date.

Suppose that you join the ongoing grizzly bear conservation effort. You must become oriented to the people, organizations, local culture, legal mandates, history of events, technical issues, and many other aspects of the issue—in a hurry. You will need to look at all existing information, determine if knowledge is lacking, decide what is included as well as missing from the reports and accounts you read and hear, assess the situation from both rational and political standpoints, and develop your own interpretation that is realistic and that can help the process function more efficiently and effectively. How should you start?

BASIC INTERDISCIPLINARY PROBLEM-SOLVING CONCEPTS

There are three key perspectives—rational, political, and moral—to any policy problem (Figure 8). The first perspective (rational) can be examined by using a problem orientation, which asks whether the policy process is reasonable. The second perspective (political) can be examined by using social and decision-process mapping. This focuses on conflicts among the participants and seeks to identify the values being used and promoted in the process. The third perspective (moral) focuses attention on the underlying assumptions or beliefs used by participants to justify their positions; it asks whether the process is moral or justified.

The Problem Orientation (e.g., Is grizzly bear policy rational?)

Problem orientation is a strategy for constructing more rational policy (Lasswell 1971; Simon 1985). In the rush to solve problems, conservation activists, politicians, and the general public have traditionally been more solution oriented than problem oriented. Yet lasting, comprehensive solutions to problems cannot be constructed unless the problems themselves are fully understood and analyzed. Being solution oriented, also called "problem blind," is often a serious problem in and of itself. Thorough problem orientation requires that five interrelated tasks be undertaken (Lasswell 1971). They are as follows:

1. *Clarify goals*—the things, events, or processes you want to achieve as the preferred outcomes.

2. *Describe trends*—historic and recent events; these should include changes relevant to the goals.

3. *Analyze conditions*—factors that shape trends, causes, motives, and policies.

4. *Make projections*—the likely future developments under various circumstances.

5. *Invent, evaluate, and select alternatives*—the possible courses of action that will likely help realize your goals.

By carrying out these five tasks fully, one can easily establish rational choices. "Choose the alternative that you expect, on the basis of trends, conditions, and projections, to be the best means of realizing your goals" (Brunner 1995:3). Best in this case may mean the most effective, efficient, and equitable alternative.

Let's continue with the grizzly bear example to illustrate these problem orientation concepts (see Table 3). Like most professionals, given your civic responsibilities, you desire to improve GYE grizzly bear conservation. You hope to accomplish this, in part, by minimizing the frequency of lethal accidents to bears (*clarify goals*). Suppose you read a report on lethal incidents from poaching, car collisions, lightning, risky scientific methods, and so on over the past decade, and how much those accidents have threatened future grizzly bear survival (*describe trends*). You conclude that the frequency and

Table 3. Problem Orientation Concepts

1. **Goals:** What outcomes do we want?

2. **Problems** (Problems are discrepancies between goals and real or likely states of affairs): What are the problems given our goal?

3. **Alternatives:** What alternatives are open to participants for solving problems?

4. **Evaluate alternatives:** Would each alternative help solve the problem?

   A. *Trends*: Did it work or not work when used in similar occasions in the past?

   B. *Conditions*: Why, or under what conditions, did it work or not work?

   C. *Projections*: Would it work satisfactorily under existing conditions?

5. **Repeat:** Repeat the above analysis on an ongoing basis within the limits of time and resources.

Sources: Based on Brunner 1995; Clark 1996; Wallace & Clark 1999.

severity of accidents is much higher than you prefer (i.e., you become aware that a problem exists).

To do something about the problem, you must understand why the frequency and severity of accidents are so high. So you take a closer look at the report and discover, perhaps, that hunter behavior, unsanitary camping conditions, and high vehicle speeds were involved in most bear deaths. On the basis of your experience as a professional, you conclude that poaching, improper solid waste management, or poorly educated backcountry users are the main causes of bear deaths (*analyze conditions*). If nothing is done or too little is done too late, the frequency and severity of accidents will remain higher than you prefer (*make projections*). On the other hand, if poaching can be reduced, if solid waste management can be made bear-proof, and if backcountry users can be educated, some progress seems possible.

To improve prospects for grizzly bear survival, you decide to promote a policy that will reduce poaching, improve waste management, and educate backpackers. In your estimation, it is the least expensive and most effective means of achieving the goal of minimizing accidental deaths (*invent, evaluate, and select a policy*), a goal that may be shared by others. Is this a rational policy? The answer depends on whether or not the number of grizzly bear deaths will actually decrease as a result of the actions you propose. Your analysis makes you confident that the policy might be rational, keeping in mind that your level of confidence might change given additional interest or time for further analysis. For example, you might try to find out what happened in other areas where one or more of these causes of accidental bear deaths were either low or absent.

In the real world, however, your policy proposal is not the end of the story. It only has your tentative support. After more reflection, you realize that you have neglected to consider something that might be key to promoting the policy successfully, if not actually improving it. Addressing the solid waste management problem or regulating backcountry users may, in fact, reduce the frequency of lethal accidents, but it may also impact the goals of other people—and perhaps other goals of your own. Consider these two examples: (1) better waste management increases costs by 20%, and (2) backcountry regulations may preclude human use of certain areas at certain times (e.g., the site you wanted to camp in next July). Given that people universally prefer to reduce their costs and increase their opportunities, it seems unlikely that your policy would receive broad public support.

Although your policy is important to you, it is far from comprehensive. It failed to consider other peoples' goals and other possible alternatives. This lack of comprehensiveness has two significant implications—political and analytical. First, your preferred policy will likely conflict with other policies supported by various interests. For example, environmentalists might oppose it because in their estimation it does not adequately protect bears. Conflicts like these are reconciled through the decision process. Mapping the grizzly bear decision process is essential, and it is described in the next section.

Second, the implication of this analysis is that a perfectly rational policy is an ideal worth striving for, but in reality there is always other relevant information to be considered. No matter how hard or long you work to solve a given problem, perfect rationality is never achieved. It requires mental capabilities well beyond human powers or resources. In addition, unanticipated events can undermine your rational consideration of a policy. For example, almost no one in the public predicted the potentially beneficial effects of the 1988 GYE fires, which burned over a million acres. As it turned out, however, the extensive fires may aid grizzly bear survival by creating more ungulate habitat (Mattson & Craighead 1994). The point is that problem solving is a series of approximations in actual practice. It is a process used to construct more rational policies, rather than an unwavering commitment to one initial policy. Thus rationality is procedural. To be rational a person must carry out certain procedures, namely, the five interrelated steps in problem orientation in the sequence listed above.

These five steps must be repeated over and over again as time permits (Dewey 1910; Lasswell 1971). Following their completion you have greater insight about the nature of the problem and the potential solutions. In each subsequent evaluation you reconsider previous findings in light of new information and changing circumstances.

The sequence of these five steps is important. It is necessary to begin by clarifying goals. Without goals, there is no rational basis for deciding on the important trends, conditions, and projections to examine. You must be selective in your analysis because you cannot possibly consider everything.

Goals should be tentative. As you consider trends, conditions, and projections, and learn more about the problem at hand, you will likely want to revisit your goals. Perhaps they will need to be changed. This iterative process (i.e., repeating the process) promotes individual and policy learning and is extremely important in real problem solving. Policy failure commonly occurs when goal clarification and trend description are downplayed in the rush to get on with the high-profile tasks that follow (e.g., alternatives).

Problem orientation not only is essential in constructing your own policy but also is useful in appraising an argument that someone else has made on behalf of their favored policy. In fact, evaluating other people's policy position is necessary to adequately construct your own. You must look for and find the goals, trends, conditions, projections, and alternatives on which other people's policy arguments are based. If any of the five elements are not explicitly stated, then you should raise important questions about them.

When one or more of the five tasks is omitted or poorly treated, there is a gap in the policy argument. Sometimes a gap is a sign of propaganda or censorship designed to manipulate viewpoints on controversial issues. For example, a promoter of an endangered species recovery program might censor all but one alternative in order to focus attention on that tactic. This, in effect, "captures" expressed or accepted goals by associating them with one alternative. Although good analysis thrives on alternative choices, politics often depends on restricting the consideration of alternatives in order to control policy outcomes.

The Social and Decision Process (The Politics of Grizzly Bear Policy)

The second perspective concerns the social and decision processes; that is, the means of reconciling conflicts and achieving agreement on policy (Lasswell 1971: Fig. 1). Politics inevitably affect policy, because people have special interests and tend to promote them to the exclusion of other alternatives. None of the interests has a completely objective view of the issue (see Brewer & Clark 1994). Yet people must reconcile their differences to find a common interest.

Politics develop as participants, with their unique perspectives, interact in complex ways and changing situations. Each participant brings certain base values (e.g., power, wealth, and enlightenment) to the table and uses them to promote his or her own interests (Table 4). The decision process produces outcomes that do or do not benefit participants. In turn, outcomes have longer-term effects.

Let's return to the grizzly bear example. Suppose that in your concern about grizzly bear population survival you advocate stronger law enforcement against poaching, better waste management, and backcountry user education. Environmentalists might advocate excluding people from bear habitat altogether. Government agencies might advocate the status quo. Hunters or energy producers might want something else entirely. It is not

Table 4. Values or "Bases of Power" Participants Use to Influence Decision Outcomes

Value Definition	Questions to Ask
Power: to give and receive support in making decisions in specific contexts.	How is power given and received in interpersonal and decision process and what are the outcomes?
Enlightenment: to give and receive information.	How is information given and received? What are the outcomes?
Wealth: to give or receive the opportunity to control resources, such as money, natural resources, and other people.	How is wealth affected (given and received) by the process? What are the outcomes?
Well-being: to give or receive the opportunity for personal safety, health, and comfort.	How is well-being, both physical and mental, affected by the decision process?
Skill: to give or receive the opportunity to develop talents into operations of all kinds including professional, vocational, and artistic skills.	How well and what kind of skills are used (or not) in problem orientation and in decision process, and with what outcomes?
Affection: to give and receive friendship, loyalty, love, and intimacy in interpersonal situations.	How are professional, friendship, and loyalty values used in decision process, and with what outcomes?
Respect: to give and receive recognition in a profession or community.	How is respect or deference used (or not) in decision process, and what are the outcomes?
Rectitude: to give and receive appraisal about responsible or ethical conduct.	What are the ethics at play in interpersonal relations and embodied in decision process outcomes?

Source: Lasswell 1971.

possible to meet all these needs at the same time. Nor is it reasonable to have several separate reserves set aside for each special interest. They all have a common interest in sharing use of the GYE and in finding a consensus on grizzly bear conservation. A policy that reflects this common interest must be found, selected, and implemented.

A selected policy is a public consensus on the rules expected to hold people accountable for behaviors that conflict with the policy. In the case of grizzly bear conservation, such rules might include regulations that (1) exclude people from bear habitat under certain circumstances, or (2) impose jail sentences for people found guilty of harming bears.

A logically inclusive and explicit set of rules specifies each of the following: (1) goals (or purposes) to be achieved through the policy, (2) rules of conduct intended to achieve those goals, (3) contingencies (or circumstances) in which the rules apply, (4) sanctions to enforce compliance with the rules in the applicable circumstances, and (5) assets to cover enforcement and other administrative expenses. (Brunner 1995:13)

Does Current Grizzly Bear Management Meet These Criteria?

A consensus in conservation does not mean that the special interests have gone away, nor does it mean that all parties agree with the rules of conduct. Essentially it means that everyone more or less expects the rules to be enforced, regardless of whether they agree with them or with the purpose of the rules. In effect, a poacher can rightly expect to be prosecuted to the full extent of the law if he is caught. "More or less" is an important qualification here. After all, consensus is never absolute and rules are seldom perfectly clear.

Grizzly bear conservation is a primary purpose (or goal) of the wildlife management agencies involved, and legal arrangements (or rules of conduct) have been made to support this goal. For the grizzly, however, the rules are contingent on where you are located. People in the Yellowstone region tend to comply with the rules, in part, out of a basic respect for the law and a concern for personal safety (benefits or positive sanctions). People also comply with the rules because they do not want to be arrested for littering or hiking in closed areas (punishments or negative sanctions). However, if enforcement by state and federal wildlife officials and other such agents becomes lax or nonexistent, people's expectations may change gradually and they will do what they like. Because enforcement and compliance are not perfect, the effective rules (as opposed to the formal rules) are more lenient.

A good policy prescription can be undone if the agency charged with carrying out the policy experiences reduced staff and operating funds. A well-enforced prescription can be undercut if the courts do not vigorously enforce the law. The goals used as criteria to define success can be retroactively redefined, or the data can be manipulated to show success. If failure is inescapable, responsibility can be deflected to scapegoats. Finally, ending a policy and moving to a new one is often difficult or impossible because special interests that benefit from the selected policy are unwilling to relinquish it, whether or not it is effective.

The discussion to this point has addressed the latter phases of the decision process and some of the political possibilities that affect outcomes. How the earlier phases unfold depends primarily on who controls planning. If an office in the federal government, such as the U.S. Fish and Wildlife Service, has a planning monopoly and is also allied closely with traditional natural resource extractors, that federal office is unlikely to comply with demands from conservationists for a viable population of grizzly bears based on the best standards of modern conservation biology. If the earlier phases of planning, debating, and setting the rules are not inclusive, open, reliable, and comprehensive, it is likely that implementation will be weak, lawsuits will proliferate, and the effort will go on with little consensus or resolution.

Knowledge of the different policy and decision phases has enabled policy researchers to distinguish patterns among successful and unsuccessful programs. Some policies have undesirable, unplanned, and often unanticipated impacts (Ascher & Healy 1990). Certain weaknesses or pitfalls characteristic of each policy or decision phase (outcome) recur time after time, regardless of the technical details of the conservation issue (Table 5). Knowing about these common pitfalls and being able to anticipate them can help you avoid them in your professional practice.

Basic Beliefs (Is Grizzly Bear Conservation Policy Morally Justified?)

The morality of a policy can be understood by examining basic beliefs. These are fundamental assumptions about the way society should function (i.e., how people should treat one another and the living world we inhabit). Human cultures and subcultures are distinguished by different belief patterns. Policy and political conflicts usually stem from differences in basic beliefs. Among other things, basic beliefs serve as guidelines on how power is used in society. The grizzly bear policy process shows high conflict at present. This is a reflection of differences in participants' basic beliefs and their notion of how power should be used.

Greater understanding of the grizzly bear policy process can be achieved by examining (1) the way in which opposing policy positions are being justified, and (2) the basic beliefs being appealed to—and by whom—in these justifications. To date there has been virtually no systematic analysis of these issues in GYE grizzly bear management. Even though participants in grizzly bear policy might agree that the bear is an important part of America's natural and cultural heritage, consensus does not flow directly into specific management decisions. Disagreement over political issues (e.g., whether to kill problem bears and to curtail development in and adjacent to bear habitat) essentially stems from opposing basic beliefs.

Social scientists recognize that basic beliefs, or premises, form the foundation of political myths. *Myth* in this sense refers to the underlying philosophy of communities or individuals rather than a fictitious story, and it should be considered synonymous with more neutral terms such as *paradigm*, *worldview*, *outlook*, or *frame of reference*.

Myths are made up of a hierarchy of three elements (Lasswell 1971). The first is doctrine, which is the part of the myth that sets out basic beliefs—aims and expectations of the community. For example, statements of doctrine can be found in preambles to constitutions. The second element is formula, which prescribes the basic rules for progress according to the com-

Table 5. Common Weaknesses or Pitfalls in Each of the Policy Phases

1. **Initiation** phase:

   Delayed sensitivity in which perception of a problem comes only after the problem has developed and harmful effects are widely felt.

   Biased initial problem definition. One interest sets out a problem definition that favors its own interests and view of the problem. This definition fails to capture the full and true nature of the emerging problem.

2. **Estimation** phase:

   Inadequate analysis of the problem. Needed data and analysis of trends, conditions, and projections are lacking or only partially carried out.

   Study the problem to buy time. This form of delaying is a common tactic of people who oppose the emerging policy picture and problem definition. These people do not accept the problem definition nor want to take action on it.

3. **Selection** phase:

   Poor coordination in government decision making. Often, complex problems are addressed by several groups simultaneously who may not be aware of each other or communicate well in developing a common understanding of the problem and what needs to be done to solve it.

   Over-control. Groups may respond to problems by automatically imposing greater controls on everyone involved. This leads to bureaucratization and sometimes paralysis, or gridlock.

4. **Implementation** phase:

   Benefit leakage. Certain socioeconomic interest groups may seek to capture and benefit more from the policy than other intended recipients.

   Limitations of state enterprises as natural resource managers. The size, slowness, political interests, conservative, and bureaucratic features of governmental organizations can all be limitations.

   Poor coordination of implementation. Often bureaucratic over-control, rivalry, and exclusion of key parties can lead to muddled policies and programs.

5. **Evaluation** phase:

   Insensitivity to criticism. Critics may try to improve a policy honestly, but government often simply ignores their input, regardless of its merits.

   Failing to learn from experience. Organizations can fail to learn and repeatedly respond to new conservation challenges using the same programs, approaches, and techniques.

6. **Termination** phase:

   Pressure to continue unsuccessful policies. Even unsuccessful or poorly performing policies and programs that have outlived their usefulness may benefit someone who then clamors for the policy to continue.

   Failure to prepare for termination. Groups may fail to appreciate and prepare for the difficulties of terminating a policy, even a bad one, early in the overall policy process.

Based on Ascher & Healy 1990.

munity's aims and expectations. Basic laws or constitutions are examples of the formula. The third element is symbol, which glorifies and legitimizes the political myth (e.g., heroes, flags, and anthems). Grizzly bears symbolize different things to different people according to their myth (e.g., conservation needs vs. conflict with development), and this can cause conflict.

Myths, or basic beliefs, are accepted as a matter of faith and often go unquestioned because they are so deeply ingrained. Basic beliefs are usually reaffirmed and redefined through their use in social interaction over time. Each generation adapts basic beliefs to its unique time and circumstances. Societies can and do change their doctrines, formula, and symbols as they confront new circumstances and changes in their own identity, expectations, and demands. However, there is considerable variation in how well individuals and societies understand these social dynamics or adapt successfully to new situations.

The grizzly bear remains a powerful symbol that connects strongly and directly with certain basic beliefs, namely, securing a national identity and sustaining a healthy environment. For example, invocations of symbols such as "grizzly bear" and "Yellowstone National Park" are often used successfully to mobilize public support for key conservation policy actions. They have been successful in reducing conflicts when they incorporate shared basic beliefs.

People use myths to reconcile differences and to grasp some understanding of a situation, because no one has a comprehensive and entirely objective view of the world. In this way science, law, and politics are made up of subcommunities with separate submyths. Many subcommunities and submyths exist within the broader human community. For example, many members of the grizzly bear research community believe that more and better research is the key to securing effective conservation. This myth about the power of science and knowledge is pursued without sufficient acknowledgment that irrespective of the nature and validity of research findings, political circumstances are often the determining factor in decisionmaking. Scientists, managers, and other natural resources professionals and community groups would benefit from becoming fully aware of their own myth and submyths (see Brunner & Ascher 1992; Brunner 1993a, 1993b). Greater appreciation of the doctrines, formula, and symbols that one follows can liberate individuals previously unaware of their own myth.

This discussion has examined three perspectives of policy process—rationality, politics, and morality. It has introduced the concepts of problem orientation, social and decision process, and basic beliefs or myths, which are the interdisciplinary conceptual tools for critical thinking and practical problem solving. However, simply knowing about these elements does not guarantee reaching consensus. It is necessary to use these concepts to gain insight into and understanding of the policy process. A successful policy integrates what is rational, politically practical, and justifiable according to basic beliefs into one set of practical actions. The three interdisciplinary concepts can help in appraising different policies. However, mastering these concepts and using them with skill require time and experience.

STANDPOINT OF PROFESSIONALS

A professional can take diverse standpoints in any policy (Clark et al. 1992). In policy process, the term *participant/observer* is used because professionals both participate in policy process and observe it at the same time from their unique vantage point. This dual role is critical to developing a greater understanding of policy process. It is important to note that all participant/observers have biases based on experience, culture, education, values, and so on (Lasswell & Kaplan 1950; Lasswell & McDougal 1992). Yet many professionals believe they are objective, neutral, and acting in the public interest. For a person to reach his full potential, he must be able to examine and understand himself and other people in the process (Lasswell 1971).

Conventional Standpoints

A traditional professional tends to see situations, events, values, and decisions in customary ways. This approach devalues or ignores alternative ways of understanding the world, even those that are empirically grounded.

Much of the conflict surrounding grizzly bear conservation arises when conventional professionals and other participants draw on differing standards, basic beliefs, and other variables to form their perspectives. In this situation, bias prevents each person from seeing the total picture. Conventional frames of reference lead to "partial blindness" about policy process because attention is restricted to the "rules" and away from the "process." Typically the values of key participants are overlooked, a hazy focus is cast over the decision process, and participants are left with an anecdotal understanding of the overall policy process at best—and confusion and misdirection at worst. These lead to unproductive conflict and power struggles.

Civic Professionalism and a Policy Orientation

A policy-oriented professional seeks to move beyond a conventional standpoint to understand the overall structure and functioning of a policy process. She seeks a clear view of herself, including an appreciation of her own biases (and myths). In carrying out her work she systematically employs the conceptual tools discussed previously. In this way, she essentially takes on an anthropological role by living in a society while simultaneously describing and analyzing that society and any decisionmaking exercises of interest to her.

Ultimately the goal of grizzly bear policy is to achieve viable, self-sustaining populations of grizzly bears in the GYE in ways that benefit from long-term public support. Exercises such as producing publications, holding meetings, performing research, or improving the political status of an organization can be advantageous but should not be the primary policy goals. Many specialists and interest groups try to justify their activities on the basis of alleged contributions to a common goal (e.g., saving the bear). However, their activities are most likely to serve grizzly bear conservation if they and other participant/observers take a functional standpoint and become knowledgeable and skillful in addressing the four elements of policy addressed to this point—rationality, politics, basic beliefs, and standpoint—rather than if they ignore or underattend to these dimensions.

To be successful, professionals should (1) provide knowledge that is useful in the policy process, and at the same time (2) have knowledge of those processes and how well they are working (Clark in press). Yet some professionals remain discipline-bound and conventionally oriented with a strictly technical focus, not policy oriented. To become policy oriented is not to give up traditional professionalism, but to add to it the benefits of commanding the conceptual tools introduced in this discussion.

Several authors discuss improving professional performance in the conservation process (e.g., Lasswell 1971; Clark & Wallace 1999). First, one must recognize that human social systems and ecological systems have co-evolved and that dealing with problems in one system will likely affect the other system (Norgaard & Dixon 1986). Many past efforts at policy improvements have had poor outcomes because they failed to recognize this relationship. Norgaard and Dixon (1986) offer guidelines for policy and management improvements. They suggest that policy success will be achieved if a group (1) sustains system productivity and diversity as the most important goal, (2) starts small and experiments with policy interventions, (3) learns from experience as quickly and thoroughly as possible, (4) maintains flexibility in new policies and programs, (5) reduces the vulnerability of policies and its consequences on both systems, and (6) avoids big plans by staying small and manageable.

Second, one must recognize that seeking ways to improve policy and management often involves pushing "conventional analytic, decision, and management means, often well beyond their effective capacities" (Brewer 1992:8). In seeking improvements, one must keep in mind that (1) science is essential in nearly all efforts to improve policy, (2) there is never enough science available at the time of decision, (3) time and space are important considerations that are often overlooked, (4) synergisms and thresholds characterize most environmental problems and need careful attention, and (5) the way the environment is affected and managed is directly related to the kinds of human institutions that are in place. Failure to attend to these issues accounts for the lack of past successes.

Third, Viederman et al. (1997:466) note that policy success rests on "how well we succeed in achieving a vision of a world that serves the needs of humans and preserves nature." They offer three operating principles: (1) be humble (the humility principle), (2) think deeply and move slowly (the precautionary principle), and (3) do not take irreversible actions (the reversibility principle). They also emphasize the need to understand the policy process as the basis for success:

There is a tendency among scientists to argue the centrality of scientific information in the policy process. Knowledge is clearly better than ignorance, but good science does not necessarily make good policy. Science may be necessary, but cannot be sufficient, because policymaking is the process of reflecting what we value in society, which is at heart a matter of ethics and values. (p. 480)

CONCLUSION

Interdisciplinary problem-solving tools should be used to better understand (1) the grizzly bear management policy process, and (2) the roles of professionals, advocates, and other interests involved in it, including government, business, and nongovernmental participants (see Figure 9). The perspectives on problem orientation, social and decision process mapping, and analyzing basic beliefs can help you evaluate and recognize good and bad policies, as well as construct better policies regardless of your role in the process (see Figure 9).

ACKNOWLEDGMENTS

Brian Miller, Richard Reading, Anne Marie Gillesberg, and Murray Rutherford critically reviewed the manuscript of this chapter. Special thanks go to Ronald Brunner, whose unpublished course notes on policy process analysis formed the basis for this chapter. He generously gave time in discussions about policy analytic theory and its practical application. Other discussions with Andrew Willard, Garry Brewer, and William Ascher also clarified essential theory. Conversations and joint analytic exercises with former students and diverse professionals from over 25 countries throughout 15 years also aided in writing this chapter. Support came from the Denver Zoological Foundation, Scott Opler Foundation, Cathy Patrick, Gil Ordway, Fanwood Foundation, and New-Land Foundation via the Northern Rockies Conservation Cooperative, and Yale University.

Figure 9. An Illustration of the Interdisciplinary Approach to Understanding and Participating in the Policy Process

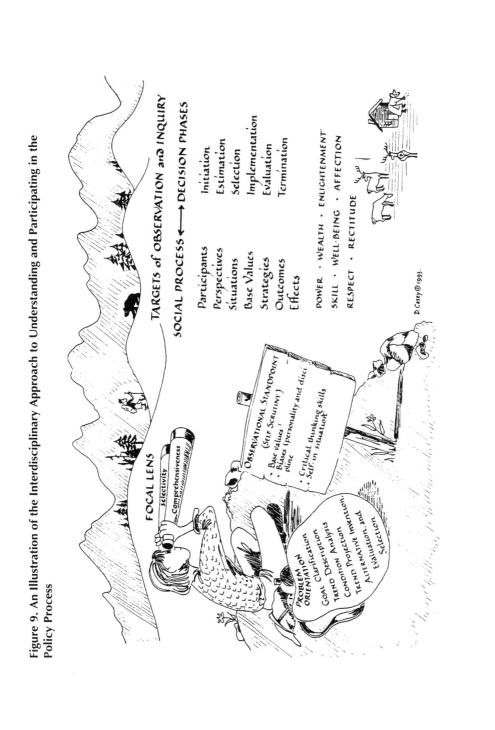

FOCAL LENS
Selectivity
Comprehensiveness

TARGETS of OBSERVATION and INQUIRY

SOCIAL PROCESS ⟷ DECISION PHASES

Participants
Perspectives
Situations
Base Values
Strategies
Outcomes
Effects

Initiation
Estimation
Selection
Implementation
Evaluation
Termination

POWER · WEALTH · ENLIGHTENMENT
SKILL · WELL-BEING · AFFECTION
RESPECT · RECTITUDE

OBSERVATIONAL STANDPOINT
(Self Scrutiny)
· Base Values !
· Biases (personality and disci
  pline
  · Critical thinking skills
  · Self in situation?

PROBLEM
ORIENTATION
Goal Clarification
Trend Description
Trend Analysis
Condition Projection.
Trend Projection Invention.
Alternative Invention, and
Evaluation, and
Selection

D. Casey © 1993.

Conclusions: Causes of Endangerment and Conflicts in Recovery

Richard P. Reading and Brian Miller

The Earth is in the midst of one of the greatest extinction episodes in its history, and the rate of extinction appears to be accelerating (Wilson 1988, 1992). To reverse this trend, it will require more effective conservation of the species that remain. Yet, as the case studies in this book illustrate, such efforts are wrought with human conflict.

Even though the effects of biodiversity loss are usually described in biological terms, the causes are firmly rooted in human values. Indeed, most people and organizations choose to maximize values that secure personal advantage, such as power or money, over species and nature conservation (see Clark this volume). Many conflicts of this type underlie the causes of species endangerment. Other conflicts prevent the development and implementation of more effective conservation programs. Examples include ineffective management of conflict among participants, use of decisionmaking processes that are poorly suited to the tasks at hand, inadequate definition of problems before forming solutions, and lack of sufficient and stable funding. It appears we may not be learning how to more effectively conserve species. Certainly we are not progressing fast enough to reverse current trends.

Collections of case studies such as this one can contribute to improved conservation in a variety of ways. First, they educate people about the range of problems facing endangered species. Second, they represent a reference for those seeking information about endangered species and associated conservation efforts. Third, they provide a rich source of case material for analysis and learning as a precursor to developing better approaches to conservation. Certain trends in the causes of species decline and the conflicts facing recovery programs become readily apparent upon review of the cases.

CAUSES OF SPECIES DECLINE

Several factors are responsible for the endangered status of the species represented in this volume, but all can ultimately trace the source of decline to human activities. People (1) destroy and degrade habitat, (2) directly kill species, (3) introduce exotic species, and (4) spread diseases. These factors, which Wilson (1992) referred to as the four "mindless horsemen of the environmental apocalypse," are all evident in this book to varying degrees (Figure 10; Table 6).

Habitat destruction and degradation involves direct habitat loss, fragmentation, and degradation. This includes the effects of pollution and heavy recreation use (particularly motorized recreation). *Direct mortality/exploitation* comprises all mortality owing to human actions, including hunting, poaching, trapping, fishing, control programs, direct and indirect poisoning, and collisions with motorized vehicles (e.g., cars and boats). *Introduced species* are exotic (alien) ones that are intentionally or unintentionally brought to areas outside of their natural ranges by people. Disease refers to introduced diseases, often brought in with exotic species.

The most common cause of species decline among the case studies presented in this book is habitat destruction and degradation, mentioned in 95.9% of the entries. Similarly, Wilcove et al. (1998), in a review of the causes of species endangerment in the United States, found that 95% of vertebrate species were imperiled by habitat loss and degradation, although the authors separated out the effects of pollution into a separate category. The case studies in this book also describe direct mortality or exploitation (81.6% of the cases), introduced species (20.4%), and disease (16.3%) as important causes of decline. These percentages differ from the findings of Wilcove et al. (1998), who reported that exploitation, alien species, and disease impacted 27%, 47%, and 11% of the vertebrate species they evaluated, respectively. Wilcove et al. (1998) did not include other sources of direct mortality in their figures on exploitation, which may account for some of the difference between their numbers and the ones presented here in Figure 10. Other differences may be related to their focus on only U.S. species.

Despite the differences between other analyses and the case studies presented in this book, it is clear that habitat destruction and degradation is the most significant cause of species decline and endangerment in the world today. It is also clear that the other three "mindless horsemen" seriously impact species as well and can act in a synergistic fashion with habitat destruction. For example, Woodroffe and Ginsberg (1998) showed that 74% of known large carnivore mortality in protected areas was human-caused and occurred near the boundaries of reserves. Habitat fragmentation increases the habitat patch's perimeter distance to area ratio, thereby increasing the number of animals that come in contact with the population sink (i.e., population in which death rates exceed birth rates), at the edge of reserves.

It appears that social, economic, and political factors lie at the root of most species' endangerment. Kellert (1985:528) suggested that "A compelling rationale and an effective strategy for protecting endangered species

Figure 10. Causes of Species Decline Cited by Case Study Authors

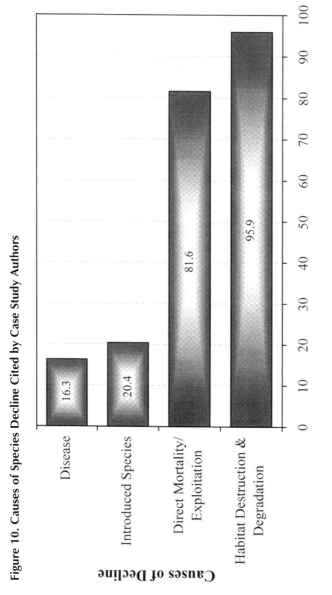

Disease — 16.3

Introduced Species — 20.4

Direct Mortality/ Exploitation — 81.6

Habitat Destruction & Degradation — 95.9

Causes of Decline

Percentage of Species Affected

Note: Percentages add to over 100% because most programs were affected by more than one factor.

Table 6. Causes of Species Decline for Each of the Case Studies

Species	Habitat Destruction & Degradation	Direct Mortality/ Exploitation	Disease	Introduced Species
African Wild Dog (*Lycaon pictus*)	X	X	X	
Altai Argali Sheep (*Ovis ammon ammon*)	X	X		
American Burying Beetle (*Nicrophorus americanus*)	X			
Andean Condor (*Vultur gryphus*)	X	X		
Anegada Iguana (*Cyclura pinguis*)	X	X		X
Aruba Island Rattlesnake (*Crotalus unicolor*)	X	X		X
Asian Elephant (*Elephas maximus*)	X	X		
Asiatic Lion (*Panthera leo persica*)	X	X		
Aye-aye (*Daubentonia madagascariensis*)	X	X		
Basking Shark (*Cetorhinus maximus*)		X		
Black-footed Ferret (*Mustela nigripes*)	X	X	X	
Boreal Toad (*Bufo boreas boreas*)	X		X	
Brown Bear (*Ursus arctos*)	X	X		
Chinese Alligator (*Alligator sinensis*)	X	X		
Dalmatian Pelican (*Pelecanus crispus*)	X	X		
Douc Langur (*Pygathrix nemaeus nemaeus*)	X	X		
Eastern Barred Bandicoot (*Perameles gunnii*)	X	X	X	X
Ethiopian Wolf (*Canis simensis*)	X	X	X	X
European Mink (*Mustela lutreola*)	X	X		X
Florida Manatee (*Trichechus manatus latirostris*)	X	X		
Giant Panda (*Ailuropoda melanoleuca*)	X	X		
Golden Lion Tamarin (*Leontopithecus rosalia*)	X	X		
Golden-rumped Elephant-shrew (*Rhynchocyon chrysopygus*)	X	X		
Gorilla (*Gorilla gorilla*)	X	X		
Grevy's Zebra (*Equus grevyi*)	X	X		
Hawaiian Goose (*Branta sandvicensis*)	X	X		X
Indiana Bat (*Myotis sodalis*)	X			
Ivory-billed Woodpecker (*Campephilus principalis*)	X	X		
Jaguar (*Panthera onca*)	X	X		
Kemp's Ridley Sea Turtle (*Lepidochelys kempii*)	X	X		
Leatherback Turtle (*Dermochelys coriacea*)	X	X		
Malagasy Freshwater Fishes	X			X
Marbled Murrelet (*Brachyramphus marmoratus*)	X	X		

Table 6. (*continued*)

Species	Habitat Destruction & Degradation	Direct Mortality/ Exploitation	Disease	Introduced Species
Mariana Crow (*Corvus kubaryi*)	X			X
Mediterranean Monk Seal (*Monachus monachus*)	X	X		
Mexican Prairie Dog (*Cynomys mexicanus*)	X	X	X	
Mexican Wolf (*Canis lupus baileyi*)	X	X		
Mountain Plover (*Charadrius montanus*)	X			
Northern Rocky Mt. Wolf (*Canis lupus nubilus*)		X		
Orinoco River Dolphin (*Inia geoffrensis*)	X	X		
Piping Plover (*Charadrius melodus*)	X	X		
Po'o-uli (*Melamprosops phaeosoma*)	X		X	X
Red-cockaded Woodpecker (*Picoides borealis*)	X			
Red Wolf (*Canis rufus*)	X	X		X
Scarlet Macaw (*Ara macao cyanoptera*)	X	X		
Snow Leopard (*Uncia uncia*)	X	X		
Southern Sea Otter (*Enhydra lutris nereis*)	X	X	X	
Sturgeons & Paddlefishes (Acipenseriformes)	X	X		
Yuma Clapper Rail (*Rallus longirostris yumanensis*)	X			
TOTAL NUMBER	47	40	8	10

will require recognition that contemporary extinction problems are the result of socioeconomic and political forces." Nevertheless, biologists continue to address conservation by means of a primarily biological approach, often becoming highly frustrated when their biological solutions do not work or are not fully utilized (Wondolleck et al. 1994; Clark & Wallace 1998). More successful conservation requires (1) recognition of the factors underlying the causes of species decline, and (2) development of interdisciplinary approaches (such as the one proposed in the previous chapter) to address those factors (see also Clark et al. in press). Addressing a conservation problem solely through the use of technical, biological approaches, while ignoring the social roots of the issue, is analogous to a physician treating the symptoms but not the disease.

CONFLICTING ISSUES

Social, economic, and political forces not only lie at the heart of species decline and endangerment, but they can also erupt into unproductive conflicts that cripple or hamper conservation efforts. Yet conflict is not always bad. Well-managed conflict can generate discussion and healthy debate, often leading to new ideas, approaches, and other innovations (Clark 1997). However, unmanaged and unproductive conflict often dominates conservation programs, as the case studies in this book demonstrate (see also Clark et al. 1994). In such instances, conflict degenerates into power struggles to control the program. Alternatively, understanding the conflict surrounding recovery programs facilitates the development of more effective responses. A careful definition of the problem, a clear statement of goals, an analysis of trends and conditions, an understanding of the perspectives of stakeholders, and an inclusive organizational process will empower programs to adequately address and rectify these difficult situations (Wallace & Clark 1999).

Although the conflicts related to endangered species are diverse, they share a number of common elements. The collection of case studies presented in this book provides an opportunity to examine trends and explore the relative success of different approaches to dealing with the conflicts. In categorizing conflicting issues, we separated them into 13 broader categories (Table 7; Figure 11). Conflicts were centered around specific topics (such as *Legislation* or *Relations among Key Actors*) or between the species in question and particular interest groups (such as *Agricultural Interests* or *Development Interests*).

Some categories require clarification. *Agricultural Interests* include farmers and ranchers. *Local Rights and Attitudes* refers to conflict surrounding the values and attitudes of local residents. These include issues of property rights, local customs, and degree of local support. *Political and Economic Instability* includes war, unstable governments, faltering economies, and social inequalities that force people to overexploit natural resources. *Development Interests* means mining, oil and gas extraction, and urban sprawl. *Fear* reflects the belief that animals harm humans or human property, including pets and livestock (e.g., large carnivores, rodents). *Poor Intelligence* indicates a lack of reliable data or inadequate infusion of data into conservation efforts.

Conflicts were common between species conservation proponents and groups that rely on natural resource extraction. Indeed, the two topics most frequently mentioned were conflicts with agricultural interests (53.1% of the cases) and consumptive uses of wildlife, such as hunting, trapping, and fishing (49.0%). Conflicts with development and forestry interests were mentioned in 30.6% and 20.4% of cases, respectively.

Factors associated with the way programs functioned (i.e., process) were also important. Two of the most common sources of conflict were program performance and organization (44.9%) and relations among key actors (42.9%). Issues of power and control within species recovery programs led to conflicts in 20.4% of the case studies. Poor intelligence was a less impor-

Table 7. Conflicting Issues Affecting Species in Each of the Case Studies

Species (See Table 6)	Agricultural Interests[1]	Forestry Interests	Hunters, Trappers, & Fishers	Development Interests[2]	Program Performance & Organization	Local Rights & Attitudes[3]	Recreational Interests	Human Fear (e.g., Pest Species)[4]	Power & Control	Relations Among Key Actors	Legislation	Political & Economic Instability[5]	Poor Intelligence[6]
African Wild Dog	X		X										
Altai Argali Sheep	X		X										
American Burying Beetle	X			X									
Andean Condor					X	X				X		X	
Anegada Iguana	X							X					
Aruba Island Rattlesnake	X		X					X	X				
Asian Elephant	X		X					X	X	X		X	
Asiatic Lion	X					X		X	X				
Aye-aye								X					
Basking Shark			X									X	X
Black-footed Ferret	X					X		X		X	X	X	
Boreal Toad											X		X
Brown Bear	X	X	X			X		X		X	X	X	
Chinese Alligator	X		X						X			X	
Dalmatian Pelican			X			X	X						
Douc Langur	X	X				X						X	
Eastern Barred Bandicoot	X				X					X			
Ethiopian Wolf	X								X			X	
European Mink			X							X	X	X	
Florida Manatee				X		X			X	X	X		X
Giant Panda	X	X				X		X		X	X	X	
Golden Lion Tamarin	X							X				X	
Golden-rumped Elephant-shrew	X	X	X	X									
Gorilla		X	X	X								X	
Grevy's Zebra	X											X	
Hawaiian Goose			X			X		X		X			
Indiana Bat			X		X				X				

Table 7. (*continued*)

Species	Agricultural Interests[1]	Forestry Interests	Hunters, Trappers, & Fishers	Development Interests[2]	Program Performance & Organization	Local Rights & Attitudes[3]	Recreational Interests	Human Fear (e.g., Pest Species)[4]	Power & Control	Relations Among Key Actors	Legislation	Political & Economic Instability[5]	Poor Intelligence[6]
Ivory-billed Woodpecker		X			X					X			
Jaguar	X	X	X		X				X	X			X
Kemp's Ridley Sea Turtle			X	X					X	X		X	
Leatherback Turtle			X	X	X	X						X	
Malagasy Freshwater Fishes			X		X						X		
Marbled Murrelet		X	X	X		X							
Mariana Crow				X		X				X			
Mediterranean Monk Seal			X							X		X	X
Mexican Prairie Dog	X				X			X					
Mexican Wolf	X		X			X			X	X			
Mountain Plover	X		X		X	X							
Northern Rocky Mountain Wolf	X		X					X					
Orinoco River Dolphin	X		X	X	X					X		X	
Piping Plover				X		X	X			X			
Po'o-uli						X							X
Red-cockaded Woodpecker		X			X	X			X	X			
Red Wolf			X			X		X		X			
Scarlet Macaw	X				X		X			X			
Snow Leopard	X		X		X					X			
Southern Sea Otter			X	X						X			
Sturgeons and Paddlefishes			X		X					X		X	X
Yuma Clapper Rail	X			X	X			X					
TOTAL NUMBER	26	10	24	15	22	20	4	12	10	21	3	17	6

[1] Agricultural interests include farmers and ranchers.

[2] Development interests means mining, oil and gas extraction, and urban sprawl.

[3] Local rights and attitude refers to conflict surrounding the values and attitudes of local residents, including issues of property rights, local customs, and degree of local support.

[4] Fear reflects the belief that animals harm humans or human property, including pets and livestock (e.g., large carnivores, rodents).

[5] Political and economic instability includes war, unstable governments, faltering economies, and social inequalities that force people to overexploit natural resources.

[6] Poor intelligence indicates a lack of reliable data or inadequate infusion of data into conservation efforts.

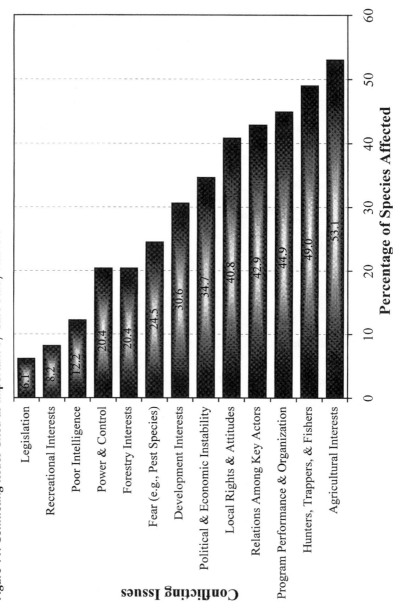

Figure 11. Conflicting Issues Cited as Important by Case Study Authors

Note: Percentages add to over 100% because most programs were affected by more than one conflict.

tant source of conflict (mentioned in 12.2% of cases), and conflicts surrounding legislation and its enforcement (or lack thereof) were even more rare (6.1%).

Finally, there were important socioeconomic considerations. Conflict surrounding local rights and attitudes affected 40.8% of the case studies, and political and economic instability affected 34.7%. Fear, or concern over pests and predators, was also important (24.5%).

It may not be surprising that as the number of threatened and endangered species grows, the amount and intensity of human conflict surrounding recovery efforts increase. Addressing conflict is no easy task, especially when it has been historically unmanaged and opposing stakeholders have become entrenched. Avoiding unproductive conflict in the first place is the best approach (Clark et al. in press); however, as the cases in this book attest, this is no longer an option for many programs. In some cases, formal dispute resolution processes provide a viable approach to reducing conflict, reaching consensus, and improving conservation (see Bacow & Wheeler 1984; Bingham 1986; Crowfoot & Wondolleck 1990). However, not all situations are amenable to such processes. To succeed, formal dispute resolution requires that all parties (1) be willing to work as co-equals, (2) have the power and ability to interact as co-equals, (3) be committed to endorsing and implementing consensus agreements, and (4) share a sense of urgency (Wondolleck et al. 1994). Situations that do not meet these criteria are more difficult to resolve, usually requiring more involved intervention such as the approach presented by Clark (this volume).

IMPROVING CONSERVATION

It is desirable to avoid and resolve unproductive conflicts in conservation programs. The extent of conflict facing the programs presented in this book varied considerably. Some case studies, such as those addressing the sturgeons and paddlefishes, black-footed ferret (*Mustela nigripes*), and Mediterranean monk seal (*Monachus monachus*), have faced a substantial amount of conflict that negatively impacted conservation efforts. In other programs, such as those addressing the golden lion tamarin (*Leontopithecus rosalia*), American burying beetle (*Nicrophorus americanus*), and red wolf (*Canis rufus*), conflicts have been relatively well managed, permitting substantial progress toward recovery. The differences in degree of success provide an opportunity to learn. The more successful programs were characterized by broader, more interdisciplinary approaches. Nevertheless, all programs can improve by incorporating processes that permit constant learning and refinement (i.e., adaptive management) (Hollings 1978; Clark 1996; Clark et al. in press).

Learning will be crucial in the effort to conserve as much of the Earth's biodiversity as possible (Clark 1996). And the reasons for protecting biodiversity are myriad and well documented. They include (1) the direct and indirect benefits to people, such as products (medicine, food, clothing), ecological services (water and air purification, nutrient cycling, pest control, pollination), recreational opportunities (hunting, fishing, bird watching, nature photography), and inspirational value (use in art, as symbols, as a source of awe); as well as (2) moral and ethical considerations, such as the intrinsic rights of species to exist (Wilson 1988; Rolston 1994; Kellert 1996; Reaka-Kudla et al. 1996; Costanza et al. 1997; Pimentel et al. 1997).

If you have taken the time to read this book, you probably already support the conservation of biological diversity and hold values that oppose humanity's role in driving species to extinction. The book provides current and future practitioners with useful information about specific endangered species conservation programs and about the conflicts typically found across programs. It might also be useful in making comparisons across programs and in developing more effective approaches to resolving conflict and improving conservation. Clark's chapter will be particularly helpful in that regard. Although the book addresses single species, it is clear that a common thread unites the causal elements—human values, attitudes, and activities. It is also clear that practitioners must expand their thinking beyond the limited scales of time and space usually encountered in human endeavors, especially the short time frames and small geographic boundaries that dominate agency or political planning. Decisions must be made over regional scales (i.e., across entire landscapes) and in the framework of evolutionary time. Although this increases the complexity of the task, these are probably the only scales that permit the development of conservation programs that will be effective at addressing present trends. The world is still rich with species, habitats, and ecological processes. It is important to act while there is still time.

Selected Conservation Organizations

The best reference for conservation activities is the *Conservation Directory*, which covers U.S. and international organizations. Information on obtaining the book can be secured by writing to the National Wildlife Federation (see address below). What follows here is a list of 25 nongovernmental organizations. It is by no means inclusive.

Africa Wildlife Foundation
P.O. Box 48177
Nairobi
KENYA

American Zoo and Aquarium Association
7970-D Old Georgetown Road
Bethesda, MD 20814
U.S.A.

BirdLife International
Wellbrook Court, Girton Road
Cambridge CB3 0NA
UNITED KINGDOM

Canadian Nature Federation
543 Sussex Drive
Ottawa, Ontario K1N 6Z4
CANADA

Canadian Wildlife Federation
2740 Queensview Drive
Ottawa, Ontario K2B 1A2
CANADA

Center for Marine Conservation
1725 DeSales Street N.W., Suite 600

Washington, DC 20036
U.S.A.

Center for Plant Conservation and Missouri Botanical Garden
P.O. Box 299
St. Louis, MO 63166–0299
U.S.A.

CITES Secretariat
15 Chemin des Anemones
Case Postale 356
1219 Chatelaine
Geneva
SWITZERLAND

Conservation International
2501 M Street N.W., Suite 200
Washington, DC 20037
U.S.A.

Defenders of Wildlife
1244 Nineteenth Street N.W.
Washington, DC 20036
U.S.A.

IUCN World Conservation Union
Avenue de Mont Blanc

CH-1196 Gland
SWITZERLAND

National Audubon Society
700 Broadway
New York, NY 10003
U.S.A.

National Wildlife Federation
1400 Sixteenth Street N.W.
Washington, DC 20036
U.S.A.

The Nature Conservancy
1815 North Lynn Street
Arlington, VA 22209
U.S.A.

Royal Botanical Gardens, Kew
Richmond
Surrey TW9 3AB
UNITED KINGDOM

Sierra Club
85 Second Street, Second Floor
San Francisco, CA 94105-3441
U.S.A.

Smithsonian Institution
Conservation and Research Center of
the National Zoological Park
1500 Remount Road
Front Royal, VA 22630
U.S.A.

**United Nations Environment
Programme (UNEP)**
P.O. Box 30552
Nairobi
KENYA

The Wilderness Society
900 Seventeenth Street N.W.
Washington, DC 20006-2598
U.S.A.

The Wildlands Project
1955 W. Grant Road, Suite 145
Tucson, AZ 85745
U.S.A.

Wildlife Conservation Society
2300 Southern Boulevard
Bronx, NY 10460
U.S.A.

World Resources Institute
1709 New York Avenue N.W.
Washington, DC 20037
U.S.A.

**World Wide Fund for Nature–
International (WWF)**
Avenue du Mont-Blanc
CH–1196, Gland
SWITZERLAND

World Wildlife Fund—Canada
90 Eglinton Avenue East, Suite 504
Toronto, Ontario M4P 2Z7
CANADA

World Wildlife Fund—U.S.
1250 Twenty-fourth Street N.W.
Washington, DC 20006
U.S.A.

Zoological Society of London
Regent's Park
London NW1 4RY
UNITED KINGDOM

Glossary

Adaptation—A structural, physiological, or behavioral trait that aids an organism in surviving in its environment.

Adaptive Management—The process of framing policy (management) decisions as experiments whereby the assumptions and predictions can be empirically tested; results can then be used to improve plans.

Biodiversity—The entire spectrum of life and functioning processes.

Carnivore—A predator that eats animals for its food source; some authors restrict their definition to include only species in the family Carnivora.

Carrying Capacity—The number of individuals of a species that a given area can support, usually through the most unfavorable time of the year.

CITES—Convention on International Trade in Endangered Species of Wild Fauna and Flora; also known as the Washington Convention.

Community—Organisms that inhabit a common environment and interact with each other in some way.

Competition—Interaction among members of the same population or of different populations to secure a mutually necessary resource.

Cryptic—Resembling or blending into habitat or background.

Demography—The study of age structure and distribution, sex ratio, reproductive rates, and mortality rates.

Deterministic—Not random or determined by chance.

Diversity—The number of species in an area and their relative abundance.

Ecology—The study of interactions among organisms and their physical environment.

Ecosystem—All organisms in a community and the nonbiological elements with which they interact.

Ecotone—The transition zone between two different communities; the edge.

Ecotourism—Ecological or nature-based tourism.

Endangered Species—According to the IUCN, a species is Critically Endangered when it has a 50% probability of becoming extinct within 10 years or three generations (whichever is longer). The species is considered Endangered if it has a 20% probability of becoming extinct within 20 years or five generations.

Endemic—A species that is restricted to a particular area or region.

Ephemeral—Lasting only a season or less.

Evolution—Any change in the gene pool of a population; in the Darwinian sense (i.e., the change is a result of natural selection acting on the natural genetic variation found in a population).

Extinction—The process whereby an entire species is eliminated from existence.

Extinction Spasm—A rapid increase in the rate of extinction over a relatively short span of (geologic) time.

Extirpation—The extinction of a species from a local area but not from its entire range.

Family—A taxonomic group of related genera (Family is one category above Genus and one category below Order).

Feral—Formally domestic (i.e., living under conditions of artificial selection), but now living wild (i.e., living under conditions of natural selection).

Forbs—Herbaceous plants other than grass, sedge, or rush; herbaceous plants with broad leaves.

Fragmentation—The process whereby available habitat is reduced in size and divided into units that are smaller and geographically isolated from other patches of similar habitat.

Gene—A unit of heredity that is located in an organism's chromosomes.

Gene Pool—All the types of genes in a population.

Genetic Drift—Random changes in the gene pool of a small population.

Genetic Isolation—Absence of genetic exchange between populations of the same species.

Genus—A taxonomic group of closely related species (Genus is one category above Species and one category below Family).

Habitat—The physical surroundings in which a species is usually found.

Habitat Fragmentation—see **Fragmentation**.

Herbivore—An organism that obtains needed energy by consuming plants.

Hibernacula—A site used by animals to hibernate (i.e., go into winter domancy, as characterized by a greatly reduced metabolic rate).

Hybrid—Offspring of parents that are genetically different or are members of different species.

Inbreeding—The mating of closely related individuals.

Inbreeding Depression—A reduction in fitness and vigor due to inbreeding.

Interdisciplinary—Not dominated by one, or a few, disciplines; instead, using all perspectives and skills of different disciplines in a multifaceted, integrated manner.

IUCN Red List—The International Union for Conservation of Nature and Natural Resources' publication of threatened species that are organized in categories reflecting their level of threat.

K-Selected Life History—A life history pattern shaped by selection under carrying capacity (K) condition (e.g., large, long lived, low reproductive rate, high parental care, etc.).

Keystone Species—A species whose impacts on its community or ecosystem are large, and much larger than would be expected from its abundance.

Landscape—An area with interacting ecosystems that are repeated in similar form throughout the region.

Manifest Destiny—The ostensibly benevolent or necessary policy of imperialistic expansion, and specifically the mid-19th century U.S. doctrine of justified expansion to the Pacific Ocean.

Metapopulation—A set of partially isolated subpopulations that interact to form a larger population.

Natural Selection—Differential reproduction and survival as a result of interactions among populations and their environments.

Niche—An organism's interaction with its physical surroundings and other organisms that share the same area.

Old Growth—The older successional stages of a forest.

Omnivore—An organism that obtains its energy by eating both animals and plants.

Order—A taxonomic group of related families (Order is one category above Family and one category below Class).

Pheromone—A chemical released by animal that influences the behavior of other individuals of the same species.

Philopatric—An affinity for an organism's native area, to which it will usually try to remain or, in the case of migratory species, return.

Population—A group of individuals of the same species that live in the same area.

Population Sink—A habitat type, or region, in which death rates exceed birth rates for a given population or species.

Population Viability—The probability that a population, or species, will persist for a given amount of time.

Precocial—Capable of a high degree of independent activity from birth.

Predator—An animal that kills and eats other animals.

Proximate Cause—The mechanics of how an adaptation arises.

Red Book—*See* IUCN Red List

Refugium—An area of relatively unaltered climate that remains a center of relict populations from which dispersal and speciation occur following strong climatic change.

Relict Population—A persistent population of a species that has been extirpated from the remained of its range.

Rut—Annual state of sexual excitement of male ungulates.

Species—A closely related group of organisms that are reproductively isolated from other groups of organisms.

Species Richness—The number of species in a given area.

Stochastic Events—Random, or chance, events that affect a population.

Succession—The progressive and orderly change in community composition over time.

Synergism—When an interaction between effects is greater than combined, independent (or additive) effects.

Take—As defined by the U.S. Endangered Species Act, means to harass, harm, pursue, hunt, shoot, wound, kill, trap, capture, or collect, or to attempt to engage in any such conduct.

Taxa—Individuals within one taxonomic group.

Taxonomy—The hierarchical classification of organisms.

Territory—An area occupied and defended by individuals or groups.

Threatened Species—According to the U.S. Endangered Species Act, a species with a high risk of becoming endangered.

Ultimate Cause—The evolutionary reason for why an adaptation arises.

Ungulate—A hooved mammal.

Viable Population—A population with sufficient numbers and distribution so as to have a high probability of survival.

Vulnerable Species—According to the IUCN, a species is Vulnerable when there is a 10% chance of it becoming extinct within 100 years.

References

INTRODUCTION

Clark, T.W. 1996. Learning as a strategy for improving endangered species conservation. *Endangered Species Update 13*: 5–6, 22–24.

Clark, T.W., and R.L. Wallace. 1998. Understanding the human factor in endangered species recovery: An introduction to human social process. *Endangered Species Update 15*: 2–9.

Clark, T.W., A.R. Willard, and C.R. Cromley, eds. In press. *Foundations of natural resources policy and management*. New Haven, CT: Yale University Press.

Gilpin, M.E., and M.E. Soulé. 1986. Minimum viable populations: Processes of species extinction. Pp. 19–34 in *Conservation biology: The science of scarcity and diversity*, ed. M.E. Soulé. Sunderland, MA: Sinauer Associates.

Gunderson, L.H., C.S. Holling, and Stephan S. Light. 1995. *Barriers and bridges to the renewal of ecosystems and institutions*. New York: Columbia University Press.

Kellert, S.R. 1996. *The value of life: Biological diversity and human society*. Washington, DC: Island Press.

Lasswell, H.D. 1971. *A pre-view of the policy sciences*. New York: American Elsevier Publishing.

Lasswell, H.D., and M.S. McDougal. 1992. *Jurisprudence for a free society: Studies in law, science, and policy*. New Haven, CT: New Haven University Press.

Murphy, D., D. Wilcove, R. Noss, J. Harte, C. Safina, J. Lubchenco, T. Root, V. Sher, L. Kaufman, M. Bean, and S. Pimm. 1994. On reauthorization of the Endangered Species Act. *Conservation Biology 8*: 1–3.

Myers, N. 1988. Tropical forests and their species: Going, going . . . ? Pp. 28–37 in *Biodiversity*, ed. E.O. Wilson. Washington DC: National Academy Press.

Noss, R.F., M.A. O'Connell, and D.D. Murphy. 1997. *The science of conservation planning*. Covelo, CA: Island Press.

Primack, R.B. 1998. *Essentials of conservation biology*, 2nd ed. Sunderland, MA: Sinauer Associates.

Quammen, D. 1998. Planet of weeds. *Harpers 297*: 57–69.

Rosenzweig, M.L. 1995. *Species diversity in space and time*. New York: Cambridge University Press.

Soulé, M. 1996. The end of evolution. *World Conservation* (originally *IUCN Bulletin*) *1/96*: 24–25.

Soulé, M., and R. Noss. 1998. Rewilding and biodiversity: Complimentary goals for continental conservation. *Wild Earth 8*: 19–28.

Terborgh, J., J. Estes, P. Paquet, K. Ralls, D. Boyd, B. Miller, and R. Noss. 1999. Role of top carnivores in regulating terrestrial ecosystems. Pp. 39–64 in *Continental conservation: Design and management principles for long-term regional conservation networks*, ed. M. Soulé and J. Terborgh. Covelo, CA: Island Press.

U.S. General Accounting Office. 1992. *Endangered Species Act: Type and number of implementing actions*. GAO/RCED-92-131BR. Washington, DC: U.S. General Accounting Office.

Vitousek, P.M., P.R. Ehrlich, A.H. Ehrlich, and P.M. Matson. 1986. Human appropriation of the productivity of photosynthesis. *BioScience 36*: 368–373.

Ward, P.D. 1997. *The call of the distant mammoths*. New York: Copernicus Press.

Wilcove, D.S., M.J. Bean, R. Bonnie, and M. McMillan. 1996. *Rebuilding the ark: Toward a more effective Endangered Species Act for private land*. Washington, DC: Environmental Defense Fund.

Wilcove, D.S., D. Rothstein, J. Dubow, A. Phillips, and E. Losos. 1998. Quantifying threats to imperiled species in the United States. *BioScience 48*: 607–615.

Wilson, E.O., ed. 1988. *Biodiversity*. Washington, DC: National Academy Press.

Wilson, E.O. 1992. *The diversity of life*. Cambridge, MA: Belknap Press of Harvard University Press.

Woodroffe, R., and J.R. Ginsberg. 1998. Edge effects and the extinction of populations inside protected areas. *Science 280*: 2126–2128.

AFRICAN WILD DOG

Alexander, K.A., and M. Appel. 1994. African wild dogs (*Lycaon pictus*) endangered by a canine distemper epizootic among domestic dogs near the Masai Mara National Reserve, Kenya. *Journal of Wildlife Diseases 30*: 481–485.

Alexander, K.A., P.W. Kat, L.A. Munsen, A. Kalake, and M.J.G. Appel. 1996. Canine distemper-related mortality among wild dogs (*Lycaon pictus*) in Chobe National Park, Botswana. *Journal of Zoo and Wildlife Medicine 27*: 426–427.

Atwell, R.I.G. 1959. The African hunting dog: A wildlife management incident. *Oryx 4*: 326–328.

Bere, R.M. 1955. The African wild dog. *Oryx 3*: 180–182.

Burrows, R. 1992. Rabies in wild dogs. *Nature 359*: 277.

Burrows, R. 1995. Demographic changes and social consequences in wild dogs 1964–1992. Pp. 400–420 in *Serengeti II: Research, management and conservation of an ecosystem*, ed. A.R.E. Sinclair and P. Arcese. Chicago: Chicago University Press.

Creel, S., and N.M. Creel. 1995. Communal hunting and pack size in African wild dogs *Lycaon pictus*. *Animal Behaviour 50*: 1325–1339.

Creel, S., and N.M. Creel. 1996. Limitation of African wild dogs by competition with larger carnivores. *Conservation Biology 10*: 1–15.

Fanshawe, J.H., J.R. Ginsberg, C. Sillero-Zubiri, and R. Woodroffe. 1997. The status and distribution of remaining wild dog populations. Pp. 11–57 in *The African wild dog—Status survey and conservation action plan*, ed. R. Woodroffe, J.R. Ginsburg, and D.W. Macdonald. Gland, Switzerland: IUCN.

Frame, L.H., J.R. Malcolm, G.W. Frame, and H. van Lawick. 1979. Social organization of African wild dogs (*Lycaon pictus*) on the Serengeti Plains. *Zeitschrift fur Tierpsychologie 50*: 225–249.

Fuller, T.K., and P.W. Kat. 1990. Movements, activity and prey relations of African wild dogs (*Lycaon pictus*) near Aitong, south-western Kenya. *African Journal of Ecology 28*: 330–350.

Fuller, T.K., P.W. Kat, J.B. Bulger, A.H. Maddock, J.R. Ginsberg, R. Burrows, J.W. McNut, and M.G.L. Mills. 1992. Population dynamics of African wild dogs. Pp. 1125–1139 in *Wildlife 2001: Populations*, ed. D.R. McCullough and H. Barrett. London: Elsevier Science Publishers.

Gascoyne, S.C., M.K. Laurenson, S. Lelo, and M. Borner. 1993. Rabies in African wild dogs (*Lycaon pictus*) in the Serengeti region, Tanzania. *Journal of Wildlife Diseases 29*: 396–402.

Girman, D.J., M.G.L. Mills, E. Geffen, and R.K. Wayne. 1997. A molecular genetic analysis of social structure, dispersal and interpack relationships of the African wild dog (*Lycaon pictus*). *Behavioural Ecology and Sociobiology 40*: 187–198.

Girman, D.J., and R.K. Wayne. 1997. Genetic perspectives on wild dog conservation. Pp. 7–10 in *The African wild dog—Status survey and conservation action plan*, ed. R. Woodroffe, J.R. Ginsburg, and D.W. Macdonald. Gland, Switzerland: IUCN.

Gorman, M.L., M.G.L. Mills, J.P. Raath, and J.R. Speakman. 1998. High hunting costs make African wild dogs vulnerable to kleptoparasitism by hyaenas. *Nature 391*: 479–481.

Kat, P.W., K.A. Alexander, J.S. Smith, and L. Munson. 1995. Rabies and African wild dogs in Kenya. *Proceedings of the Royal Society of London B 262*: 229–233.

Malcolm, J.R., and K. Marten. 1982. Natural selection and the communal rearing of pups in African wild dogs (*Lycaon pictus*). *Behavioural Ecology and Sociobiology 10*: 1–13.

McNutt, J.W. 1996. Sex-biased dispersal in African wild dogs, *Lycaon pictus*. *Animal Behaviour 52*: 1067–1077.

Mills, M.G.L. 1993. Social systems and behaviour of the African wild dog *Lycaon pictus* and the spotted hyaena *Crocuta crocuta* with special reference to rabies. *Onderstepoort Journal of Veterinary Science 60*: 405–409.

Mills, M.G.L., S. Ellis, R. Woodroffe, A. Maddock, P. Stander, A. Pole, G. Rasmussen, P. Fletcher, M. Bruford, D. Wildt, D. Macdonald, and U. Seal, eds. 1998. *Population and habitat viability assessment for the African wild dog* (Lycaon pictus) *in southern Africa*. Final Workshop Report. Apple Valley, MN: IUCN/SSC Conservation Breeding Specialist Group.

Mills, M.G.L., and M.L. Gorman. 1997. Factors affecting the density and distribution of wild dogs *Lycaon pictus* in the Kruger National Park. *Conservation Biology 11*: 1397–1406.

Rasmussen, G.S.A. 1996. Highly endangered painted hunting dogs used as an excuse for stock loss. *The Farmer 66*: 30–31.

Schaller, G.B. 1972. *The Serengeti lion*. Chicago: Chicago University Press.

Scheepers, J.L., and K.A.E. Venzke. 1995. Attempts to reintroduce African wild dogs *Lycaon pictus* into Etosha National Park, Namibia. *South African Journal of Wildlife Research 25*: 138–140.

Stevenson-Hamilton, J. 1947. *Wild life in South Africa*. London: Cassell.

van Heerden, J., M.G.L. Mills, M.J. van Vuuren, P.J. Kelly, and M.J. Dyer. 1995. An investigation into the health status and diseases of wild dogs (*Lycaon pictus*) in the Kruger National Park. *Journal of the South African Veterinary Association 66*: 18–27.

Woodroffe, R., and J.R. Ginsberg. 1997a. Measures for the conservation and management of free-ranging wild dog populations. Pp. 88–99 in *The African wild dog—Status survey and conservation action plan*, ed. R. Woodroffe, J.R. Ginsburg, and D.W. Macdonald. Gland, Switzerland: IUCN.

Woodroffe, R., and J.R. Ginsberg. 1997b. Past and future causes of wild dog's population decline. Pp. 58–74 in *The African wild dog—Status survey and conservation action plan*, ed. R. Woodroffe, J.R. Ginsburg, and D.W. Macdonald. Gland, Switzerland: IUCN.

ALTAI ARGALI SHEEP

Baillie, J., and B. Groombridge. 1996. *1996 IUCN red list of threatened animals*. Gland, Switzerland: IUCN.

Luschekina, A. 1994. The status of argali in Kirgizstan, Tadjikistan, and Mongolia. Unpublished report to the U.S. Fish and Wildlife Service, Washington, DC. 44 pp.

Mallon, D.P., A. Bold, S. Dulamtseren, R.P. Reading, and S. Amgalanbaatar. 1997. Mongolia.

Pp. 193–201 in *Wild sheep and goats and their relatives: Status survey and conservation action plan for Caprinae*, ed. D. Schackleton. Gland, Switzerland: IUCN.

MNE. 1996. *Biodiversity conservation action plan for Mongolia.* The Ministry for Nature and the Environment and United Nations Development Programme's Mongolia Biodiversity Project. Ulaanbaatar, Mongolia: Ministry for Nature and the Environment.

Reading, R.P., S. Amgalanbaatar, H. Mix. 1998. Recent conservation activities for Argali (*Ovis ammon*) in Mongolia, Part 1. *Caprinae* (August): 1–3.

Reading, R.P., S. Amgalanbaatar, and H. Mix. 1999. Recent conservation activities for Argali (*Ovis ammon*) in Mongolia, Part 2. *Caprinae* (January): 1–4.

Reading, R.P., S. Amgalanbaatar, H. Mix, and B. Lhagvasuren. 1997. Argali *Ovis ammon* surveys in Mongolia's South Gobi. *Oryx 31*: 285–294.

Reading, R.P., H. Mix, B. Lhagvasuren, and D. Tseveenmyadag. 1998. The commercial harvest of wildlife in Dornod Aimag, Mongolia. *Journal of Wildlife Management 62*: 59–71.

Schackleton, D., ed. 1997. *Wild sheep and goats and their relatives: Status survey and conservation action plan for Caprinae.* Gland, Switzerland: IUCN.

Schaller, G.B. 1977. *Mountain monarchs: Wild sheep and goats of the Himalaya.* Chicago: University of Chicago Press.

Schaller, G.B. 1998. *Wildlife of the Tibetan steppe.* Chicago: University of Chicago Press.

Shiirevdamba, Ts., O. Shagdarsuren, G. Erdenjav., Ts. Amgalan, and Ts. Tsetsegma, eds. 1997. *Mongolian red book.* Ulaanbaatar, Mongolia: Ministry for Nature and the Environment of Mongolia. (In Mongolian, with English summaries.)

Valdez, R. 1982. *The wild sheep of the world.* Mesa, NM: Wild Sheep and Goat International.

Valdez, R., and M. Frisina. 1993. *Wild sheep surveys in eastern and central Gobi desert and Altai mountains, Mongolia.* Unpublished report to the Mongolia Biodiversity Project, Ulaanbaatar, Mongolia. 9 pp.

AMERICAN BURYING BEETLE

Amaral, M., A. Kozol, and T. French. 1997. Conservation status and reintroduction of the endangered American burying beetle. *Northeastern Naturalist 4*: 121–132.

Anderson, R.S. 1982. On the decreasing abundance of *Nicrophorus americanus* Olivier (Coleoptera: Silphidae) in eastern North America. *Coleopterists Bulletin 36*: 362–365.

Backlund, D.C., and G.M. Marrone. 1997. New records of the endangered American burying beetle, *Nicrophorus americanus* Olivier (Coleoptera: Silphidae) in South Dakota. *Coleopterists Bulletin 51*: 53–58.

Bartlett, J. 1987. Evidence for a sex attractant in burying beetles. *Ecological Entomology 12*: 179–183.

Creighton, J.C., and G.D. Schnell. 1998. Short-term movement patterns of the endangered American burying beetle *Nicrophorus americanus*. *Biological Conservation 81*: 281–287.

Eggert, A-K., and J.K. Muller. 1989. Pheromone-mediated attraction in burying beetles. *Ecological Entomology 14*: 235–237.

Holloway, A.K., and G.D. Schnell. 1997. Relationship between numbers of the endangered American burying beetle *Nicrophorus americanus* Olivier (Coleoptera: Silphidae) and available food resources. *Biological Conservation 81*: 145–152.

Kozol, A.J., M.P. Scott, and J.F.A. Traniello. 1988. The American burying beetle: Studies on the natural history of an endangered species. *Psyche 95*: 167–176.

Lomolino, M.V., and J.C. Creighton. 1996. Habitat selection, breeding success and conservation of the endangered American burying beetle *Nicrophorus americanus*. *Biological Conservation 77*: 235–241.

Lomolino, M.V., J.C. Creighton, G.D. Schnell, and D.L. Certain. 1995. Ecology and conservation of the endangered American burying beetle (*Nicrophorus americanus*). (*Conservation Biology 9*: 605–614.

Muths, E. 1992. Substrate discrimination in burying beetles, *Nicrophorus orbicollis* (Coleoptera: Silphidae). *Journal of the Kansas Entomological Society 64*: 447–450.

Pukowski, E. 1933. Okologische Untersuchungen an Necrophorus F. *Zeitschrift fur Morphologie und Ökologie der Tiere 27*: 518–586.
Ratcliffe, B.C. 1996. The carrion beetles (Coleoptera: Silphidae) of Nebraska. *Bulletin of the University of Nebraska State Museum 13*: 1–100.
Scott, M.P. 1998. The ecology and behavior of burying beetles. *Annual Review Entomology 43*: 595–618.
USFWS. 1991. American burying beetle (*Nicrophorus americana*) recovery plan. Newton Corner, MA: U.S. Fish and Wildlife Service.
Wells, S.M., R.M. Pyle, and N.M. Collins. 1983. *The IUCN red data book*. Gland, Switzerland: IUCN.
Wilson, D.S., and J. Fudge. 1984. Burying beetles: Intraspecific interactions and reproductive success in the field. *Ecological Entomology 9*: 195–203.

ANDEAN CONDOR

Rees, M.D. 1989. Andean condors released in experiment to aid the California condor. *Endangered Species Technical Bulletin 14*: 8–9.
Toone, W.D., and A.C. Risser Jr. 1987. Captive management of the California condor. *International Zoo Yearbook 10*: 15–17.
Wallace, M.P. 1987. *A proposal for the experimental release of Andean condors in California.* Unpublished presentation to the California Condor Recovery Team, Los Angeles Zoo.
Wallace, M.P., and S.A. Temple. 1983. An evaluation of techniques for releasing hand-reared vultures to the wild. Pp. 400–423 in *Vulture biology and management*, ed. S.R. Wilbur and J.A. Jackson. Berkeley: University of California Press.
Wallace, M.P., and S.A. Temple. 1987. Competitive interactions within and between species in a guild of avian scavengers. *The Auk 104*: 290–295.
Wallace, M.P., and S.A. Temple. 1989. Andean condor experimental releases to enhance California condor recovery. *Endangered Species Updated 6*: 1–4.
Zonfrillo, B. 1977. Re-discovery of the Andean condor *Vultur gryphus* in Venezuela. *Bulletin of the British Ornithology Club 97*: 17–18.

ANEGADA IGUANA

Carey, W.M. 1975. The rock iguana, *Cyclura pinguis*, on Anegada, British Virgin Islands, with notes on *Cyclura ricordi* and *Cyclura cornuta* on Hispanola. *Bulletin of the Florida State Museum, Biological Sciences 19*: 189–233.
Geoghegan, T., A.D. Putney, and N.V. Clarke. 1986. *A parks and protected areas system plan for the British Virgin Islands.* Report prepared by the BVI National Parks Trust and the Eastern Caribbean Natural Area Management Program, with assistance from the Ministry of Natural Resources and the Town and Country Planning Office. 90 pp.
Goodyear, N.C. 1992. Flamingos return to Anegada: Status update. *National Parks Trust News, British Virgin Islands* (August): 1.
Goodyear, N.C., and R. DeRavariere. 1993. *Anegada National Park: Revenue through conservation.* Unpublished proposal from the National Parks Trust and the Conservation Agency to Town and Country Planning, Ministry of Natural Resources, British Virgin Islands. 9 pp.
Goodyear, N.C., and J.D. Lazell. 1994. Status of a relocated population of endangered *Iguana pinguis* on Guana Island, British Virgin Islands. *Restoration Ecology 2*: 43–50.
Government of the British Virgin Islands. 1993. *Anegada development plan.* Town and Country Planning Department, Office of the Chief Minister.
Mitchell, N.C. 1999. Effect of introduced ungulates on density, dietary preferences, home range, and physical condition of the iguana (*Cyclura pinguis*) on Anegada. *Herpetologica 55*:7–17.
Mitchell, N.C. In Press. Species accounts: *Cyclura pinguis.* Pp. 60–62 in *West Indian iguanas:*

Status survey and conservation action plan, ed. A. Alberts. Gland, Switzerland: IUCN SSC.

Pregill, G. 1981. Late pleistocene herpetofaunas from Puerto Rico. *University of Kansas Museum of Natural History Miscellaneous Publication 71*: 1–72.

Renwick, J.D.B. 1987. *Report of the Anegada Lands Commission.* Report commissioned by Governor J.M.A. Hurdman. British Virgin Islands. 20 pp. plus 3 appendices.

Schwartz, A., and R.W. Henderson. 1991. *Amphibians and reptiles of the West Indies: Descriptions, distributions, and natural history.* Gainesville: University of Florida Press.

Sites, J.W., S.K. Davis, T. Guerra, J. Iverson, and H.L. Snell. 1996. Character congruence and phylogenetic signal in molecular and morphological data sets: A case study in the living iguanas (Squamata, Iguanidae). *Molecular Biology and Evolution 13*: 1087–1105.

ARUBA ISLAND RATTLESNAKE

Arikok, Parke Nacional. 1998. Web site. www.arubanationalparks.com.

Campbell, J.A., and W.W. Lamar. 1989. *The venomous snakes of Latin America.* Ithaca, NY: Comstock Publishing.

Carpenter, C.C. 1980. An ethological approach to reproductive success in reptiles. Pp. 33–48 in *Reproductive biology and diseases of captive reptiles*, ed. J.B. Murphy and J.T. Collins. Oxford, OH: Society for the Study of Amphibians and Reptiles.

Chiszar, D., J.B. Murphy, and H.M. Smith. 1993. In search of zoo-academic collaborations: A research agenda for the 1990's. *Herpetologica 49*: 488–500.

Chiszar, D., B. O'Connell, R. Greenlee, B. Demeter, T. Walsh, J. Chiszar, K. Moran, and H.M. Smith. 1985. Duration of strike-induced chemosensory searching in long-term captive rattlesnakes at National Zoo, Audubon Zoo, and San Diego Zoo. *Zoo Biology 4*: 291–294.

Chiszar, D., H.M. Smith, and C.C. Carpenter. 1994. An ethological approach to reproductive success in reptiles. Pp. 147–173 in *Captive management and conservation of amphibians and reptiles*, ed. J.B. Murphy, K. Adler, and J.T. Collins. Oxford, OH: Society for the Study of Amphibians and Reptiles.

Chiszar, D., H.M. Smith, and C.W. Radcliffe. 1993. Zoo and laboratory experiments on the behavior of snakes: Assessments of competence in captive-raised animals. *American Zoologist 33*: 109–116.

Crews, D., ed. 1987. *Psychobiology of reproductive behavior: An evolutionary perspective.* Englewood Cliffs, NJ: Prentice-Hall.

Dodd, C.K., Jr., and R.A. Seigel. 1991. Relocation, repatriation and translocation of amphibians and reptiles: Are they conservation strategies that work? *Herpetologica 47*: 336–350.

Fitch, H.S. 1970. Reproductive cycles of lizards and snakes. *University of Kansas Museum of Natural History Miscellaneous Publication 52*: 1–247.

Goode, M., C.W. Radcliffe, K. Estep, A. Odum, and D. Chiszar. 1990. Field observations on feeding behavior in an Aruba Island rattlesnake (*Crotalus durissus unicolor*): Strike-induced chemosensory searching and trail following. *Bulletin of the Psychonomic Society 28*: 312–314.

IUCN. 1996. *1996 IUCN red list of threatened animals*, eds. J. Baillie and B. Groombridge. Gland, Switzerland, and Cambridge, United Kingdom: IUCN.

Klauber, L.M. 1972. *Rattlesnakes: Their habits, life history, and influence on mankind.* Berkeley and Los Angeles: University of California Press.

O'Connell, B., R. Greenlee, J. Bacon, and D. Chiszar. 1982. Strike-induced chemosensory searching in old-world vipers and new-world pit vipers at San Diego Zoo. *Zoo Biology 1*: 287–294.

Odum, R.A. 1996. *Aruba Island rattlesnake-species survival plan master plan.* Toledo, OH: Toledo Zoological Society.

Reinert, H.K. 1991. Translocations as a conservation strategy for amphibians and reptiles. Some comments, concerns and observations. *Herpetologica 47*: 357–363.

Reinert, H.K., L.M. Bushar, and R.A. Odum. 1995. Recommendations for the conservation and management of the Aruba Island rattlesnake on Aruba. Section 2, pp.1–5 in *Aruba Island rattlesnake conservation action plan*, ed. R.A. Odum. Toledo, OH: Toledo Zoological Society.

Wilson, E.O. 1998. *Consilience: The unity of knowledge*. New York: Alfred A. Knopf.

ASIAN ELEPHANT

Baskaran, N., M. Balasubramanian, S. Swaminathan, and A.A. Desai. 1995. Home range of elephants in the Nilgiri Biosphere Reserve, South India. Pp. 296–313 in *A week with elephants*, ed. J.C. Daniel and H. Datye. Bombay: Bombay Natural History Society, and New Delhi: Oxford University Press.

Blair, J.A.S., G.L. Boon, and N.M. Noor. 1979. Conservation or cultivation: The confrontation between the Asian elephant and land development in peninsular Malaysia. *Land Development Digest 2*: 25–58.

de Silva, M. 1998. Status and conservation of the elephant (*Elephas maximus*) and the alleviation of man-elephant conflict in Sri Lanka. *Gajah 19*: 1–68.

Eisenberg, J.F., G.M. McKay, and M.R. Jainudeen. 1971. Reproductive behaviour of the Asiatic elephant (*Elephas maximus maximus* L.). *Behaviour 38*: 193–225.

McKay, G.M. 1973. The ecology and behavior of the Asiatic elephant in southeastern Ceylon. *Smithsonian Contributions to Zoology 125*: 1–113.

Nath, C.D., and R. Sukumar. 1998. *Elephant-human conflict in Kodagu, southern India: Distribution patterns, people's perceptions and mitigation methods*. Bangalore, India: Asian Elephant Research & Conservation Centre, Indian Institute of Science.

Owen-Smith, R.N. 1988. *Megaherbivores: The influence of very large body size on ecology*. Cambridge, U.K.: Cambridge University Press.

Rasmussen, L.E.L. 1997. Chemical communication: An integral part of functional Asian elephant (*Elephas maximus*) society. *Ecoscience 5*: 411–429.

Santiapillai, C., and P. Jackson. 1990. *The Asian elephant: An action plan for its conservation*. Gland, Switzerland: IUCN/SSC Asian Elephant Specialist Group, IUCN.

Santiapillai, C., and W.S. Ramono. 1990. Sumatran elephant database. *IUCN/SSC Asian Elephant Specialist Group Newsletter 5*: 3–35.

Shoshani, J., and J.F. Eisenberg. 1982. *Elephas maximus*. *Mammalian Species 182*: 1–8.

Sukumar, R. 1989. *The Asian elephant: Ecology and management*. Cambridge, U.K.: Cambridge University Press.

Sukumar, R. 1991. The management of large mammals in relation to male strategies and conflict with people. *Biological Conservation 55*: 93–102.

Sukumar, R. 1994. *Elephant days and nights: Ten years with the Indian elephant*. New Delhi, India: Oxford University Press.

Sukumar, R. 1995. Elephant raiders and rogues. *Natural History 104*: 52–60.

Sukumar, R., and M. Gadgil. 1988. Male-female differences in foraging on crops by Asian elephants. *Animal Behaviour 36*: 1233–1235.

Sukumar, R., and C. Santiapillai. 1996. *Elephas maximus*: Status and distribution. Pp. 327–331 in *The Proboscidea: Evolution and palaeoecology of elephants and their relatives*, ed. J. Shoshani and P. Tassy. Oxford: Oxford University Press.

Williams, A.C., and A.J.T. Johnsingh. 1997. *A status survey of elephants* (Elephas maximus), *their habitats and an assessment of the elephant-human conflict in Garo Hills, Meghalaya*. Dehra Dun, India: Wildlife Institute of India.

ASIATIC LION

Joslin, P. 1973. The Asiatic lion: A study of ecology and behaviour. Ph.D. thesis, University of Edinburgh.

Ravi Chellam. 1993. Ecology of the Asiatic lion, *Panthera leo persica*. Ph.D. thesis, Saurashtra University, Rajkot.

Ravi Chellam and A.J.T. Johnsingh. 1993. Management of Asiatic lions in the Gir forest, India. *Symposia of the Zoological Society of London 65*: 409–424.

Ravi Chellam, J. Joshua, C.A. Williams, and A.J.T. Johnsingh. 1995. *Survey of the potential sites for the re-introduction of Asiatic lion.* Project report submitted to the Wildlife Institute of India, P.O. Box 18, Dehra Dun, 248 001.

Saberwal, V.K., J. Gibbs, Ravi Chellam, and A.J.T. Johnsingh. 1994. Lion-human conflict in the Gir Forest, India. *Conservation Biology 8*: 501–507.

AYE-AYE

Adriamampianina, J. 1984. Nature reserves and nature conservation in Madagascar. In *Key environments: Madagascar*, ed. A. Jolly, P. Oberle, and R. Albignac. Oxford: Pergamon Press.

Carroll, J.B., and D.M. Haring. 1994. Maintenance and breeding of aye-ayes (*Daubentonia madagascariensis*) in captivity: A review. *Folia Primatologica 62*: 54–62.

Curtis, D.J., and A.T.C. Feistner. 1994. Positional behaviour in captive aye-ayes (*Daubentonia madagascariensis*). *Folia Primatologica 62*: 155–159.

Dubois, C., and M.K. Izard. 1990. Social and sexual behaviors in captive aye-ayes (*Daubentonia madagascarienses*). *Journal of Psychology and the Behavioral Sciences 5*: 1–10.

Erickson, C.J., S. Nowicki, L. Dollar, and N. Goehring. 1998. Percussive foraging: Stimuli for prey location by aye-ayes (*Daubentonia madagascariensis*). *International Journal of Primatology 19*: 111–122.

Feistner, A.T.C., and C. Ashbourne. 1994. Infant development in a captive-bred aye-aye (*Daubentonia madagascariensis*) over the first year of life. *Folia Primatologica 62*: 74–92.

Feistner, A.T.C., and J.B. Carroll. 1993. Breeding aye-ayes: An aid to preserving biodiversity. *Biodiversity and Conservation 2*: 283–289.

Feistner, A.T.C., J.B. Carroll, and J.C. Beattie. 1996. *International studybook for the aye-aye* Daubentonia madagascariensis, *Number One, 1986–1995.* Jersey: Jersey Wildlife Preservation Trust.

Feistner, A.T.C., E.C. Price, and G. Milliken. 1994. Preliminary observations on hand preference for tapping, digit-feeding and food-holding in captive aye-ayes (*Daubentonia madagascariensis*). *Folia Primatologica 62*: 136–141.

Feistner, A.T.C., and E.J. Sterling, eds. 1994. The aye-aye: Madagascar's most puzzling primate. *Folia Primatologica 62*: 1–3.

Feistner, A.T.C., and E.J. Sterling. 1995. Body mass and sexual dimorphism in the aye-aye. *Dodo, Journal of the Wildlife Preservation Trust 31*: 73–76.

Grandidier, G. 1908. Les moeurs de l'aye-aye. *La Nature 1812*: 161–162.

Green, G.M., and R.W. Sussman. 1990. Deforestation history of the eastern rain forests of Madagascar from satellite images. *Science 248*: 212–215.

Lamberton, C. 1910. Contributions à l'étude des moeurs du aye-aye. *Bulletin de l'Academie Malgache 8*: 129–140.

Lamberton, C. 1934. Contribution à la connaissance de la faune subfossile de Madagascar. *Mémoires de l'Academie Malgache 17*: 40–46.

MacPhee, R.D.E., and E.M. Raholimavo. 1988. Modified subfossil aye-aye incisors from southwestern Madagascar: Species allocation and paleoecological significance. *Folia Primatologica 51*: 126–142.

Milliken, G.W., J.P. Ward, and C.J. Erickson. 1991. Independent digit control in foraging by the aye-aye (*Daubentonia madagascariensis*). *Folia Primatologica 56*: 219–224.

Mittermeier, R.A., I. Tattersall, W.R. Konstant, D.M. Meyers, and R.B. Mast. 1994. *Lemurs of Madagascar.* Washington, DC: Conservation International.

Petter, J.J. 1977. The aye-aye. Pp. 37–57 in *Primate conservation*, ed. Prince Rainier III and G.H. Bourne. New York: Academic Press.

Petter, J.J., and A. Peyrieras. 1970. Nouvelle contribution à l'étude d'un lémurien malgache, le aye-aye (*Daubentonia madagascariensis* E. Geoffroy). *Mammalia 34*: 167–193.

Price, E.C., and A.T.C. Feistner. 1994. Responses of captive aye-ayes (*Daubentonia madagascariensis*) to the scent of conspecifics: A preliminary investigation. *Folia Primatologica 62*: 170–174.

Richard, A.F., S.J. Goldstein, and R.E. Dewar. 1989. Weed macaques: The evolutionary implications of macaque feeding ecology. *International Journal of Primatology 10*: 569–594.

Simons, E.L. 1994. The giant aye-aye, *Daubentonia robusta*. *Folia Primatologica 62*: 14–21.

Sterling, E.J. 1993. *The behavioral ecology of the aye-aye on Nosy Mangabe, Madagascar*. Ph.D. thesis, Yale University.

Sterling, E.J. 1994a. Aye-ayes: Specialists on structurally defended resources. *Folia Primatologica 62*: 142–154.

Sterling, E.J. 1994b. Evidence for non-seasonal reproduction in aye-ayes (*Daubentonia madagascariensis*) in the wild. *Folia Primatologica 62*: 46–53.

Sterling, E.J., E.S. Dierenfeld, C.J. Ashbourne, and A.T.C. Feistner. 1994. Dietary intake, food composition and nutrient intake in wild and captive populations of *Daubentonia madagascariensis: Folia Primatologica 62*: 115–124.

Winn, R.M. 1989. The aye-ayes, *Daubentonia madagascariensis*, at the Paris Zoological Gardens: Maintenance and preliminary behavioural observations. *Folia Primatologica 52*: 109–123.

Winn, R.M. 1994a. Development of behaviour in a young aye-aye (*Daubentonia madagascariensis*) in captivity. *Folia Primatologica 62*: 93–107.

Winn, R.M. 1994b. Preliminary study of the sexual behaviour of three aye-ayes (*Daubentonia madagascarienses*) in captivity. *Folia Primatologica 62*: 63–73.

BASKING SHARK

Anderson, E.D. 1990. Fishery models as applied to elasmobranch fisheries. Pp. 473–484 in *Elasmobranchs as living resources: Advances in the biology, ecology, systematics, and the status of fisheries*, ed. H.L. Pratt, S.H. Gruber, and T. Taniuchi. U.S. Department of Commerce, NOAA Technical Report NMFS 90. Washington, D.C., U.S. Department of Commerce, NOAA.

Anonymous. 1999. Proposal to include the basking shark (*Cetorhinus maximus*) on Appendix II of the Convention on International Trade in Endangered Species (CITES). Submitted by the U.K. government. Global Wildlife Division, Department of the Environment, Transport and Regions, Bristol, U.K. November 1999.

Anonymous. 1993. *Fishery management plan for sharks of the Atlantic Ocean*. National Marine Fisheries Service, National Oceanic and Atmospheric Administration. Washington, DC: U.S. Department of Commerce.

Bonfil, R. 1994. *Overview of world elasmobranch fisheries*. FAO Fisheries Technical Paper No. 341. Rome, Italy: FAO.

Camhi, M., S.L. Fowler, J. Musick, A. Brautigam, and S. Fordham. 1998. The implications of biology for the conservation and management of sharks and their relatives. IUCN Occasional Paper No. 20. Gland, Switzerland: IUCN/SSC Shark Specialist Group.

Clemens, W.A., and G.V. Wilby. 1961. *Fishes of the Pacific coast of Canada*, 2nd ed. Bulletin of the Fisheries Research Board of Canada, No. 86.

Compagno, L.J.V. 1984. *Sharks of the world. Hexanchiformes to Lamniformes*. FAO Fisheries Synopsis No. 124, Vol. 4, Pt.1. Rome, Italy: UNDP and FAO.

Darling, J.D., and K.E. Keogh. 1994. Observations of basking sharks *Cetorhinus maximus* in Clayoquot Sound, British Columbia. *Canadian Field Naturalist 108*: 199–210.

Earll, R.C. 1990. The basking shark: Its fishery and conservation. *British Wildlife 1* (3): 121–129.

Fairfax, D. 1998. *The basking shark in Scotland.* East Linton, Scotland: Tuckwell Press.

Fleming, E.H., and P. Papageorgiou. 1996. European regional overview of elasmobranch fisheries and trade in selected Atlantic and Mediterranean countries. Brussels, Belgium: TRAFFIC Europe.

Gauld, J.A. 1989. *Records of Porbeagles landed in Scotland, with observations on the biology, distribution and exploitation of the species.* Scottish Fisheries Research Report No. 45. Aberdeen, Scotland: Department of Agriculture and Fisheries for Scotland.

Kunzlik, P.A. 1988. *The basking shark.* Scottish Fisheries Information Pamphlet No. 14. Aberdeen, Scotland: Department of Agriculture and Fisheries for Scotland.

Lien, J., and L. Aldrich. 1982. *The basking shark* (Cetorhinus maximus) *in Newfoundland.* Report to the Department of Fisheries, Government of Newfoundland and Labrador, St. Johns, Canada.

Lien, J., and L. Fawcett. 1986. Distribution of basking sharks *Cetorhinus maximus* incidentally caught in inshore fishing gear in Newfoundland. *Canadian Field Naturalist 100*: 246–252.

Lum, M. 1996. Every mouthful of shark's fin in high demand. *Singapore Sunday Times* (Straits Times), May 19, 1996.

Matthews, L.H. 1950. Reproduction in the basking shark. *Philosophical Transactions of the Royal Society of London, Series B 234*: 247–316.

McNally, K. 1976. *The sun-fish hunt.* Belfast, Ireland: Blackstaff Press.

Olsen, A.M. 1959. The status of the school shark fishery in south-eastern waters. *Australian Journal of Marine and Freshwater Research 10*: 150–176.

Owen, R.E. 1984. *Distribution and ecology of the basking shark* Cetorhinus maximus *(Gunnerus 1765).* Master's thesis, University of Rhode Island, Providence.

Parker, H.W., and M. Boeseman. 1954. The basking shark *Cetorhinus maximus* in winter. *Proceedings of the Zoological Society of London 124*: 185–194.

Pauly, D. In press. Growth and mortality of the basking shark *Cetorhinus maximus* and their implications for management of whale sharks *Rhincodon typus.* In *Proceedings of International Seminar and Workshop on Elasmobranch Biodiversity, Conservation and Management* (Sabah, July 1997), ed. S.L. Fowler. Gland, Switzerland: IUCN/SSC Shark Specialist Group.

Phillips, J.B. 1947. Basking shark fishery revived in California. California Department of Fish and Game, *Fisheries Bulletin 61*: 11–23.

Roedel, P.M., and W.M.E. Ripley. 1950. California sharks and rays. California Department of Fish and Game, *Fisheries Bulletin 64*: 7–37.

Rose, D. 1996. *An overview of world trade in sharks and other cartilaginous fishes.* Cambridge, UK: TRAFFIC International.

Scott, G., P.J. Phares, and B. Slater. 1996. *Recreational catch, average size and effort information for sharks in US Atlantic and Gulf of Mexico waters.* 1996 NMFS Stock Evaluation Workshop document, SB-111–5. Miami, FL: NMFS Southeast Fisheries Scientific Center.

Sund, O. 1943. Et brugdelbarsel. *Naturen 67*: 285–286.

Walker, T.I. 1993. Conserving the shark stocks of Southern Australia. Pp. 33–40 in *Shark conservation*, ed. J. Pepperell, J. West, and P. Woon. Sydney, NSW, Australia: Zoological Parks Board of NSW.

Watkins, A. 1958. *The sea my hunting ground.* London: Heinemann.

BLACK-FOOTED FERRET

Anderson, E., S.C. Forrest, T.W. Clark, and L. Richardson. 1986. Paleobiology, biogeography, and systematics of the black-footed ferret, *Mustela nigripes* (Audubon and Bachman, 1851). *Great Basin Naturalist Memoirs 8*: 11–62.

Biggins, D.E., B.J. Miller, T.W. Clark, and R.P. Reading. 1997. Management of an endangered species: The black-footed ferret. Pp. 420–426 in *Principles of conservation biology*, 2nd ed., ed. G.K. Meffe, C.R. Carroll, and contributors. Sunderland, MA: Sinauer Associates.

Clark, T.W. 1989. *Conservation biology of the black-footed ferret*, Mustela nigripes. Wildlife Preservation Trust Special Scientific Report #3. Philadelphia: Wildlife Preservation Trust International.

Clark, T.W. 1997. *Averting extinction: Reconstructing endangered species recovery.* New Haven: Yale University Press.

Forrest, S.C., D.E. Biggins, L. Richardson, T.W. Clark, T.M. Campbell, K.A. Fagerstone, and E.T. Thorne. 1988. Black-footed ferret (*Mustela nigripes*) attributes at Meeteetse, Wyoming, 1981–1985. *Journal of Mammalogy 69*: 261–273.

Miller, B., R.P. Reading, and S. Forrest. 1996. *Prairie night: Black-footed ferrets and the recovery of endangered species.* Washington, DC: Smithsonian Institution Press.

Reading, R.P. 1993. Toward an endangered species reintroduction paradigm: A case study of the black-footed ferret. Ph.D. thesis, Yale University, New Haven, CT.

Reading, R.P., T.W. Clark, A. Vargas, L.R. Hanebury, B.J. Miller, D.E. Biggins, and P.E. Marinari. 1997. Black-footed ferret (*Mustela nigripes*): Conservation update. *Small Carnivore Conservation 17*: 1–6.

Reading, R.P., and S.R. Kellert. 1993. Attitudes toward a proposed reintroduction of black-footed ferrets (*Mustela nigripes*). *Conservation Biology 7*: 569–580.

Wilcox, B.A., and D.D. Murphy. 1985. Conservation strategy: The effects of fragmentation on extinction. *American Naturalist 125*: 879–887.

BOREAL TOAD

Arnold, S.J., and R.J. Wassersug. 1978. Differential predation on metamorphic anurans by gartersnakes (*Thanophis*): Social behavior as a possible defense. *Ecology 59*: 1014–1022.

Beiswenger, R.E. 1981. Predation by gray jays on aggregating tadpoles of the boreal toad (*Bufo boreas*). *Copeia 1981*: 459–460.

Blaustein, A.R., P.D. Hoffman, D.G. Hokit, J.M. Kiesecker, S.C. Walls, and J.B. Hays. 1994. UV repair and resistance to solar UV-B in amphibian eggs: A link to population declines. *Proceedings of the National Academy of Sciences 91*: 1791–1795.

Burke, R.L. 1991. Relocations, repatriations, and translocations of amphibians and reptiles: Taking a broader view. *Herpetologica 47*: 350–357.

Campbell, J.B. 1970a. Hibernacula of a population of *Bufo boreas boreas* in the Colorado Front Range. *Herpetologica 25*: 278–282.

Campbell, J.B. 1970b. Life history of *Bufo boreas boreas* in the Colorado Front Range. Ph.D. thesis, University of Colorado, Boulder.

Carey, C. 1993. Hypothesis concerning the causes of the disappearance of boreal toads from the mountains of Colorado. *Conservation Biology 7*: 355–362.

Corn, P.S. 1993. *Bufo boreas* (boreal toad) predation. *Herpetological Review 24* (2): 57.

Corn, P.S. 1994. What we know and don't know about amphibian declines in the west. *U.S. Forest Service Rocky Mountain General Technical Report RM-247*: 59–67.

Corn, P.S. 1998. Effects of ultraviolet radiation on boreal toads in Colorado. *Ecological Applications 8*: 18–26.

Corn, P.S., M.L. Jennings, and E. Muths. 1997. Survey and assessment of amphibian populations in Rocky Mountain National Park. *Northwestern Naturalist 78*: 34–55.

Corn, P.S., and F.A. Vertucci. 1992. Descriptive risk assessment of the effects of acidic deposition on Rocky Mountain amphibians. *Journal of Herpetology 26*: 361–369.

Dodd, C.K., Jr., and R.A. Seigel. 1991. Relocation, repatriation and translocation of amphibians and reptiles: Are they conservation strategies that work? *Herpetologica 47*: 336–350.

Goebel, A.M. 1996. Systematics and conservation of bufonids in North America and in the *Bufo boreas* species group. Ph.D. thesis, University of Colorado, Boulder.

Goebel, A.M. 1998. Molecular genetic analyses of the endangered boreal toad in Colorado and southeast Wyoming. Pp. 147–171 in *CDOW boreal toad research progress report 1995–1997*, ed. M.S. Jones, J.P. Goettl, K.L. Scherff-Norris, S. Brinkman, L.J. Livo, and A.M. Goebel. Denver: Colorado Division of Wildlife.

Griffith, B., J.M. Scott, J.W. Carpenter, and C. Reed. 1989. Translocation as a species conservation tool: Status and strategy. *Science 245*: 477–480.

Hammerson, G.A. 1992. *Field surveys of amphibians in the mountains of Colorado, 1991.* Report funded by the USFWS, USFS, CDOW, and the Colorado Office of the Nature Conservancy. Denver: Colorado Division of Wildlife.

Jones, M. 1997. Studies of boreal toads in the Henderson Mine area. Pp. 25–26 in *Report on the status and conservation of the boreal toad in the Southern Rocky Mountains*, ed. C. Loeffler. Denver: Colorado Division of Wildlife.

Jones, M.S., and J.P. Goettl. 1998. Henderson/Urad boreal toad studies. Pp. 21–82 in *CDOW boreal toad research progress report 1995–1997*, ed. M.S. Jones, J.P. Goettl, K.L. Scherff-Norris, S. Brinkman, L.J. Livo, and A.M. Goebel. Denver: Colorado Division of Wildlife.

Kiesecker, J.M., and A.R. Blaustein. 1995. Synergism between UV-B radiation and a pathogen magnifies amphibian embryo mortality in nature. *Proceedings of the National Academy for the Sciences 92*: 11049–11052.

Livo, L.J. 1998. Predators of larval *Bufo boreas. Colorado-Wyoming Academy of Science 38*(1): 32.

Livo, L.J., and C. Fetkavich. 1998. Late-season boreal toad tadpoles. *Northwestern Naturalist 79*: 120–121.

Loeffler, C. 1998. Conservation plan and agreement for the management and recovery of the southern Rocky Mountain population of the boreal toad (*Bufo boreas boreas*). Denver: State of Colorado Department of Natural Resources, Colorado Division of Wildlife. (Unpublished report).

Olson, D.H. 1989. Predation on breeding western toads (*Bufo boreas*). *Copeia 1989*: 391–397.

Reinert, H.K. 1991. Translocation as a conservation strategy for amphibians and reptiles: Some comments, concerns, and observations. *Herpetologica 47*: 357–363.

Stebbins, R.C. 1954. *Amphibians and reptiles of western North America.* New York: McGraw-Hill.

Stebbins, R.C. 1985. *A field guide to western reptiles and amphibians*, 2nd ed., rev. Boston: Houghton Mifflin.

Stuart, J.N., and C.W. Painter. 1994. A review of the distribution and status of the boreal toad, *Bufo boreas boreas* in New Mexico. *Bulletin of the Chicago Herpetological Society 29*: 113–116.

Thomas, B.W., and T.H. Whitaker. 1994. Translocation of the Fiordland skink *Leiolopisma acrinasum* to Hawea island, Breaksea Sound, Fiordland, New Zealand. In *Reintroduction biology of Australian and New Zealand fauna*, ed. M. Serena. Chipping Norton, New South Wales, Australia: Surrey Beatty and Sons Party Limited.

USFWS. 1995. Endangered and threatened wildlife and plants: 12 month finding for a petition to list the southern Rocky Mountain population of the boreal toad as endangered. *Federal Register 60*: 15282–15283.

Vertucci, F.A., and P.S. Corn. 1996. Evaluation of episodic acidification and amphibian declines in the Rocky Mountains. *Ecological Applications 6*: 449–457.

BROWN BEAR

Brown, D.E. 1985. *The Grizzly in the Southwest.* Norman: University of Oklahoma Press.

Chestin, I.E., Y.P. Gubar, V.E. Sokolov, and V.S. Lobachev. 1992. The brown bear (*Ursus arctos* L.) in the USSR: Numbers, hunting and systematics. *Acta Zologica Fennici 29*: 57–68.

Clark, T.W., and D. Casey. 1992. *Tales of the Grizzly.* Moose, WY: Homestead Publishing.

Couturier, M. 1954. *L'ours brun.* Grenoble, France: 904 pp. (In French).

Craighead, J.J., and J.A. Mitchell. 1982. Grizzly bear. Pp. 515–556 in *Wild mammals of North America—biology, management, economics*, ed. J.A. Chapman and G.A. Feldhamer. Baltimore: Johns Hopkins University Press.

Elgmork, K. 1996. The brown bear (*Ursus arctos* L.) in Norway: Assessment of status around 1990. *Biological Conservation 78*: 223–237.

Highley, K., and S.C. Highley. 1994. *Bear farming and trade in China and Taiwan*. Washington, DC: Humane Society of the United States and Humane Society International.

Kellert, S.R., M. Black, C.R. Rush, and A.J. Bath. 1996. Human culture and large carnivore conservation in North America. *Conservation Biology 10*: 977–990.

Kohn, M., F. Knauer, A. Stoffella, W. Schröder, and S. Pääbo. 1995. Conservation genetics of the European brown bear—a study using excremental PCR of nuclear and mitochondrial sequences. *Molecular Ecology 4*: 95–103.

Mattson, D.J. 1990. Human impacts on bear habitat use. *International Conference on Bear Research and Management 8*: 33–56.

Mattson, D.J. 1998. Diet and morphology of extant and recently extinct northern bears. *International Conference on Bear Research and Management 10*: 479–496.

Mattson, D.J., and J.J. Craighead. 1994. The Yellowstone grizzly bear recovery program. Pp. 101–129 in *Endangered species recovery: Finding the lessons, improving the process*, ed. T.W. Clark, R.P. Reading, and A.L. Clarke. Washington, DC: Island Press.

Mattson, D.J., S. Herrero, R.G. Wright, and C.M. Pease. 1996. Science and management of Rocky Mountain grizzly bears. *Conservation Biology 10*: 1013–1025.

Mattson, D.J., R.G. Wright, K.C. Kendall, and C.J. Martinka. 1995. Grizzly bears. Pp. 103–105 in *Our living resources*, ed. E.T. LaRoe, G.S. Farris, C.E. Puckett, P.D. Doran, and M.J. Mac. Washington, DC: U.S. Department of the Interior, National Biological Service.

Miller, S.D., G.C. White, S.A. Sellers, H.V. Reynolds, J.W. Schoen, K. Titus, V.G. Barnes Jr., R.B. Smith, R.R. Nelson, W.B. Ballard, and C.C. Schwartz. 1997. Brown and black bear density estimation in Alaska using radiotelemetry and replicated mark-resight techniques. *Wildlife Monographs 133*: 1–55.

Nelson, R.A., G.E. Flok Jr., E.W. Pfeiffer, J.J. Craighead, C.J. Jonkel, and D.L. Steiger. 1983. Behavior, biochemistry, and hibernation in black, grizzly and polar bears. *International Conference on Bear Research and Management 5*: 284–290.

Parde, J.M. 1997. The brown bear in the central Pyrenees: Its decline and present situation. *International Conference on Bear Research and Management 9*: 45–52.

Primm, S.A. 1992. Grizzly conservation in Greater Yellowstone. Master's thesis, University of Colorado, Boulder.

Primm, S.A. 1996. A pragmatic approach to grizzly bear conservation. *Conservation Biology 10*: 1026–1035.

Servheen, C. 1990. The status and conservation of the bears of the world. *International Conference on Bear Research and Management, Monograph Series 2*: 1–32.

Waits, L.P., S.L. Talbot, R.H. Ward, and G.F. Shields. 1998. Mitochondrial DNA phylogeography of the North American brown bear and implications for conservation. *Conservation Biology 12*: 408–417.

CHINESE ALLIGATOR

Anonymous. 1992. *Registration of the Anhui Research Centre of Chinese Alligator Reproduction for* Alligator sinensis. Unpublished proposal to CITES.

Behler, J. 1977. A propagation program for Chinese alligators (*Alligator sinensis*) in captivity. *Herpetological Review 84*: 124–125.

Behler, J. 1993. Chinese alligator (*Alligator sinensis*). *AAZPA Annual Report on Conservation and Science 1992–1993*: 227–229.

Brazaitis, P. 1973. The identification of living crocodilians. *Zoologica 58*: 59–101.

Chen, B. 1985. *Chinese alligator*. Hefei, Anhui, China: Anhui Science and Technology Press.

Chen, B. 1991. Chinese alligator. Pp. 361–365 in *The amphibian and reptilian fauna of Anhui*, ed. B. Chen. Hefei, Anhui, China: Anhui Publishing House of Science and Technology.

Chen, B., and B. Li. 1979. Initial observations on ecology of the Chinese alligator. *Journal of Anhui Teacher's College 1*: 69–73.

Chen, B.C. 1990. The past and present situation of the Chinese alligator. *Asiatic Herpetological Research 3*: 129–136.

Chu-cheng, K. 1957. Observations on the life history of the Chinese alligator. *Acta Zoologica Sinica 92*: 129–143.

Hsiao, S.D. 1935. Natural history notes on the Yangtze alligator. *Peking Natural History Bulletin 9*: 283–293.

Huang, C. 1982. The ecology of the Chinese alligator and changes in its geographical distribution. Pp. 54–62 in *Crocodiles. Proceedings of the 5th Working Meeting of the IUCN/SSC Crocodile Specialist Group*. Gland, Switzerland: IUCN-World Conservation Union.

Lang, J.W. and H.V. Andrews. 1994. Temperature-dependent sex determination in crocodilians. *Journal of Experimental Zoology 270*: 28–45.

Wan, Z., C. Gu, X. Wang, and C. Wang. 1999. Conservation, management and farming of crocodiles in China. Pp. 80–100 in *Crocodiles. Proceedings of the 14th Working Meeting of the Crocodile Specialist Group*. Gland, Switzerland: IUCN-World Conservation Union.

Watanabe, M.E. 1982. The Chinese alligator: Is farming the last hope? *Oryx 17*: 176–181.

Webb, G.J.W., and B. Vernon. 1992. Crocodilian management in the People's Republic of China. A review with recommendations. Pp. 1–27 in *Crocodile conservation action*. A special publication of the Crocodile Specialist Group of the Species Survival Commission of the IUCN-World Conservation Union. Gland, Switzerland: IUCN.

DALMATIAN PELICAN

Anonymous. 1998. Disturbance for Dalmatian pelican. *World Birdwatch 20*(2): 5.

Bräunlich, A. 1995. Report on the first WWF expedition to the Great Lakes Basin, western Mongolia, May–July 1995, and preliminary recommendations for the establishment of a new protected area. Ulaanbaatar, Mongolia: WWF Mongolia.

Collar, N.J., M.J. Crosby, and A.J. Stattersfield. 1994. *Birds to watch 2: The world list of threatened birds*. Cambridge, UK: BirdLife International.

Cramp, S., ed. 1977. *Handbook of the birds of Europe, the Middle East and North Africa*. Vol. 1, *Ostrich to ducks*. Oxford: Oxford University Press.

Crivelli, A.J. 1994a. Dalmatian pelican *Pelecanus crispus*. Pp. 86–87 in *Birds in Europe: Their conservation status*, ed. G.M. Tucker and M.F. Heath. BirdLife Conservation Series No. 3. Cambridge, UK: BirdLife International.

Crivelli, A.J. 1994b. The importance of the former USSR for the conservation of pelican populations nesting in the Palearctic. Pp. 1–4 in *Pelicans in the former USSR*, ed. A.J. Crivelli, V.G. Krivenko, and V.G. Vinogradov. IWRB Publication 27. Slimbridge, UK: IWRB.

Crivelli, A.J. 1996. Action plan for the Dalmatian pelican (*Pelecanus crispus*) in Europe. Pp. 53–66 in *Globally threatened birds in Europe. Action plans*, ed. B. Heredia, L. Rose, and M. Painter. Strasbourg, Germany: Council of Europe.

Crivelli, A.J., V.G. Krivenko, and V.G. Vinogradov, eds. 1994. *Pelicans in the former USSR*. IWRB Publication 27. Slimbridge, UK: IWRB.

Crivelli, A.J., and T. Michev. 1997. Dalmatian pelican *Pelecanus crispus*. P. 33 in *The EBCC atlas of European breeding birds*, ed. W.J.M. Hagemeijer and M.J. Blair. London: T&AD Poyser.

Dementiev, G.P., and N.A. Gladkov. 1951. *Ptitsy Sovietskogo Soyuza 1, 2*. Moscow: Sovietskaya Nauka. (In Russian)

Dokulil, M.T. 1994. Anthropogenic impacts to lakes. Are shallow lakes more vulnerable than deep lakes? Pp. 81–97 in *Proceedings of the International Symposium Wuxi March 27 to April 1, 1993*. Nanjing, China: China Science and Technology Press.

IUCN. 1996. *1996 IUCN red list of threatened animals*. Gland, Switzerland: IUCN.

Krivenko, V.G., A.J. Crivelli, and V.G. Vinogradov. 1994. Historical changes and present status of pelicans in the former USSR: A synthesis with recommendations for their conservation. Pp. 132–151 in *Pelicans in the former USSR*, ed. A.J. Crivelli, V.G. Krivenko, and V.G. Vinogradov. IWRB Publication 27. Slimbridge, UK: IWRB.

Martin, E.B., and L. Vigne. 1995. Agate replaces rhino horn in Yemen's new dagger handles. *Oryx 29*: 154.

MNE (Mongolian Ministry for Nature and Environment). 1996. *Biodiversity conservation action plan for Mongolia*. Ulaanbaatar, Mongolia: Ministry for Nature and the Environment.

Perennou, C., T. Mundkur, D.A. Scott, A. Follestad, and L. Kvenild. 1994. The Asian waterfowl census 1987–91: Distribution and status of Asian waterfowl. *AWB Publication* No. 86, *IWRB Publication* No. 24. Kuala Lumpur: AWB; Slimbridge, UK: IWRB.

Rose, P.M., and D.A. Scott 1997. *Waterfowl Population Estimates*, 2nd ed. Wetlands International Publication 44. Wageningen, The Netherlands: Wetlands International.

Shiirevdamba, Ts., ed. 1997. *Mongolian red book*. Ulaanbaatar, Mongolia: Ministry for Nature and the Environment of Mongolia. (In Mongolian with English summaries)

Zhatkanbaev, A.Zh., and A.E. Gavrilov. 1994. Ringing and migration of pelicans in Kazakhstan. Pp. 124–131 in *Pelicans in the former USSR*, eds. A.J. Crivelli, V.G. Krivenko, and V.G. Vinogradov. IWRB Publication 27. Slimbridge, UK: IWRB.

DOUC LANGUR

Bennett, E., and G. Davies. 1994. The ecology of Asian colobines. Pp. 129–171 in *Colobine monkeys: Their ecology, behaviour and evolution*, ed. G. Davies and J. Oates. Cambridge, UK: Cambridge University Press.

Davies, A. 1994. Colobine populations. Pp. 285–310 in *Colobine monkeys: Their ecology, behaviour and evolution*, eds. G. Davies and J. Oates. Cambridge UK: Cambridge University Press.

Eames, J., and C. Robson. 1993. Threatened primates in southern Vietnam. *Oryx 27*: 146–154.

Eudey, A. 1987. *Action plan for Asian primate conservation 1987–1991*. IUCN/SSC Primate Specialist Group. Gland, Switzerland: IUCN.

Eudey, A. 1991. Human population profiles of Asian primate habitat countries in 1990. *Asian Primates 1*(2): 3–5.

Gochfeld, M. 1974. Douc langurs. *Nature 247*: 167.

Groves, C. 1970. The forgotten leaf-eaters and the phylogeny of the colobinae. Pp. 555–588 in *Old World Monkeys*, ed. J.R. Napier and P.H. Napier. New York: Academic Press.

Groves, C. 1993. Order primates. Pp. 243–277 in *Mammal species of the world: A taxonomic and geographic reference*, 2nd ed., ed. D. Wilson and D. Reeder. Washington, DC: Smithsonian Institution Press.

Jablonski, N. 1995. The phyletic position and systematics of the Douc langurs of southeast Asia. *American Journal of Primatology 35*: 185–205.

Kirkpatrick, C. 1998. Ecology and behavior in snub-nosed and douc langurs. Pp. 155–190 in *The natural history of the douc and snub-nosed monkeys*, ed. N. Jablonski. Singapore: World Scientific.

Lippold, L. 1977. The Douc langur: A time for conservation. Pp. 513–537 in *Primate conservation*, eds. H.S.H. Prince Rainier III of Monaco and G.H. Bourne. New York: Academic Press.

Lippold, L. 1979. Uta and Jack: Adoption in Douc langurs. *Zoonooz 52*(6): 7–9.

Lippold, L. 1981. Monitoring female reproductive status in the Douc langur at San Diego Zoo. *International Zoo Yearbook 21*: 184–187.

Lippold, L. 1989. Reproduction and survivorship in Douc langurs, *Pygathrix nemaeus* in zoos. *International Zoo Yearbook 28*: 252–255.

Lippold, L. 1995. Distribution and conservation status of Douc langurs in Vietnam. *Asian Primates* 4(4): 4–6.

Lippold, L. 1998. Natural history of Douc langurs. Pp. 191–206 in *The natural history of the douc and snub-nosed monkeys*, ed. N. Jablonski. Singapore: World Scientific.

Lippold, L., and V.N. Thanh. 1995. Douc langur variety in the central highlands of Vietnam. *Asian Primates* 5(1–2): 6–8.

Lippold, L. and V.N. Thanh. 1998. Primate conservation in Vietnam. Pp. 293–300 in *The natural history of the douc and snub-nosed monkeys*, ed. N. Jablonski. Singapore: World Scientific.

Mittermeier, R., and D.L. Cheney. 1987. Conservation of primates and their habitats. Pp. 477–490 in *Primate societies*, ed. B. Smuts, D. Cheney, R. Seyfarth, R. Wrangham, and T. Struhsaker. Chicago: University of Chicago Press.

Nadler, T. 1995. Douc langur (*Pygathrix nemaeus* ssp.) and Francois' langur (*Trachypithecus francoisi* ssp.) with questionable taxonomic status in the Endangered Primate Rescue Center, Vietnam. *Asian Primates* 5(1–2): 8–9.

Napier, J.P., and P.H. Napier. 1967. *A handbook of living primates*. London: Academic Press.

Napier, J.P., and P.H. Napier. 1985. *The natural history of the primates*. Cambridge, MA: MIT Press.

Oates, J., and G. Davies. 1994a. What are the Colobines? Pp. 1–9 in *Colobine monkeys: Their ecology, behaviour and evolution*, ed. G. Davies and J. Oates. Cambridge, UK: Cambridge University Press.

Oates, J., and G. Davies. 1994b. Conclusions: The past, present and future of the Colobines. Pp. 347–358 in *Colobine monkeys : Their ecology, behaviour and evolution*, eds. G. Davies and J. Oates. Cambridge, UK: Cambridge University Press.

Orions, G., and E. Pfeiffer. 1970. Ecological effects of the war in Vietnam. *Science 168*: 544–554.

Rowe, N. 1996. *The pictorial guide to the living primates*. New York: Pogonias Press.

Traitel, D. 1996. Building a safe haven for Douc langurs. *Zoonooz 59*(10): 10–15.

Wang, S., and G. Quan. 1986. Primate status and conservation in China. Pp. 213–220 in *Primates: The road to self-sustaining populations*, ed. K. Benirschke. New York: Springer-Verlag.

Wirth, R., H. Adler, and N.Q. Thang. 1991. Douc langurs: How many species are there? *Zoonooz 64*: 12–13.

Wolfheim, J. 1983. *Primates of the world: Distribution, abundance and conservation*. Seattle: University of Washington Press.

EASTERN BARRED BANDICOOT

Atlas of Victoria Wildlife. 1998. Atlas of Victoria Wildlife Database. Heidelberg, Victoria: Arthur Rylah Institute for Environmental Research.

Backhouse, G.N. 1992. Recovery plan for the eastern barred bandicoot *Perameles gunnii*. Melbourne, Victoria: Department of Conservation and Environment, Victoria and Zoological Board of Victoria.

Backhouse, G.N., T.W. Clark, and R.P. Reading. 1994. The Australian eastern barred bandicoot recovery program: Evaluation and reorganization. Pp. 251–271 in *Saving endangered species: Professional and organizational lessons for improvement*, ed. T.W. Clark, R.P. Reading, and A. Clarke. Washington, DC: Island Press.

Brown, P.R. 1989. *Management plan for the conservation of the eastern barred bandicoot*, Perameles gunnii, *in Victoria*. Arthur Rylah Institute for Environmental Research Technical Report Series No. 63. Melbourne, Victoria: National Parks and Wildlife Division, Department of Conservation, Forests and Lands.

Dreissen, M.M., and G.J. Hocking. 1991. *The eastern barred bandicoot recovery plan for Tasmania: Research phase*. Hobart, Tasmania: Tasmanian Department of Parks, Wildlife and Heritage.

Dufty, A.C. 1988. The distribution, population abundance, status, movement and activity of the eastern barred bandicoot, *Perameles gunnii*, at Hamilton. Bachelor's thesis, La Trobe University, Melbourne, Victoria.

Dufty, A.C. 1991. Some population characteristics of *Perameles gunnii* in Victoria. *Wildlife Research 18*: 355–366.

Dufty, A.C. 1994a. Field observations of the behaviour of free-ranging eastern barred bandicoots, *Perameles gunnii*, at Hamilton, Victoria. *Victorian Naturalist 111*: 54–59.

Dufty, A.C. 1994b. Habitat and spatial requirements of the eastern barred bandicoot (*Perameles gunnii*) at Hamilton, Victoria. *Wildlife Research 21*: 459–472.

Dufty, A.C. 1995. The growth and development of the eastern barred bandicoot *Perameles gunnii* in Victoria. *Victorian Naturalist 112*: 79–85.

Harper, F. 1945. Extinct and vanishing mammals of the old world. Baltimore: Lord Baltimore Press.

IUCN. 1996. *1996 IUCN red list of threatened animals*. Gland, Switzerland: IUCN.

Lacy, R.C., and T.W. Clark. 1990. Population viability assessment of the eastern barred bandicoot in Victoria. Pp. 131–146 in *Management and conservation of small populations*, ed. T.W. Clark and J.H. Seebeck. Brookfield, IL: Chicago Zoological Society.

Maxwell, S., A.A. Burbidge, and K. Morris. 1996. *The 1996 action plan for Australian marsupials and monotremes*. Canberra, Australia: Wildlife Australia.

Reading, R.P., T.W. Clark, and A. Arnold. 1995. Attitudes towards the endangered eastern barred bandicoot. *Anthrozoos 7*: 255–269.

Reading, R.P., T.W. Clark, P.W. Goldstraw, A.J. Watson, and J.H. Seebeck. 1991. *An overview of eastern barred bandicoot reintroduction programs in Victoria, Australia: With recommendations for future reintroductions*. Melbourne, Victoria: Department of Conservation and Environment.

Robinson, N.A., N.D. Murray, and W.B. Sherwin. 1993. VNTR loci reveal differentiation between and structure within populations of the eastern barred bandicoot *Perameles gunnii*. *Molecular Biology 2*: 195–207.

Seebeck, J.H. 1979. Status of the barred bandicoot, *Perameles gunnii*, in Victoria: With a note on husbandry of a captive colony. *Australian Wildlife Research 6*: 255–264.

Seebeck, J.H. 1995. Eastern barred bandicoot *Perameles gunnii* Gray 1838. Pp. 75–77 in *Mammals of Victoria. Distribution, ecology and conservation*, ed. P.W. Menkhorst. Melbourne, Victoria: Oxford University Press.

ETHIOPIAN WOLF

Gottelli, D., and C. Sillero-Zubiri. 1992. The Ethiopian wolf—an endangered endemic canid. *Oryx 26*: 205–214.

Gottelli, D., C. Sillero-Zubiri, G.D. Applebaum, D. Girman, M. Roy, J. Garcia-Moreno, E. Ostrander, and R.K. Wayne. 1994. Molecular genetics of the most endangered canid: The Ethiopian wolf, *Canis simensis. Molecular Ecology 3*: 301–312.

Kingdon, J. 1990. *Island Africa*. London: Collins.

Laurenson, K., F. Shiferaw, and C. Sillero-Zubiri. 1997. Disease, domestic dogs and the Ethiopian wolf: The current situation. Pp. 32–42 in *The Ethiopian wolf: Status survey and conservation action plan*, ed. C. Sillero-Zubiri and D.W. Macdonald. Gland, Switzerland: IUCN-World Conservation Union.

Laurenson, K., C. Sillero-Zubiri, H. Thompson, F. Shiferaw, S. Thirgood, and J.R. Malcolm. 1998. Disease threats to endangered species: Patterns of infection by canine pathogens in Ethiopian wolves (*Canis simensis*) and sympatric domestic dogs. *Animal Conservation 1*: 273–280.

Malcolm, J.R., and C. Sillero-Zubiri. 1997. The Ethiopian wolf: Distribution and population status. Pp. 12–31 in *The Ethiopian wolf: Status survey and conservation action plan*, ed. C. Sillero-Zubiri and D.W. Macdonald. Gland, Switzerland: IUCN-World Conservation Union.

Sillero-Zubiri, C. 1994. Behavioural ecology of the Ethiopian wolf, *Canis simensis*. D.Phil. thesis, Oxford University.

Sillero-Zubiri, C., and D. Gottelli. 1994. *Canis simensis*. *Mammalian Species 485*: 1–6.

Sillero-Zubiri, C., and D. Gottelli. 1995a. Diet and feeding behavior of Ethiopian wolves. (*Canis simensis*). *Journal of Mammalogy 76*: 531–541.

Sillero-Zubiri, C., and D. Gottelli. 1995b. Spatial organization in the Ethiopian wolf *Canis simensis*: Large packs and small stable home ranges. *Journal of Zoology 237*: 65–81.

Sillero-Zubiri, C., D. Gottelli, and D.W. Macdonald. 1996. Male philopatry, extra-pack copulations and inbreeding avoidance in the Ethiopian wolf (*Canis simensis*). *Behavioural Ecology and Sociobiology 38*: 331–340.

Sillero-Zubiri, C., A.A. King, and D.W. Macdonald. 1996. Rabies and mortality in Ethiopian wolves (*Canis simensis*). *Journal of Wildlife Diseases 32*: 80–86.

Sillero-Zubiri, C., and D.W. Macdonald, eds. 1997. *The Ethiopian wolf: Status survey and conservation action plan*. Gland, Switzerland: IUCN-World Conservation Union.

Sillero-Zubiri, C., and D.W. Macdonald. 1998. Scent-marking and territoriality behavior of Ethiopian wolves *Canis simensis*. *Journal of Zoology 245*: 351–361.

Sillero-Zubiri, C., F.H. Tattersall, and D.W. Macdonald. 1995. Habitat selection and daily activity of giant molerats (*Tachyoryctes macrocephalus*): Significance to the Ethiopian wolf (*Canis simensis*) in the Afroalpine ecosystem. *Biological Conservation 72*: 77–84.

Yalden, D.W., and M.J. Largen. 1992. The endemic mammals of Ethiopia. *Mammal Review 22*: 115–150.

EUROPEAN MINK

Danilov, P.I., and I.L. Tumanov. 1976. The ecology of the European and American mink in the Northwest of the USSR. Pp. 118–143 in *Ecology of birds and mammals in the Northwest of the USSR*. Petrozavodsk; Akademia Nauk Karelski filial, Instut Biologij. (In Russian)

Maran, T. 1991. Distribution of the European mink, *Mustela lutreola*, in Estonia: A historical review. *Folia Theriologica Estonica 1*: 1–17.

Maran, T. 1994. On the status and the management of the European mink *Mustela lutreola*. Seminar on the management of small populations of threatened mammals. *Environmental Encounters* (Council of Europe Press) *17*: 84–90.

Maran, T. 1996a. Erhaltung des europäischen Nerzes. Zookunft, 1996. Pp. 79–92 in *Zoos in harmony with the World Zoo Conservation Strategy*. Münster: Quantum Conservation e.V.

Maran, T. 1996b. Ex situ and in situ conservation of the European mink. *International Zoo News 43*: 399–407.

Maran, T., and H. Henttonen. 1995. Why is the European mink, *Mustela lutreola*, disappearing? A review of the process and hypotheses. *Annales Zoologici Fennici 32*: 47–54.

Maran, T., D. MacDonald, H. Kruuk, V. Sidorovich, and V.V. Rozhnov. 1998. The continuing decline of the European mink, *Mustela lutreola*: Evidence for the intra-guild competition hypothesis. *Symposium of the Zoological Society of London 71*: 297–325.

Maran, T., and P. Robinson. 1996. European mink, *Mustela lutreola*, captive breeding and husbandry protocol. Vol. 1. Tallinn, Estonia: European Mink Conservation and Breeding Committee.

Novikov, G.A. 1939. *The European mink*. Leningrad: Izdatelstvo Leningradkogo Gosudarstvennovo Universiteta, Russia. (In Russian.)

Ognev, S.I. 1931. *Animals of eastern Europe and northern Asia. Carnivores. II*. Moscow-Leningrad: Izdatelstvo Akademij Nauk, USSR.

Schreiber, A.R., R. Wirth, M. Riffel, and H. van Rompaey. 1989. *Weasels, civets, mongooses, and their relatives. An action plan for the conservation of mustelids and viverrids*. Gland, Switzerland: IUCN/SSC Mustelid and Viverrid Specialist Group.

Shvarts, E.A., and M.A. Vaisfeld. 1993. Problem of saving vanishing species and the islands (discussion of the introduction of the European mink *Mustela lutreola* on Kunashir Island). *Uspehhi Sovremennoi Biologii 113*: 46–59. (In Russian.)

Sidorovich, V., H. Kruuk, D.W. MacDonald, and T. Maran. 1998. Diets of semi-aquatic carnivores in Northern Belarus, with implications for population changes. *Symposium of the Zoological Society of London 71*: 117–191.

Sidorovich, V.E., and A.V. Kozhulin. 1994. Preliminary data on the status of the European mink's (*Mustela lutreola*) abundance in the centre of the eastern part of its present range. *Small Carnivore Conservation 9*: 10–11.

Sidorovich V.E., V.V. Savchenko, and V.B. Bundy. 1995. Some data about the European mink *Mustela lutreola* distribution in the Lovat River Basin in Russia and Belarus: Current status and retrospective analysis. *Small Carnivore Conservation 12*: 14–18.

Youngman, P.M. 1982. Distribution and the systematics of the European mink *Mustela lutreola* Linnaeus 1761. *Acta Zoologica Fennica 166*: 1–48.

Youngman, P.M. 1990. *Mustela lutreola. Mammalian Species 362*: 1–3.

FLORIDA MANATEE

Ackerman, B.B., S.D. Wright, R.K. Bonde, D.K. Odell, and D.J. Banowetz. 1995. Trends and patterns in mortality of manatees in Florida, 1974–1992. Pp. 223–258 in *Population biology of the Florida manatee*, ed. T.J. O'Shea, B.B. Ackerman, and H.F. Percival. Information and Technology Report 1. Washington, DC: U.S. National Biological Service.

Buffett, J. 1996. Club marks 15th anniversary. *Save the Manatee Club Newsletter* (September): 1–2.

Clark, T.W., and J.R. Cragun. 1994. Organizational and managerial guidelines for endangered species restoration programs and recovery teams. Pp. 9–33 in *Restoration of endangered species*, ed. M.L. Bowles and C.J. Whelan. Cambridge, UK: Cambridge University Press.

Frohlich, R.K. 1998. Power plants: Good or bad for manatees? *Sirenews—Newsletter of the IUCN/SSC Sirenia Specialist Group 29* (April): 5–6.

Hartman, D.S. 1979. *Ecology and behavior of the manatee* (Trichechus manatus) *in Florida*. American Society of Mammalogists, Special Publication No. 5.

Irvine, A.B. 1983. Manatee metabolism and its influence on distribution in Florida. *Biological Conservation 25*: 315–334.

Lefebvre, L.W., T.J. O'Shea, G.B. Rathbun, and R.C. Best. 1989. Distribution, status, and biogeography of the West Indian manatee. Pp. 567–610 in *Biogeography of the West Indies*, ed. C.A. Woods. Gainesville, FL: Sandhill Crane Press.

O'Shea, T.J. 1988. The past, present, and future of manatees in the southeastern United States: Realities, misunderstandings, and enigmas. Pp. 184–204 in *Proceedings of the Third Southeastern Nongame and Endangered Wildlife Symposium*, ed. R.R. Odom, K.A. Riddleberger, and J.C. Osier. Social Circle, GA: Georgia Department of Natural Resources, Game and Fish Division.

O'Shea, T.J. 1995. Waterborne recreation and the Florida manatee. Pp. 297–311 in *Wildlife and recreationists: Coexistence through management and research*, ed. R.L. Knight and K.J. Gutzwiller. Covelo, CA: Island Press.

O'Shea, T.J., and B.B. Ackerman. 1995. Population biology of the Florida manatee: An overview. Pp. 280–287 in *Population biology of the Florida manatee*, ed. T.J. O'Shea, B.B. Ackerman, and H.F. Percival. Information and Technology Report 1. Washington, DC: U.S. National Biological Service.

Packard, J.M., R.K. Frohlich, J.E. Reynolds III, and J.R. Wilcox. 1989. Manatee response to interruption of a thermal effluent. *Journal of Wildlife Management 53*: 692–700.

Rathbun, G.B., J.P. Reid, R.K. Bonde, and J.A. Powell. 1995. Reproduction in free-ranging Florida manatees. Pp. 135–156 in *Population biology of the Florida manatee*, ed. T.J. O'Shea, B.B. Ackerman, and H.F. Percival. Information and Technology Report 1. Washington, DC: U.S. National Biological Service.

Rathbun, G.B., J.P. Reid, and G. Carowan. 1990. Distribution and movement patterns of manatees (*Trichechus manatus*) in northwestern peninsular Florida. *Florida Marine Research Publications 48*: 1–33.

Reid, J.P., G.B. Rathbun, and J.R. Wilcox. 1991. Distribution patterns of individually identifiable West Indian manatees (*Trichechus manatus*) in Florida. *Marine Mammal Science 7*: 180–190.

Reynolds, J.E., III, and D.K. Odell. 1991. *Manatees and dugongs*. New York: Facts on File.

Rose, P.M. 1997. Manatees and the future of electric utilities deregulation in Florida. *Sirenews— Newsletter of the IUCN/SSC Sirenia Specialist Group 28* (October): 1–3.

U.S. Fish and Wildlife Service. 1996. *Florida manatee recovery plan*, 2nd rev. Atlanta, GA: U.S. Fish and Wildlife Service.

Westrum, R. 1994. An organizational perspective: Designing teams from the inside out. Pp. 327–350 in *Endangered species recovery: Finding the lessons, improving the process*, ed. T.W. Clark, R.P. Reading, and A.L. Clarke. Covelo, CA: Island Press.

Wright, S.D., B.B. Ackerman, R.K. Bonde, C.A. Beck, and D.J. Banowetz. 1995. Analysis of watercraft-related mortality of manatees in Florida, 1979–1991. Pp. 259–268 in *Population biology of the Florida manatee*, ed. T.J. O'Shea, B.B. Ackerman, and H.F. Percival. Information and Technology Report 1. Washington, DC: U.S. National Biological Service.

GIANT PANDA

Hu J. 1998. Re-introduction and conservation of giant panda. In *Proceedings of giant panda reintroduction workshop, 24–29 September, 1997, Wolong, China*, ed. S. Mainka. Beijing: Ministry of Forestry China Protecting Giant Panda Project Office and WWF-China Programme.

Johnson, K., G. Schaller, and Hu J. 1988. Responses of giant pandas to a bamboo die-off. *National Geographic Research 4*: 161–177.

Johnson, K.G., Yin Y., Chengxia Y., Senshan Y., and Zhangmin S. 1996. Human/carnivore interactions: Conservation and management implications from China. Pp. 337–370 in *Carnivore behavior, ecology, and evolution*, Vol. 2, ed. J.L. Gittleman. Ithaca, NY: Cornell University Press.

Li Z. and Zhou Z. 1998. Update on activities to implement the National Conservation Project for the giant panda and its habitat, and current status of giant panda conservation. In *Proceedings of giant panda reintroduction workshop, 24–29 September, 1997, Wolong, China*, ed. S. Mainka. Beijing: Ministry of Forestry China Protecting Giant Panda Project Office and WWF-China Programme.

Lu Z. 1991. *Movement patterns, population dynamics, and social behaviour of the giant panda in Qinling*. Ph.D. thesis, Peking University, Beijing. (In Chinese)

Lu Z. 1999. *WWF's current panda action plan*. Beijing: WWF China Programme.

Lu Z., Pan W., Zhu X., Wang D., and Wang H. In press. What has the panda taught us? In *Future priorities for the conservation of mammalian diversity*, ed. A. Entwistle and N. Dunstone. Cambridge: Cambridge University Press.

MacKinnon, J.J., Bi F.Z., Qiu M.J., Fu C.D., Wang H.B., Yuan S.J., Tian A.S., and Li J.G. 1989. *National conservation management plan of the giant panda and its habitat*. Gland, Switzerland: WWF and the Ministry of Forestry.

Mainka, S., ed. 1998. *Proceedings of giant panda reintroduction workshop, 24–29 September, 1997, Wolong, China*. Beijing: Ministry of Forestry China Protecting Giant Panda Project office and WWF-China Programme.

Nowak, R.M. 1991. *Walker's mammals of the world*, 5th ed., Vol. 2. Baltimore: Johns Hopkins University Press.

Pan W. 1998. Ecological study on the giant pandas in Qinling and some thought about the giant panda re-introduction in Wolong. In *Proceedings of giant panda reintroduction workshop, 24–29 September, 1997, Wolong, China*, ed. S. Mainka. Beijing: Ministry of Forestry China Protecting Giant Panda Project Office and WWF-China Programme.

Pan W., Gao Z., and Lu Z. 1988. *The giant panda's natural refuge in the Qinling Mountains*. Beijing: Peking University Press. (In Chinese with English summaries)

Schaller, G.B. 1993. *The last panda*. Chicago: University of Chicago Press.

Schaller, G.B. 1998. Giant panda biology and its relevance to re-introduction efforts. In *Proceedings of giant panda reintroduction workshop, 24–29 September, 1997, Wolong, China*, ed. S. Mainka. Beijing: Ministry of Forestry China Protecting Giant Panda Project Office and WWF-China Programme.

Schaller, G.B., Hu J., Pan W., and Zhu J. 1985. *The giant pandas of Wolong*. Chicago: University of Chicago Press.

Schaller, G., Teng Q., K. Johnson, Wang X., Shen H., and Hu J. 1989. The feeding ecology of giant pandas and Asiatic black bears in the Tangjiahe Reserve, China. Pp. 212–241 in *Carnivore behaviour, ecology, and evolution*, ed. J. Gittleman. Ithaca: Cornell University Press.

Wong J. 1991. The panda in peril. *Globe and Mail*, August 24: D10.

Yang J., Zhang H., Tan Y., Wei R., Zhou S., Huang J., He T., and S.A. Mainka. 1998. Population monitoring of wild pandas and other wildlife at Wuyipeng. In *Proceedings of giant panda reintroduction workshop, 24–29 September, 1997, Wolong, China*, ed. S. Mainka. Beijing: Ministry of Forestry China Protecting Giant Panda Project Office and WWF-China Programme.

GOLDEN LION TAMARIN

Baker, A.J., J.M. Dietz, and D.G. Kleiman. 1993. Behavioural evidence for monopolization of paternity in multi-male groups of golden lion tamarins. *Animal Behaviour 46*: 1091–1103.

Ballou, J.D., R.C. Lacy, D. Kleiman, A. Rylands, and S. Ellis. 1998. Leontopithecus *II: The second population and habitat viability assessment for lion tamarins* (Leontopithecus). Apple Valley, MN: IUCN Conservation Breeding Specialist Group/SSC.

Ballou, J.D., and A. Sherr. 1996. *1995 international studbook, golden lion tamarin*. Washington, DC: National Zoological Park, Smithsonian Institution.

Beck, B.B., M.I. Castro, T.S. Stoinski, and J.D. Ballou. In press. The effects of pre-release environments on survivorship in golden lion tamarins. In *The lion tamarins: Twenty-five years of research and conservation*, ed. D.G. Kleiman and A. Rylands. Washington, DC: Smithsonian Institution Press.

Beck, B.B., D.G. Kleiman, J.M. Dietz, I. Castro, C. Carvalho, A. Martins, and B. Rettberg-Beck. 1991. Losses and reproduction in reintroduced golden lion tamarins *Leontopithecus rosalia*. *Dodo 27*: 50–61.

Beck, B.B., and A.F. Martins. 1997. *Golden lion tamarin reintroduction, annual report*. Unpublished report. Golden Lion Tamarin Association, Washington, DC.

Coimbra-Filho, A.F. 1969. Mico-leão, *Leontideus rosalia* (Linnaeus, 1766), situaço atual da espécie no Brasil (Callitrichidae-Primates). *Annais da Academia Brasileira de Ciências 41* (supplement): 29–52.

Coimbra-Filho, A.F. 1977. Natural shelters of *Leontopithecus rosalia* and some ecological implications (Callitrichidae: Primates). Pp. 79–90 in *The biology and conservation of the Callitrichidae*, ed. D.G. Kleiman. Washington, DC: Smithsonian Institution Press.

Dietz, J.M., and A.J. Baker. 1993. Polygyny and female reproductive success in golden lion tamarins, *Leontopithecus rosalia*. *Animal Behaviour 46*: 1067–1078.

Dietz, J.M., A.J. Baker, and D. Miglioretti. 1994. Seasonal variation in reproduction, juvenile growth, and adult body mass in golden lion tamarins (*Leontopithecus rosalia*). *American Journal of Primatology 34*: 115–132.

Dietz, J.M., L.A. Dietz, and E. Nagagata. 1994. The effective use of flagship species for conservation of biodiversity: The example of lion tamarins in Brazil. Pp. 33–49 in *Creative conservation: Interactive management of wild and captive animals*, ed. P.J.S. Olney, G.M. Mace, and A.T.C. Feistner. London: Chapman and Hall.

Dietz, L.A., and E. Nagagata. 1995. Golden lion tamarin conservation program: A community effort for forest conservation, Rio de Janeiro State, Brazil. Pp. 95–124 in *Conserving wildlife: International education/communication approaches*, ed. S.K. Jacobson. New York: Columbia University Press.

Hershkovitz, P. 1977. *Living new world monkeys (Platyrrhini)*, Vol. 1. Chicago: University of Chicago Press.

Kierulff, M.C., and P.P. Oliveira. 1994. Habitat preservation and the translocation of threatened groups of golden lion tamarins, *Leontopithecus rosalia*. *Neotropical Primates 2* (supplement): 15–18.

Kleiman, D.G. 1977. Characteristics of reproduction and sociosexual interactions in pairs of lion tamarins (*Leontopithecus rosalia*) during the reproductive cycle. Pp. 181–190 in *The biology and conservation of the Callitrichidae*, ed. D.G. Kleiman. Washington, DC: Smithsonian Institution Press.

Kleiman, D.G., and J.J.C. Mallinson. 1997. Recovery and management committees for lion tamarins: Partnerships in conservation planning and implementation. *Conservation Biology 12*: 27–38.

Mallinson, J.J.C. 1996. The history of golden lion tamarin management and propagation outside of Brazil and current management practices. *Zoologische Garten N.F. 66*: 197–217.

Mori, S.A., B.M. Boom, and G.T. Prance. 1981. Distribution patterns and conservation of eastern Brazil coastal tree species. *Brittonia 33*: 233–245.

Seal, U.S., J.D. Ballou, and C. Valladares Pádua, eds. 1990. Leontopithecus *population viability analysis workshop report*. Apple Valley, MN: Captive Breeding Specialist Group, World Conservation Union/Species Survival Commission.

GOLDEN-RUMPED ELEPHANT-SHREW

Burgess, N., C. FitzGibbon, and P. Clarke. 1996. Coastal forests. Pp. 329–359 in *East African ecosystems and their conservation*, ed. T.R. McClanahan and T.P. Young. New York: Oxford University Press.

Corbet, G.B., and J. Hanks. 1968. A revision of the elephant-shrews, Family Macroscelididae. *Bulletin of the British Museum (Natural History) 16*: 47–111.

Cunningham, A.B. 1998. Kenya's carvings—the ecological footprint of the wooden rhino. *Africa—Environment and Wildlife 6*(2): 43–50.

FitzGibbon, C.D. 1994. The distribution and abundance of the golden-rumped elephant-shrew *Rhynchocyon chyrsopygus* in Kenyan coastal forests. *Biological Conservation 67*: 153–160.

FitzGibbon, C.D. 1997. The adaptive significance of monogamy in the golden-rumped elephant-shrew. *Journal of Zoology* (London) *242*: 167–177.

FitzGibbon, C.D., H. Mogaka, and J.H. Fanshawe. 1995. Subsistence hunting in Arabuko-Sokoke Forest, Kenya, and its effects on mammal populations. *Conservation Biology 9*: 1116–1126.

FitzGibbon, C.D., H. Mogaka, and J.H. Fanshawe. 1996. Subsistence hunting and mammal conservation in a Kenyan coastal forest: Resolving a conflict. Pp. 147–159 in *The exploitation of mammals*, ed. N. Dunstone and V. Taylor. New York: Chapman and Hall.

Hawthorne, W.D. 1993. East African coastal forest botany. Pp. 57–99 in *Biogeography and ecology of the rain forests of eastern Africa*, ed. J.C. Lovett and S.K. Wasser. Cambridge: Cambridge University Press.

Kyalo, S.N. 1997. The ecological impact of the Kenya woodcarving industry, a preliminary survey of the biota of *Brachylaena huillensis* (muhuhu) trees. Unpublished report, People and Plants Initiative, WWF/UNESCO.

Marshall, N.T., and M. Jenkins. 1994. *Hard times for hardwood: The indigenous timber and the timber trade in Kenya*. Cambridge: TRAFFIC International.

Mogaka, H. 1991. Local utilisation of Arabuko Sokoke Forest Reserve. Unpublished report, Kenya Indigenous Forest Conservation Programme, Natural Resources Institute, Chatham Maritime, Kent, England.

Nicoll, M.E., and G.B. Rathbun. 1990. *African Insectivora and elephant-shrews: An action plan for their conservation*. Gland, Switzerland: IUCN.

Rathbun, G.B. 1979a. *Rhynchocyon chrysopygus. Mammalian Species. 117*: 1–4.

Rathbun, G.B. 1979b. The social structure and ecology of elephant-shrews. *Advances in Ethology, Supplement to Journal of Comparative Ethology 20*: 1–77.

Rodgers, W.A. 1993. The conservation of the forest resources of eastern Africa; past influences, present practices and future needs. Pp. 283–331 in *Biogeography and ecology of the rain forests of eastern Africa*, ed. J.C. Lovett and S.K. Wasser. Cambridge: Cambridge University Press.

Springer, M.S., G.C. Cleven, O. Madsen, W.W. de Jong, V.G. Waddell, H.M. Amrine, and M.J. Stanhope. 1997. Endemic African mammals shake the phylogenetic tree. *Nature* (London) *388*: 61–64.

Turner, I.M., and R.T. Corlett. 1996. The conservation value of small, isolated fragments of lowland tropical rain forest. *Trends in Ecology and Evolution 11*: 330–333.

GORILLA

Hall, J., K. Saltonstall, B. Inogwabini, and I. Omari, 1998. Distribution, abundance, and conservation status of Grauer's gorilla (*Gorilla gorilla graueri*). *Oryx 32*: 122–130.

Harcourt, A.H., K.J. Stewart, and I.M. Inahoro. 1989. Nigeria's gorillas: A survey and recommendations. *Primate Conservation 10*: 73–76.

Tutin, C.E.G., and M. Fernandez. 1984. Nationwide census of gorilla (*Gorilla g. gorilla*) and chimpanzee (*Pan t. troglodytes*) populations in Gabon. *American Journal of Primatology 6*: 313–336.

Vedder, A.L. 1984. Movement patterns of a group of free-ranging mountain gorillas (*Gorilla gorilla beringei*) and their relationship to food availability. *African Journal of Primatology 7*: 73–88.

Vedder, A.L. 1989. Feeding ecology and conservation of the mountain gorilla (*Gorilla gorilla beringei*). Ph.D. thesis, University of Wisconsin, Madison.

Vedder, A.L., and W. Weber. 1990. The Mountain Gorilla Project (Volcanoes National Park)—Rwanda. In *Living with wildlife*, ed. A. Kiss. World Bank Technical Paper No. 130. Washington, DC: World Bank.

Watts, D.P. 1984. Composition and variability of mountain gorilla diets in the central Virungas. *American Journal of Primatology 7*: 323–356.

Weber, A.W. 1989. Conservation and development on the Zaire-Nile divide: An analysis of value conflicts and convergence in the management of afromontane forests in Rwanda. Ph.D. thesis, University of Wisconsin, Madison.

Weber, A.W., and A.L. Vedder. 1983. Population dynamics of the Virunga gorilla: 1959–1978. *Biological Conservation 26*: 341–366.

Weber, W. 1987. Socioecological factors in the conservation of afromontane forest reserves. Pp. 205–229 in *Primate conservation in the tropical rainforest*, ed. C.W. Marsh and R.A. Mittermeier. New York: Alan R. Liss.

White, L.J.T., and C.E.G. Tutin. In press. Why chimpanzees and gorillas respond differently to logging: A cautionary tale from Gabon. In *African rain forest ecology and conservation*, ed. W. Weber, L.J.T. White, A. Vedder, and L. Naughton-Treves. New Haven: Yale University Press.

Wilkie, D.S., J.G. Sidle, and G.C. Boundzanga. 1992. Mechanized logging, market hunting, and a bank loan in Congo. *Conservation Biology 6*: 570–580.

GREVY'S ZEBRA

Becker, C.D., and J.R. Ginsberg. 1990. Mother-infant behavior of wild Grevy's zebra: Adaptations for survival in semi-desert east Africa. *Animal Behaviour 40*: 1111–1118.

Dirschl, H.J., and S.P. Wetmore. 1978. *Grevy's zebra abundance and distribution in Kenya, 1997*. Nairobi: Kenya Rangeland Ecological Monitoring Unit.

Ginsberg, J.R. 1988. *Social organization and mating strategies of an arid-adapted equid: The Grevy's zebra*. Princeton: Princeton University.

Ginsberg, J.R. 1989. The ecology of female behaviour and male reproductive success in Grevy's zebra, *Equus grevyi*. *Symposia of the Zoological Society* (London) *61*: 89–110.

Grunblatt, J., M.Y. Said, and J.K. Nutira. 1989. *Livestock and wildlife summary 1987–1988 for Kenya Rangelands*. Nairobi: Department of Resource Surveys and Remote Sensing, Ministry of Planning and National Development.

Herlocker, D.J. 1992. Vegetation type. In *Range management handbook of Kenya*, Vol. 2, 2: *Samburu District*, ed. S.B. Shaabani, M. Walsh, D.J. Herlocker, and D. Walther. Nairobi: Republic of Kenya, Ministry of Livestock Development (MOLD).

Herlocker, D.J. 1993. Vegetation types. In *Range management handbook of Kenya*, Vol. 2, 5: *Isiolo District*, ed. D.J. Herlocker, S.B. Shaabani, and S. Wilkes. Nairobi: Republic of Kenya, Ministry of Agriculture, Livestock Development and Marketing.

IUCN. 1996. *1996 IUCN red list of threatened animals*. Gland, Switzerland: IUCN Species Survival Commission.

Klingel, H. 1974. Social organisation and behaviour of the Grevy's zebra. *Zeitscrift für Tierpsychologie 36*: 36–70.

Klingel, H. 1980. *Survey of African Equidae*. Gland, Switzerland: IUCN Survival Service Commission.

Rowen, M. 1992. Mother-infant behavior and ecology of Grevy's zebra, *Equus grevyi*. Ph.D. thesis, Yale University.

Rowen, M., and J.R. Ginsberg. 1992. Grevy's zebra (*Equus grevyi* Oustalet). In *Equid action plan*, ed. P. Duncan. Gland, Switzerland: IUCN.

Rubenstein, D.I. 1989. Life history and social organization in arid adapted ungulates. *Journal of Arid Environments 17*: 145–156.

Thouless, C.R. 1995a. *Aerial surveys for wildlife in eastern Ethiopia*. London: Ecosystem Consultants.

Thouless, C.R. 1995b. *Aerial surveys for wildlife in Omo Valley, Chew Bahir and Borana areas of southern Ethiopia*. London: Ecosystem Consultants.

Williams, S.D. 1998. Grevy's zebra: Ecology in a heterogeneous environment. Ph.D. thesis, University College London, London.

Williams, S.D. In press. Grevy's zebra. In *Zebras, asses, and horses: An action plan for the conservation of wild equids*, ed. P. Moehlman. Gland, Switzerland: IUCN.

Wisbey, J. 1995. The population status of Grevy's zebra (*Equus grevyi*) in selected areas of northern Kenya. Master's thesis, University College London, London.

HAWAIIAN GOOSE

Baldwin, P.H. 1945. The Hawaiian goose, its distribution and reduction in numbers. *Condor 47*: 27–37.

Baldwin, P.H. 1947. Foods of the Hawaiian goose. *Condor 49*: 108–120.

Banko, P.C. 1988. Breeding biology and conservation of the nene, Hawaiian goose (*Branta sandvicensis*). Ph.D. thesis, University of Washington, Seattle.

Banko, P.C. 1992. Constraints on productivity of wild nene or Hawaiian geese *Branta sandvicensis*. Wildfowl 43: 99–106.

Banko, P.C., J.M. Black, and W.E. Banko. 1999. Hawaiian goose (*Branta sandvicensis*). In *Birds of North America* No. 434, ed. A. Poole and F. Gill. Philadelphia: Academy of Natural Sciences; Washington, DC: American Ornithologists' Union.

Beatly, T. 1994. *Habitat conservation planning.* Austin: University of Texas Press.

Berger, A.J. 1978. Reintroduction of Hawaiian geese. Pp. 339–344 in *Endangered birds: Management techniques for bird recovery programs,* ed. S.A. Temple. Madison: University of Wisconsin Press.

Black, J.M. 1998. Threatened waterfowl: Recovery priorities and reintroduction potential with special reference to the Hawaiian goose. Pp. 125–140 in *Avian conservation: Research and management,* ed. J.M. Marzluff and R. Sallabanks. Washington, DC: Island Press.

Black, J.M., and P.C. Banko. 1994. Is the Hawaiian goose (*Branta sandvicensis*) saved from extinction? Pp. 394–410 in *Creative conservation: Interactive management of wild and captive animals,* ed. P.J. Olney, G. Mace, and A. Feistner. London: Chapman & Hall.

Black, J.M., A.P. Marshall, A. Gilbum, N. Santos, H. Hoshide, J. Medeiros, J. Mello, C. Natividad Hodges, and L. Katahira. 1997. Survival, movements, and breeding of released Hawaiian geese: An assessment of the reintroduction program. *Journal of Wildlife Management 61*: 1161–1173.

Black, J.M., J. Prop, J.M. Hunter, F. Woog, A.P. Marshall, and J.M. Bowler. 1994. Foraging behavior and energetics of the Hawaiian goose *Branta sandvicensis. Wildfowl 45*: 65–109.

Elder, W.H., and D.H.W. Woodside. 1958. Biology and management of the Hawaiian goose. *Transactions of the North American Wildlife Conference 23*: 198–215.

Henshaw, H.W. 1902. Complete list of the birds of the Hawaiian Possessions with notes on their habits. *Thrum's Hawaiian Almanac and Annual 1904*: 113–145.

Johnsgard, P.A. 1965. *Handbook of waterfowl behavior.* Ithaca: Cornell University Press.

Kear, J., and A. Berger. 1980. *The Hawaiian goose: An experiment in conservation.* Vermillion, SD: Buteo Books.

Kirch, P.V. 1985. *Feathered gods and fishhooks: An introduction to Hawaiian archaeology and prehistory.* Honolulu: University of Hawaii Press.

Marshall, A.P., and J.M. Black. 1992. The effect of rearing experience on subsequent behavioural traits in Hawaiian geese (*Branta sanvicensis*): Implications for the recovery programme. *Bird Conservation International 2*: 131–147.

Olson, S.L., and H.F. James. 1982. Prodromus of the fossil avifauna of Hawaiian Islands. *Smithsonian Contributions to Zoology 365.*

Olson, S.L., and H.F. James. 1991. Description of 32 of birds from the Hawaiian Islands. Part 1. Non-passeriformes. *Ornithological Monographs 45*: 1–88.

Perkins, R.C.L. 1903. Fauna Hawaiiensis: Vertebrata (Aves). Pp. 365–466 in *Zoology of the Sandwich (Hawaiian) Islands,* Vol. 1, Part 4, ed. D. Sharp. Cambridge: Cambridge University Press.

Stone, C.P., H.M. Hoshide, and P.C. Banko. 1983. Productivity, mortality, and movements of nene in the Ka'u Desert, Hawaii Volcanoes National Park, 1981–1982. *Pacific Science 38*: 301–311.

Stone, C.P., R.L. Walker, J.M. Scott, and P.C. Banko. 1983. Hawaiian goose management and research—Where do we go from here? *Elepaio 44*: 11–15.

Tomich, P.Q. 1969. *Mammals in Hawaii.* Bernice P. Bishops Museum Special Publication 57. Honolulu: Bishop Museum Press.

U.S. Fish and Wildlife Service. 1983. *Nene: Hawaiian goose recovery plan.* Portland, OR: U.S. Fish and Wildlife Service.

Wilson, S.B., and A.H. Evans. 1890–1899. *Aves Hawaiiensis. The birds of the Sandwich Islands.* Reprint edition (1974) of the 1890–1899 edition published by R.H. Porter, London.

INDIANA BAT

Barbour, R.W., and W.H. Davis. 1969. *Bats of America.* Lexington: University Press of Kentucky.

Brack, V., and R.K. LaVal. 1985. Food habits of the Indiana bat in Missouri. *Journal of Mammalogy 66*: 308–315.

Clawson, R. 1987. Indiana bats: Down for the count. *Endangered Species Technical Bulletin* 22: 9–11.

Cope, J.B., and S.R. Humphrey. 1977. Spring and autumn swarming behavior in the Indiana bat, *Myotis sodalis. Journal of Mammalogy 58*: 93–95.

Fenton, B.M. 1983. *Just bats.* Toronto: University of Toronto Press.

Fenton, B.M. 1992. *Bats.* New York: Facts on File.

Hall, E.R. 1981. *The mammals of North America.* New York: John Wiley & Sons.

Hill, H.E., and J.D. Smith. 1984. *Bats: A natural history.* Austin: University of Texas Press.

Humphrey, S.R. 1978. Status, winter habitat and management of the endangered Indiana bat, *Myotis sodalis. Florida Scientist 41*: 65–76.

Humphrey, S.R., A.R. Richter, and J.B. Cope. 1977. Summer habitat and ecology of the endangered Indiana bat, *Myotis sodalis. Journal of Mammalogy 58*: 334–346.

Indiana Bat Recovery Team. 1996. *Indiana bat* (Myotis sodalis) *recovery plan.* Technical draft. Washington, DC: U.S. Fish and Wildlife Service, Department of the Interior.

Kurta, A., D. King, J.A. Teramino, J.M. Stribley, and K.J. Williams. 1993. Summer roosts of the endangered Indiana bat (*Myotis sodalis*) on the northern edge of its range. *American Midland Naturalist 129*: 132–138.

Kurta, A., and J.O. Whitaker Jr. 1998. Diet of the endangered Indiana bat (*Myotis sodalis*) on the northern edge of its range. *American Midland Naturalist 140*: 280–286.

Mumford, R.E., and J.O. Whitaker. 1982. *Mammals of Indiana.* Bloomington: Indiana University Press.

Pierson, E.D. 1998. Tall trees, deep holes, and scarred landscapes: Conservation biology of North American bats. Pp. 309–325 in *Bat biology and conservation*, ed. T.H. Kunz and P.A. Racey. Washington, DC: Smithsonian Institution Press.

U.S. Fish and Wildlife Service. 1983. *Recovery plan for the Indiana bat.* Washington, DC: Department of the Interior.

Wilson, D.E. 1997. *Bats in question.* Washington, DC: Smithsonian Institution Press.

IVORY-BILLED WOODPECKER

Allen, A.A., and P.P. Kellogg. 1937. Recent observations on the ivory-billed woodpecker. *Auk 54*: 164–184.

Andrews, R.C. 1951. *Nature's ways.* New York: Crown.

Anonymous. 1985. Status review on ivory-billed woodpecker. *Endangered Species Technical Bulletin 10*(5): 7.

Audubon, J.J., and J.B. Chevalier. 1840–1844. *The birds of America.* New York: Dover.

Baker, J.H. 1942. The director reports to you. *Audubon Magazine 44*: 367–376.

Baker, J.H. 1950. News of wildlife and conservation; ivory-bills now have sanctuary. *Audubon Magazine 52*: 391–392.

Barbour, T. 1944. *Vanishing Eden.* Boston: Little, Brown.

Bird, A.R. 1932. Ivory-bill is still king! *American Forests 38*: 634–635, 667.

Cahalane, V.H., C. Cottam, W.L. Finley, and A. Leopold. 1941. Report of the Committee on Bird Protection, 1940. *Auk 58*: 292–298.

Catesby, M. 1731. *Natural history of Carolina, Florida and the Bahama Islands.* Vol. 1. London: N.p.

Croker, T.C., Jr. 1979. Longleaf pine. *Journal of Forest History 23*: 32–43.

Dennis, J.V. 1948. A last remnant of ivory-billed woodpeckers in Cuba. *Auk 65*: 497–507.

Dennis, J.V. 1988. *The great cypress swamps.* Baton Rouge: Louisiana State University Press.

Diamond, J. 1987. Extant unless proven extinct? Or, extinct unless proven extant? *Conservation Biology 1*: 77–79.

Eastman, W. 1958. Ten-year search for the ivory-billed woodpecker. *Atlantic Naturalist 13*: 216–228.

Edge, R. 1943. The Singer Tract and the ivory-billed woodpecker. P. 22 in *Conservation for victory*. Publication No. 88, Annual Report 1942. New York: Emergency Conservation Committee.

Jackson, J.A. 1989. *Past history, habitats, and present status of the ivory-billed woodpecker* (Campephilus principalis) *in North America*. Final Report. Atlanta: U.S. Fish and Wildlife Service.

Jackson, J.A. 1991. Will-o'-the-wisp. *Living Bird Quarterly 10*(1): 29–32.

Jackson, J.A. 1996. Ivory-billed woodpecker. Pp. 103–112 in *Rare and endangered biota of Florida*, ed. J.A. Rodgers Jr., H.W. Kale II, and H.T. Smith. Gainesville: University Press of Florida.

Lamb, G.R. 1957. *The ivory-billed woodpecker in Cuba*. Research Report No. 1. New York: Pan-American Section, International Committee for Bird Preservation.

Lamb, G.R. 1958. Excerpts from a report on the ivory-billed woodpecker (*Campephilus principalis bairdi*) in Cuba. *Bulletin of the International Committee for Bird Preservation 7*: 139–144.

Lammertink, J.M. 1992. Search for ivory-billed woodpecker in Cuba. *Dutch Birding 14*: 170–177.

Lammertink, J.M. 1995. No more hope for the ivory-billed woodpecker *Campephilus principalis*. *Cotinga 3*: 45–47.

Lammertink, J.M., and A.R. Estrada. 1995. Status of the ivory-billed woodpecker *Campephilus principalis* in Cuba: Almost certainly extinct. *Bird Conservation International 5*: 53–59.

Lillard, R.G. 1947. *The great forest*. New York: Alfred A. Knopf.

Reynard, G. 1988. *Bird songs in Cuba*. Ithaca: Cornell Laboratory of Ornithology.

Short, L.L. 1982. *Woodpeckers of the world*. Greenville: Delaware Museum of Natural History.

Short, L.L. 1985. Last chance for the ivory-bill. *Natural History 94*: 66–68.

Short, L.L., and J.F.M. Horne. 1986. The ivory-bill still lives. *Natural History 95*(7): 26–28.

Skinner, A. 1926. Ethnology of the Ioway Indians. *Bulletin of the Public Museum of Milwaukee 5*(4): 181–354.

Sprunt, A., Jr., and E.B. Chamberlain. 1970. *South Carolina bird life*. Columbia: University of South Carolina Press.

Tanner, J.T. 1941. Three years with the ivory-billed woodpecker, America's rarest bird. *Audubon Magazine 43*(1): 4–14.

Tanner, J.T. 1942a. The ivory-billed woodpecker. Research Report No. 1. New York: National Audubon Society.

Tanner, J.T. 1942b. Present status of the ivory-billed woodpecker. *Wilson Bulletin 54*: 57–58.

Wayne, A.T. 1893. Additional notes on the birds of the Suwanee River. *Auk 10*: 336–338.

JAGUAR

Austin, W., and B. Palmer. 1997. *Final rule, U.S. population of jaguar*. Albuquerque, NM: U.S. Fish and Wildlife Service.

Caughley, G. 1993. Elephants and economics. *Conservation Biology 7*: 943–945.

Conway, W. 1992. *The role of zoos and aquariums in biological conservation: Past, present, and future*. Paper presented at AAZPA Annual Conference, September 17.

Crawshaw, P.G., and H.B. Quigley. 1991. Jaguar spacing, activity, and habitat use in a seasonally flooded environment in Brazil. *Journal of Zoology 223*: 357–370.

Frazier, J. 1990. International resource conservation: Thoughts on a challenge. *Transactions of the North American Wildlife and Natural Resources Conference 55*: 384–395.

Hoogesteijn, R., A. Hoogesteijn, and E. Mondolfi. 1993. Jaguar predation and conservation: Cattle mortality caused by felines on three ranches in the Venezuelan Llanos. *Symposium of the Zoological Society of London 65*: 391–407.

Larson, S.E. 1997. Taxonomic reevaluation of the jaguar. *Zoo Biology 16*: 107–120.

Mares, M.A. 1991. How scientists can impede the development of their discipline: Egocentrism, small pool size, and the evolution of sapismo. Pp. 57–75 in *Latin American mammalogy: History, biodiversity, and conservation*, ed. M.A. Mares and D.J. Schmidley. Norman: Oklahoma Museum of Natural History.

Middleton, N., P. O'Keefe, and S. Moyo. 1993. *Tears of the crocodile: From Rio to reality in the developing world*. Boulder, CO: Pluto Press.

Nowell, K., and P. Jackson. 1996. *Wild cats*. Gland, Switzerland: IUCN.

Rabinowitz, A.R., and B.G. Nottingham. 1986. Ecology and behavior of the jaguar (*Panthera onca*) in Belize, Central America. *Journal of the Zoological Society of London 210*: 149–159.

Seymour, K.L. 1989. *Panthera onca*. *Mammalian Species 340*: 1–9.

Swank, W.G., and J.G. Teer. 1989. Status of the jaguar—1987. *Oryx 23*: 14–21.

Woodroffe, R., and J.R. Ginsberg. 1998. Edge effects and the extinction of populations inside protected areas. *Science 280*: 2126–2128.

KEMP'S RIDLEY SEA TURTLE

Bjorndal, K.A. 1997. Foraging ecology and nutrition of sea turtles. Pp. 199–231 in *The biology of sea turtles*, ed. P.L. Lutz and J.A. Musick. New York: CRC Press.

Bookchin, M. 1994. *Which way for the ecology movement?* San Francisco: AK Press.

Bowen, B.W., and S.A. Karl. 1997. Population genetics, phylogeography, and molecular evolution. Pp. 29–50 in *The biology of sea turtles*, ed. P.L. Lutz and J.A. Musick. New York: CRC Press.

Brongersma, L. 1972. European Atlantic turtles. *Zoologische Verhandelingen. Leiden 121*: 1–318.

Brulle, R.J. 1996. Environmental discourse and social movement organizations: A historical and rhetorical perspective on the development of U.S. environmental organizations. *Social Inquiry 66*: 58–83.

Burchfield, P.M., L. Dierauf, and R.A. Byles. 1997. *Report on the Mexico/United States of American population restoration project for Kemp's ridley sea turtle*, Lepidochelys kempi, *on the coasts of Tamaulipas and Veracruz, Mexico, 1997*. Albuquerque, NM: U.S. Department of the Interior, Fish and Wildlife Service.

Caillouet, E.W., and A.M. Landry, eds. 1989. *Proceedings of the first international symposium on Kemp's ridley sea turtle biology, conservation and management*. Galveston: Texas A&M University. (TAMU-SG-89-105.)

Eckert, S.A., D. Crouse, L.B. Crowder, M. Maceina, and A. Shah. 1994. *Review of the Kemp's ridley sea turtle headstart program*. NOAA Technical Memorandum NMFS-OPR-3. Washington, DC: U.S. Department of Commerce, National Oceanic and Atmospheric Administration, National Marine Fisheries Service.

Frazier, J. 1997. Sustainable development: Modern elixir or sack dress? *Environmental Conservation 24*: 182–193.

Margavio, A.V., and C.J. Forsyth. 1996. *Caught in the net: The conflict between shrimpers and conservationists*. College Station: Texas A&M University Press.

Márquez M.R. 1994. *Synopsis of biological data on the Kemp's ridley turtle*, Lepidochelys kempi *(Garman, 1880)*. NOAA Technical Memorandum NMFS-SEFSC-343. Miami: U.S. Department of Commerce, National Oceanic and Atmospheric Administration, National Marine Fisheries Service, Southeast Fisheries Science Center.

Millett, B.L. 1998. *Kemp's ridley recovery: From myopic management to hope for the new millenium* [*sic*]. Washington, DC: PEAT Institute and Darden Environmental Trust.

Musick, J.A., and C.J. Limpus. 1997. Habitat utilization and migration in juvenile sea turtles. Pp. 137–163 in *The biology of sea turtles*, ed. P.L. Lutz and J.A. Musick. New York: CRC Press.

NRC (National Research Council). 1990. *Decline of the sea turtles: Causes and prevention.* Washington, DC: National Academy Press.

Owens, D.W. (comp.), and M. Evans. 1999. Sharing the Gulf: A challenge for us all. Conference Proceedings, June 10–12, 1998. College Station: Sea Grant, Texas A&M University.

Pulliam, H.R. 1998. The political education of a biologist: Part II. *Wildlife Society Bulletin 26*: 499–503.

Ross, J.P., S. Beavers, D. Mundell, and M. Airth-Kindree. 1989. *The status of Kemp's ridley: A report to the Center for Marine Conservation from the Caribbean Conservation Corporation.* Washington, DC: Center for Marine Conservation.

Shaver, D.J., and C.W. Caillouet. 1998. More Kemp's ridley turtles return to South Texas to nest. *Marine Turtle Newsletter 82*: 1–5.

TEWG (Turtle Expert Working Group). 1998. *An assessment of Kemp's ridley* (Lepidochelys kempii) *and loggerhead* (Caretta caretta) *sea turtle populations in the western North Atlantic.* Washington, DC: NOAA Technical Memorandum NMFS-SEFSC-409.

USFWS and NMFS (U.S. Fish and Wildlife Service and National Marine Fisheries Service). 1992. *Recovery plan for the Kemp's ridley sea turtle* (Lepidochelys kempii). St. Petersburg, FL: National Marine Fisheries Service.

Weber, M. 1996. Book review: Caught in the net: The conflict between shrimpers and conservationists. *Marine Turtle Newsletter 75*: 31–32.

Weber, M., D. Crouse, R. Irvin, and S. Iudicello. 1995. *Delay and denial: A political history of sea turtles and shrimp fishing.* Washington, DC: Center for Marine Conservation.

Wood, J.R., and F. Wood. 1988. Captive reproduction of the Kemp's ridley, *Lepidochelys kempi. Herpetological Journal 1*: 247–249.

Zug, G.R., H.J. Kalb, and S.J. Luzar. 1996. Age and growth in wild Kemp's ridley sea turtles (*Lepidochelys kempii*) from skeletochronological data. *Biological Conservation 80*: 261–268.

LEATHERBACK TURTLE

Boulon, R.H., Jr., P.H. Dutton, and D.L. McDonald. 1996. Leatherback turtles (*Dermochelys coriacea*) on St. Croix, U.S. Virgin Islands: Fifteen years of conservation. *Chelonian Conservation and Biology 2*: 141–147.

Chan, Eng-Heng, and H.-C. Liew. 1996. Decline of the leatherback population in Terengganu, Malaysia, 1956–1995. *Chelonian Conservation and Biology 2*: 196–203.

Deraniyagala, P.E.P. 1939. *The tetrapod reptiles of Ceylon*, Vol. 1. *Testudiates and Crocodilians.* Colombo, Sri Lanka: Colombo Museum.

Duguy, R., P. Moriniere, and C. Le Milinaire. 1998. Factors of mortality of marine turtles in the Bay of Biscay. *Oceanologica Acta 21*: 383–388.

Dutton, D.L., P.H. Dutton, and R. Boulon. In press. Recruitment and mortality estimates for female leatherbacks, *Dermochelys coriacea*, nesting on St. Croix, U.S. Virgin Islands. In *Proceedings of the 19th Annual Symposium on the Biology and Conservation of Sea Turtles*, comp. H. Kalb and T. Wibbles. NOAA Technical Memorandum NMFA-SEFSC. Miami: U.S. Department of Commerce, National Oceanic and Atmospheric Administration, National Marine Fisheries Service, Southeast Fisheries Science Center.

Dutton, P.H., B.W. Bowen, D.W. Owens, A.R. Barragan, and S.K. Davis. 1999. Global phylogeography of the leatherback turtle, *Dermochelys coriacea*: Shallow phylogenetic history in an ancient organismal lineage. *Journal of Zoology 248*: 397–407.

Dutton, P.H., C. Whitmore, and N. Mrosovsky. 1985. Masculinization of leatherback turtle, *Dermochelys coriacea*, hatchlings from eggs incubated in styrofoam boxes. *Biological Conservation 31*: 249–264.

Eckert, S.A., H.-C. Liew, K.L. Eckert, and E.-H. Chan. 1996. Shallow water diving by leatherback turtles in the South China Sea. *Chelonian Conservation and Biology 2*: 237–243.

Eckert, S.A., and L. Sarti M. 1997. Distant fisheries implicated in the loss of the world's largest leatherback nesting population. *Marine Turtle Newsletter 78*: 2–7.

Fairlie, S., ed. 1995. Overfishing: Its causes and consequences. *The Ecologist 25*: 41–127.

FAO (Food and Agriculture Organization of the United Nations). 1995. *Code of conduct for responsible fisheries*. Rome: FAO.

Frazier, J. 1987. Semantics and the leathery turtle, *Dermochelys coriacea*. *Journal of Herpetology 21*: 240–242.

Frazier, J. 1997. Sustainable development: Modern elixir or sack dress? *Environmental Conservation 24*: 182–193.

Frazier, J. 1999. Guest editorial: Update on the Inter-American Convention for the Protection and Conservation of Sea Turtles. *Marine Turtle Newsletter 84*: 1–3.

Fretey, J. 1998. *Marine turtles of the Atlantic coast of Africa*. UNEP/CMS Technical Publication 1. Bonn, Germany: UNEP/CMS.

Girondot, M., and J. Fretey. 1996. Leatherback turtles, *Dermochelys coriacea*, nesting in French Guiana, 1978–1995. *Chelonian Conservation and Biology 2*: 204–208.

Hughes, G.R. 1996. Nesting of the leatherback turtle (*Dermochelys coriacea*) in Tongaland, KwaZulu-Natal, South Africa, 1963–1995. *Chelonian Conservation and Biology 2*: 153–158.

Hunter, J.R. 1997. *Simple things won't save the Earth*. Austin: University of Texas Press.

Juárez Cerón, J.A., A.R. Barragán, and H. Gómez Ruíz. In press. Contamination by phthalate ester plasticizers in the yolk of two marine turtle species. In *Proceedings of the 18th Annual Symposium on the Biology and Conservation of Sea Turtles*, comp. A. Abreu, R. Briseño, R. Márquez, and L. Sarti. NOAA Technical Memorandum NMFS-SEFSC. Miami: U.S. Department of Commerce, National Oceanic and Atmospheric Administration, National Marine Fisheries Service, Southeast Fisheries Science Center.

Korten, D.C. 1995. *When corporations rule the world*. West Hartford, CT, and San Francisco: Kumarian Press and Berrett-Koehler Publishers.

Lutcavage, M.E., and P.E. Lutz. 1997. Diving physiology. Pp. 277–296 in *The biology of sea turtles*, ed. P.L. Lutz and J.A. Musick. New York: CRC Press.

Lutcavage, M.E., P. Plotkin, B. Witherington, and P.E. Lutz. 1997. Human impacts on sea turtle survival. Pp. 387–409 in *The biology of sea turtles*, ed. P.L. Lutz and J.A. Musick. New York: CRC Press.

McDonald, D.L., and P.H. Dutton. 1996. Use of PIT tags and photoidentification to revise remigration estimates of leatherback turtles (*Dermochelys coriacea*) nesting in St. Croix, U.S. Virgin Islands, 1979–1995. *Chelonian Conservation and Biology 2*: 148–152.

McGoodwin, J.R. 1990. *Crisis in the world's fisheries: People, problems, and politics*. Stanford: Stanford University Press.

Miller, J.A. 1997. Reproduction in sea turtles. Pp. 51–81 in *The biology of sea turtles*, ed. P.L. Lutz and J.A. Musick. New York: CRC Press.

Morgan, P.J. 1989. Occurrence of leatherback turtles (*Dermochelys coriacea*) in the British Isles in 1988, with reference to a record specimen. Pp. 119–120 in *Proceedings of the 9th annual symposium on the biology and conservation of sea turtles*, comp. S.A. Eckert, K.L. Eckert, and T.H. Richardson. NOAA Technical Memorandum NMFA-SEFC 232. Miami: U.S. Department of Commerce, National Oceanic and Atmospheric Administration, National Marine Fisheries Service, Southeast Fisheries Center.

NRC (National Research Council). 1995. *Understanding marine biodiversity: A research agenda for the nation*. Washington, DC: National Academy Press.

Paladino, F.V., J.R. Spotila, M.P. O'Connor, and R.E. Gatten Jr. 1996. Respiratory physiology of adult leatherback turtles (*Dermochelys coriacea*) while nesting on land. *Chelonian Conservation and Biology 2*: 223–229.

Pritchard, P.C.H. 1996. Are leatherbacks really threatened with extinction? *Chelonian Conservation and Biology 2*: 303–305.

Pritchard, P.C.H. 1997. Evolution, phylogeny, and current status. Pp. 1–28 in *The biology of sea turtles*, ed. P.L. Lutz and J.A. Musick. New York: CRC Press.

Rhodin, J.A.G., A.G.J. Rhodin, and J.R. Spotila. 1996. Electron microscopic analysis of vascular cartilage canals in the humeral epiphysis of hatchling leatherback turtles, *Dermochelys coriacea*. *Chelonian Conservation and Biology 2*: 250–260.

Sarti, L., L. Flores, and A. Aguayo. 1994. Evidence of predation of killer whale (*Oricinus orca*) on a leatherback sea turtle (*Dermochelys coraica*) in Michoacan, Mexico. *Revista de la Investigación Científica 2* (Numero especial de SOMEMA 2): 23–26.

Sarti, L., S. Eckert, P. Dutton, A. Barragán, and N. García-T. In press. The current situation of the leatherback population on the Pacific coast of México and Central America, abundance and distribution of the nestlings: An update. In *Proceedings of the 19th annual symposium on the biology and conservation of sea turtles*, comp. H. Kalb and T. Wibbels. NOAA Technical Memorandum NMFS-SEFSC. Miami: U.S. Department of Commerce, National Oceanic and Atmospheric Administration, National Marine Fisheries Service, Southeast Fisheries Science Center.

Spotila, J.R., A.E. Dunham, A.J. Leslie, A.C. Steyermark, P.T. Plotkin, and F.V. Paladino. 1996. Worldwide population decline of *Dermochelys coriacea*: Are leatherback turtles going extinct? *Chelonian Conservation and Biology 2*: 209–222.

Spotila, J.R., M.P. O'Connor, and F.V. Paladino. 1997. Thermal biology. Pp. 297–314 in *The biology of sea turtles*, ed. P.L. Lutz and J.A. Musick. New York: CRC Press.

Steyermark, A.C., K. Williams, J.R. Spotila, F.V. Paladino, D.C. Rostral, S.J. Morreale, M.T. Koberg, and R. Arauz. 1996. Nesting leatherback turtles at Las Baulas National Park, Costa Rica. *Chelonian Conservation and Biology 2*: 173–183.

Suarez, A., and C.H. Starbird. 1996. Subsistence hunting of leatherback turtles, *Dermochelys coriacea*, in the Kai Islands, Indonesia. *Chelonian Conservation and Biology 2*: 190–195.

Wetherall, J.A., G.A. Balazs, R.A. Tokunaga, and M.Y.Y. Yong. 1993. Bycatch of marine turtles in North Pacific high-seas driftnet fisheries and impacts on the stocks. *International North Pacific Fisheries Commission Bulletin 53*: 519–538.

Witzell, W.N. 1999. Distribution and relative abundance of sea turtles caught incidentally by the longline fleet in the western north Atlantic Ocean 1992–1995. *Fisheries Bulletin 97*: 200–211.

Zug, G.R., and J.F. Parham. 1996. Age and growth in leatherback turtles, *Dermochelys coriacea* (Testudines: Dermochelyidae): A skeletonchronological analysis. *Chelonian Conservation and Biology 2*: 244–249.

MALAGASY FRESHWATER FISHES

Arnoult, J. 1959. Poissons des eaux douces. Faune De Madagascar 10, I.R.S.M., Tananarive.

Cann, C. 1996. *Naturalized fishes of the world*. San Diego: Academic Press.

De Rham, P. 1997. Main results of October 1997 collecting trip to Madagascar. *Aquatic Survival Bulletin 6*: 1–11.

Haeffner, R. 1998. Conservation of Malagasy freshwater fishes. Pp. 51–54 in *American Zoo and Aquarium Regional Conference Proceedings*. Bethesda, MD: American Zoo and Aquarium Association.

Jolly, A., P. Oberle, and R. Albignac, eds. 1984. *Madagascar*. Oxford: Pergamon Press.

Kiener, A. 1963. Poissons, peche et pisciculture à Madagascar. *Centre Technique Forestier Tropical 24*: 1–224.

Kiener, A., and M. Mauge. 1966. Contribution à l'etude systematique et ecologique des poissons Cichlidae endemiques de Madagascar. *Memoirs de Museum Nationale d'Histoire 40*(2): 51–99.

Loiselle, P. 1994. The Cichlids of Madagascar: Going . . . going . . . gone? *Buntbarsche Bulletin 161*: 1–7.

Mittermeier, R.A. 1988. Primate diversity and the tropical forest: Case studies from Brazil and Madagascar and the importance of the megadiversity countries. Pp. 145–154 in *Biodiversity*, ed. E.O. Wilson. Washington, DC: National Academy Press.

Preston-Mafham, K. 1991. *Madagascar: A natural history*. New York: Facts on File.

Reinthal, P.N., and M.L.J. Stiassny. 1991. The freshwater fishes of Madagascar: A study of an endangered fauna with recommendations for a conservation strategy. *Conservation Biology 5*: 231–243.

Stiassny, M.L.J. 1990. Notes on the anatomy and relationships of the bedotiid fishes of Madagascar with a taxonomic revision of the genus *Rheocles* (Atherinomorpha: Bedotiidae). *American Museum of Novitates 2979*: 1–33.

Stiassny, M.L.J., and N. Raminosoa. 1994. Biological diversity in African fresh- and brackish water fishes. Geographical overviews. *Annales du Musée Royal de l'Afrique Centrale, Zoologie 275*: 133–149.

MARBLED MURRELET

American Ornithologists' Union. 1997. Forty-first supplement to the American Ornithologists' Union check-list of North American birds. *Auk 114*: 542–552.

Beissinger, S.R. 1995. Population trends of the marbled murrelet from demographic analyses. Pp. 385–393 in *Ecology and conservation of the marbled murrelet*, ed. C.J. Ralph, G.L. Hunt Jr., M.G. Raphael, and J.F. Piatt. Albany, CA: U.S. Department of Agriculture, Forest Service, Southwest Research Station General Technical Report PSW-GTR-152.

Beissinger, S.R., and N. Nur. 1997. Appendix B: Population trends of the marbled murrelet projected from demographic analysis. Pp. B1–B35 in *Recovery plan for the marbled murrelet* (Brachyramphus marmoratus) *in Washington, Oregon, and California*. Portland, OR: U.S. Fish and Wildlife Service.

Binford, L.C., B.G. Elliott, and S.W. Singer. 1975. Discovery of a nest and the downy young of the marbled murrelet. *Wilson Bulletin 87*: 303–319.

Booth, D.E. 1991. Estimating prelogging old-growth in the Pacific Northwest. *Journal of Forestry 89*: 25–29.

Carter, H.R., M.L.C. McAllister, and M.E.P. Isleib. 1995. Mortality of marbled murrelets in gil¹ nets in North America. Pp. 271–283 in *Ecology and conservation of the marbled murrelet*, ed. C.J. Ralph, G.L. Hunt Jr., M.G. Raphael, and J.F. Piatt. Albany, CA: U.S. Department of Agriculture, Forest Service, Southwest Research Station General Technical Report PSW-GTR-152.

De Santo, T.L., and S.K. Nelson. 1995. Comparative reproductive ecology of the auks (Family Alcidae) with emphasis on the marbled murrelet. Pp. 33–47 in *Ecology and conservation of the marbled murrelet*, ed. C.J. Ralph, G.L. Hunt Jr., M.G. Raphael, and J.F. Piatt. Albany, CA: U.S. Department of Agriculture, Forest Service, Southwest Research Station General Technical Report PSW-GTR-152.

Hamer, T.E., B.A. Cooper, and C.J. Ralph. 1995. Use of radar to study the movements of marbled murrelets at inland sites. *Northwestern Naturalist 76*: 73–78.

Marshall, D.B. 1988. Status of the marbled murrelet in North America: With special emphasis on populations in California, Oregon and Washington. *USDI Fish and Wildlife Service Biological Report 88*(30): 1–19.

Nelson, S.K., and S.G. Sealy. 1995. Symposium: Biology of the marbled murrelet: Inland and at sea. *Northwestern Naturalist 76*: 1–119.

Norse, E.A. 1990. *Ancient forests of the Pacific Northwest*. Washington, DC: Island Press.

Noss, R.F., and A.Y. Cooperrider. 1994. *Saving nature's legacy: Protecting and restoring biodiversity*. Washington, DC: Island Press.

Piatt, J.F., and C.J. Lensink. 1989. *Valdez* bird toll. *Nature 342*: 865–866.

Ralph, C.J., G.L. Hunt Jr., M.G. Raphael, and J.F. Piatt. 1995. *Ecology and conservation of the marbled murrelet*. Albany, CA: U.S. Department of Agriculture, Forest Service, Southwest Research Station General Technical Report PSW-GTR-152.

U.S. Department of Agriculture and U.S. Department of the Interior. 1994. *Supplemental environmental impact statement on management of habitat for late-successional and old-*

growth forest related species within the range of the northern spotted owl. Portland, OR: U.S.D.A. Forest Service and U.S.D.I. Bureau of Land Management.

U.S. Fish and Wildlife Service. 1997. *Recovery plan for the threatened marbled murrelet* (Brachyramphus marmoratus) *in Washington, Oregon, and California.* Portland, OR.

MARIANA CROW

Aguon, C.F, R.E. Beck Jr., and M.W. Ritter. 1998. A method for protecting nests of the Mariana crow from brown tree snake predation. In *Problem snake management: Habu and brown tree snake examples,* ed. G.H. Rodda, Y. Sawai, D. Chizar, and H. Tanaka. Ithaca: Cornell University Press.

Baker, R.H. 1951. The avifauna of Micronesia, its origin, evolution, and distribution. *University of Kansas Publications of the Museum of Natural History 3*: 1–359.

Beck, R.E., Jr., and J.A. Savidge. 1990. *Native forest birds of Guam and Rota of the Commonwealth of the Northern Mariana Islands recovery plan.* Portland, OR: U.S. Fish & Wildlife Service.

Collar, N.J., M.J. Crosby, and A.J. Stattersfield. 1994. *Birds to watch 2: The world list of threatened birds.* BirdLife Conservation Series No. 4., BirdLife International. Washington, DC: Smithsonian Institution Press.

DAWR (Division of Aquatic and Wildlife Resources). 1985–1995, 1997. *Survey and inventory of native land birds on Guam.* Annual reports. Mangilao: Division of Aquatic and Wildlife Resources, Guam Department of Agriculture.

Engbring, J., and F.L. Ramsey. 1984. *Distribution and abundance of the forest birds of Guam: Results of a 1981 survey.* FWS/OBS-84/20. Washington, DC: U.S. Fish and Wildlife Service.

Engbring, J., F.L. Ramsey, and V.J. Wildman. 1986. *Micronesian forest bird survey, 1982: Saipan, Tinian, Agiguan, and Rota.* Honolulu: U.S. Fish and Wildlife Service.

Fritts, T.H. 1988. The brown tree snake, *Boiga irregularis,* a threat to Pacific islands. *U.S. Fish and Wildlife Service Biological Bulletin 88*(31): 1–36.

Grout, D.J, M. Lusk, and S.G. Fancey. 1996. *Results of the 1995 Mariana crow survey on Rota.* Final report. Honolulu: U.S. Fish and Wildlife Service.

Jenkins, J.M. 1983. *The native forest birds of Guam.* Ornithological Monographs No. 31. Washington, DC: American Ornithologists' Union.

Marshall, J.T., Jr. 1949. The endemic avifauna of Saipan, Tinian, Guam and Palau. *Condor 51*: 200–221.

Michael, G.A. 1987. Notes on the breeding biology and ecology of the Mariana or Guam crow. *Aviculture Magazine 93*: 73–82.

Morton, J.M. 1996. The effects of aircraft overflights on endangered Mariana crows and Mariana fruit bats at Anderson Air Force Base, Guam. Pearl Harbor: Department of the Navy, Pacific Division, Naval Facilities Engineering Command.

National Research Council. 1997. *The scientific basis for preservation of the Mariana crow.* Washington, DC: National Academy Press.

Savidge, J.A. 1987. Extinction of an island forest avifauna by an introduced snake. *Ecology 68*: 660–668.

Strophet, J.J. 1946. Birds of Guam. *Auk 63*: 534–540.

Tarr, C.L., and R.C. Fleischer. 1999. Molecular assessment of genetic variability and population differentiation in the endangered Mariana crow (*Corvus kubaryi*). *Molecular Ecology 8*: 941–950.

Tomback, D.F. 1986. Observations on the behavior and ecology of the Mariana crow. *Condor 88*: 398–401.

USFWS (U.S. Fish and Wildlife Service). 1984. Endangered and threatened wildlife and plants: Determination of endangered status for seven birds and two bats of Guam and the Northern Mariana Islands. *Federal Register 49*(167): 33881–33885.

MEDITERRANEAN MONK SEAL

Aguilar, A. 1997. *Current situation of the monk seal die-off in Western Sahara-Mauritania.* Report in Internet-Marmam Research and Conservation Discussion, 17 July 1997. Marmamed@uvic.ca, MARMAM Editors, CA.

Aguilar, A. 1998. *Current status of Mediterranean Monk sea* (Monachus monachus). Regional Activity Center for Specially Protected Areas UNEP. RAC/SPA, B.P. 337, 1080. Tunis, Tunisia: CEDEX.

Forcada, J., P. Hammond, and A. Aguilar. 1998. Population size of the monk seal of Cabo Blanco colony. P.15 in *Workshop biology and conservation of endangered monk seals, Monaco, 19–20 January.* Lawrence, KS: Society for Marine Mammalogy.

González, L.M., A. Aguilar, L.F. López-Jurado, and E. Grau. 1997. Status and conservation of Mediterranean monk seal (*Monachus monachus*) on the Cabo Blanco peninsula (Western Sahara-Mauritania) in 1993–95. *Biological Conservation 80*: 225–233.

González, L.M., L.F. López-Jurado, and P. Lopez. 1997. *Report on the release of two juvenile Mediterranean monk seals in Cape Blanc peninsula.* Report in Internet-Marmam Research and Conservation Discussion, 31 December 1997. Marmamed@uvic.ca, MARMAM Editors, CA.

Harwood, J., D. Lavigne, and P. Reijnders. 1998. *Workshop on the causes and consequences of the 1997 mass mortality of Mediterranean monk seals in the Western Sahara.* IBN Scientific Contributions 11. Wageningen, The Netherlands: DLO Institute for Forestry and Nature Research.

Hernández, M., I. Robinson, A. Aguilar, L.M. González, L.F. López-Jurado, M.I. Reyero, E. Cacho, J.M. Franco, V. López-Rodas, and E. Costas. 1998. Did algal toxins cause monk seal mortality? *Nature 393*: 28–29.

King, J.E. 1955. The monk seals (genus *Monachus*). *Bulletin of the British Museum (Natural History) Zoology 3*: 201–256.

Marchessaux, D. 1989. *Recherches sur la biologie, écologie et le status du phoque moine.* Marseille, France: GIS Posidonie.

Muizon, de, C. 1982. Phocid phylogeny and dispersal. *Annals of the South African Museum 89*: 175–213.

Osterhaus, A., J. Groen, H. Niesters, L. Vedder, H. Van Egmond, B.A. Sidi, and M.E.O. Bartham. 1997. Morbillivirus in monk seal mass mortality? *Nature 388*: 838–839.

Reijnders, P.J.H. 1998. The Mediterranean monk seal: Present status and conservation efforts to remedy threats. P. 29 in *Workshop biology and conservation of endangered monk seals, Monaco, 19–20 January.* Lawrence, KS: Society for Marine Mammalogy.

SRRC (Seal Research and Rehabilitation Center). 1997a. *Note sur la situation actuelle des phoques et analyse du plan d'action d'urgence.* Interim Report to the European Commission, July 1997. DGXI/B2/CE. Brussels, Belgium: European Commission.

SRRC (Seal Research and Rehabilitation Center). 1997b. *Update with respect to the mass-mortality of the monk seal population on the Atlantic coast of the Cap-Blanc peninsula.* Report in Internet-Marmam Research and Conservation Discussion, 29 June 1997. Marmamed@uvic.ca, MARMAM Editors, CA.

Universidad de Barcelona. 1997. *Actuaciones y viabilidad de la colonia de focas monje de la península de Cabo Blanco, 1996–97.* D.G.XI/B2CE. Brussels, Belgium: E.U.-LIFE.

Universidad de Las Palmas de Gran Canaria. 1997. *Actuaciones y viabilidad de la colonia de focas monje de la península de Cabo Blanco, 1996–97.* D.G.XI/B2CE. Brussels, Belgium: E.U.-LIFE.

Vedder, E. 1998. *Status report of the presently released monk seals off the coast of Mauritania.* Unpublished interim report, January 8, 1998. Seal Research and Rehabilitation Center (SRRC), 9968 AG Pieterburen, the Netherlands.

Vedder, E., E. Androukaki, S. Abou, L. Hart, J. Azza, S. Kotomatas, and A. Osterhaus. 1998. Rehabilitation program for orphaned mediterranean monk seal pups. P. 32 in *Workshop biology and conservation of the endangered monk seals, Monaco, 19–20 January.* Lawrence, KS: Society for Marine Mammalogy.

MEXICAN PRAIRIE DOG

Baker, R.H. 1956. Mammals of Coahuila, Mexico. University of Kansas Publications. *Museum of Natural History* 9: 125–335.

Ceballos, G., and D.E. Wilson. 1985. *Cynomys mexicanus. Mammalian Species 248*: 1–3.

Chesser, R.K. 1983. Genetic variability within and among populations of the black-tailed prairie dog. *Evolution 37*: 320–331.

CITES (Convention on International Trade in Endangered Species of Wild Fauna and Flora). 1992. *Appendix I and II.* Châtelaine-Genève, Switzerland: United Nations Environmental Program.

Clark, T.W. 1977. Ecology and ethology of the white-tailed prairie dog (*Cynomys leucurus*). Milwaukee Publications. *Museum Publications on Biology and Geology 9*: 1–97.

Garret, M.G. 1982. *Dispersal of black-tailed prairie dogs* (Cynomys ludovicianus) *in Wind Cave National Park, South Dakota.* Master's thesis, Iowa State University, Ames.

Hall, E.R. 1981. *The mammals of North America*, 2nd ed. New York: John Wiley & Sons.

Hoogland, J.L. 1981a. The evolution of coloniality in white-tailed and black-tailed prairie dogs (Sciuridae: *Cynomys leucurus* and *C. ludovicianus*). *Ecology 62*: 252–272.

Hoogland, J.L. 1981b. Nepotism and cooperative breeding in black-tailed prairie dog (Sciuridae: *Cynomys ludovicianus*). Pp. 283–310 in *Natural selection and social behavior*, ed. R.D. Alexander and D.W. Twinkle. New York: Chiron Press.

Hoogland, J.L. 1995. The black-tailed prairie dog: Social life of a burrowing mammal. Chicago: University of Chicago Press.

IUCN (International Union for the Conservation of Nature). 1990. *Red list of threatened animals.* Gland, Switzerland: Conservation Monitoring Center, International Union for the Conservation of Nature—The World Conservation Union.

Jiménez-Guzmán, A. 1966. *Mammals of Nuevo Leon, Mexico.* Master's thesis, University of Kansas, Lawrence.

King, J.A. 1955. Social behavior, social organization and population dynamics in a black-tailed prairie dog town in the Black Hills of South Dakota. *Contributions from the Laboratory of Vertebrate Biology, University of Michigan, Ann Arbor 67*: 1–123.

McCullough, D.A., and R.K. Chesser. 1987. Genetic variation among populations of the Mexican prairie dog. *Journal of Mammalogy 68*: 555–560.

Medina-T., J.G., and J.A. de la Cruz. 1976. Ecología y control del perrito de las praderas *Cynomys mexicanus* Merriam en el Norte de México. *Monografía Técnica Científica, Universidad Autónoma Agraria Antonio Narro 2*: 365–418.

Mellink, E. 1989. La erosión del suelo como una amenaza para las colonias de perro llanero, en el norte de San Luis Potosí. Pp. 68–76 in *Memorias del Simposio sobre Fauna Silvestre*, ed. M.A. Roa R. and L. Palazuelos. Ciudad de México, Distrito Federal 7: Universidad Nacional Autónoma de México.

Oldemeyer, J.L., D.E. Biggins, and B.J. Miller. 1993. Proceedings of the symposium on the management of prairie dog complexes for the reintroduction of the black-footed ferret. *United States Fish and Wildlife Service Biological Report 93*: 1–97.

Scott-M., K.M. 1984. *Taxonomía y relación con los cultivos, de roedores y lagomorfos, en el Ejido El Tokio, Galeana, Nuevo León, México.* Thesis, Facultad de Ciencias Biológicas, Universidad Autónoma de Nuevo León, Monterrey, Nuevo León, México.

SEDESOL (Secretaría de Desarrollo Social). 1994. Norma Oficial Mexicana NOM-059-ECOL-1994, que determina las especies y subespecies de flora y fauna silvestres terrestres y acuáticas en peligro de extinción, amenazadas, raras y las sujetas a protección especial, y que establece especificaciones para su protección. *Diario Oficial de la Federación 438*: 1–60.

Treviño-Villarreal, J. 1988. *Area protegida del altiplano mexicano en el municipio de Galeana, Nuevo León, México*. Proposal presented to the Gobierno del Estado de Nuevo León, Secretaría de Fomento Agropecuario, Dirección de Planeación.

Treviño-Villarreal, J. 1990. The annual cycle of the Mexican prairie dog (*Cynomys mexicanus*). *Occasional Papers of the Museum of Natural History, University of Kansas 139*: 1–27.

Treviño-Villarreal, J., I.M. Berk, A. Aguirre, and W.E. Grant. 1998. Survey for sylvatic plague in the Mexican prairie dog (*Cynomys mexicanus*). *Southwestern Naturalist 43*: 147–154.

Treviño-Villarreal, J., I.M. Berk, and E.C. Andrade-Limas. 1996. The fate of the Mexican prairie dog (*Cynomys mexicanus*) in Coahuila, Nuevo León, and San Luis Potosí: A case study of the human induced changes in the landscape of northern Mexico. Pp. 44–51 in *Proceedings of the symposium on the ecology of our landscape: The botany of where we live*, ed. M. Hackett and S.H. Sohmer. Fort Worth: Botanical Research Institute of Texas.

Treviño-Villarreal, J., and W.E. Grant. 1998. Geographic range of the endangered Mexican prairie dog (*Cynomys mexicanus*). *Journal of Mammalogy 79*: 1273–1287.

Treviño-Villarreal, J., W.E. Grant, and A. Cardona-Estrada. 1997. Characterization of soil texture in Mexican prairie dog (*Cynomys mexicanus*) colonies. *Texas Journal of Science 49*: 207–214.

Treviño-Villarreal, J., C. Gutiérrez, A. Mora, A. Cardona, A. García, L. Corral, J.L. Mora, and A.M. Delgado. 1992. *Informe adicional sobre el perrito mexicano de las praderas* Cynomys mexicanus *al estudio de impacto ambiental del rancho Santa Ana ubicado en el Municipio de Vanegas, San Luis Potosí*. Instituto de Ecología y Alimentos. Cd. Victoria, Tamaulipas, México: Universidad Autónoma de Tamaulipas.

USFWS (U.S. Fish and Wildlife Service). 1991. *Endangered and threatened wildlife and plants*. Washington, DC: Department of the Interior, United States Fish and Wildlife Service.

MEXICAN WOLF

Barsness, J. 1998. Killing machines. *Field and Stream CIII*: 57–64.

Bass, R. 1998. Halfway home. *Audubon* (March-April): 61–67, 102.

Benke, R. 1998. Wolf killings by saboteurs? *Daily News* (Albuquerque, NM), 11 November.

Bernal, J.F., and J.M. Packard. 1997. Differences in winter activity, courtship and social behavior of two captive family groups of Mexican wolves *Canis lupus baileyi*. *Zoo Biology 16*: 435–443.

Bernal-Stoopen, J.F. 1999. *Binational cooperation in endangered species recovery: Case study of Mexican wolf*. Ph.D. thesis, Texas A&M University, College Station.

Bowden, C. 1992. Lonesome lobo. *Wildlife Conservation* (January/February): 46–73.

Brown, D., ed. 1983. *The wolf in the Southwest: The making of an endangered species*. Tucson: University of Arizona Press.

Burbank, J.C. 1990. *Vanishing lobo*. Boulder, Co: Johnson Publishing.

Crane, C. 1989. The last of the lobos. *Animals* (July/August): 18–24.

Dale, D. 1992. Chances with wolves. *Beef* (February): 88–89.

Dinon, J. 1998. Mexican gray wolves return to the wild. *CHAT 3*: 8–10.

Groebner, D.J., A.L. Girmendonk, and T.B. Johnson. 1995. *A proposed cooperative reintroduction plan for the Mexican wolf in Arizona*. Technical Report 56. Phoenix: Nongame and Endangered Wildlife Program, Arizona Game and Fish Department.

Hedrick, P., P. Miller, E. Geffen, and R. Wayne. 1997. Genetic evaluation of the three captive Mexican wolf lineages. *Zoo Biology 161*: 47–69.

Leopold, A. 1959. *Wildlife of Mexico*. Berkeley: University of California.

Packard, J.M. In press. Social behavior of wolves: Reproduction and development in family groups. In *The ecology and behavior of the wolf*, ed. L. Mech and L. Boitani. Chicago: University of Chicago Press.

Page, J. 1992. New Mexico's Gray Ranch: The natural preserve as big as all outdoors. *Smithsonian 22*: 30–43.

Savage, H. 1995. Waiting for el lobo. *Defenders 70*: 8–15.

Servín-Martínez, J. 1991. Algunos aspectos de la conducta social del lobo mexicano (*Canis lupus baileyi*) en cautiverio. *Acta Zoologico Mexico 45*: 1–43.

Servín-Martínez, J. 1993. Lobo . . . ¿Estás ahí? *Ciencias 32*: 3–10.

Servín-Martínez, J. 1997. El periodo de apareamiento, nacimiento y crecimiento del lobo Mexicano (*Canis lupus baileyi*). *Acta Zoologico Mexico 71*: 45–56.

Stolzenberg, W. 1998. Above the rim. *Nature Conservation* (September/October): 12–18.

U.S. Fish and Wildlife Service. 1997. *Reintroduction of the Mexican wolf within its historic range in the southwestern United States: Final environmental impact statement.* Washington, DC: U.S. Department of the Interior.

Young, S., and E. Goldman. 1944. *The wolves of North America.* Washington, DC: American Wildlife Institute.

MOUNTAIN PLOVER

Bureau of Land Management. 1992. *The Judith-Valley-Phillips resource management plan.* Billings, MT: Bureau of Land Management.

Graul, W.D. 1974. Vocalizations of the mountain plover. *Wilson Bulletin 86*: 221–229.

Graul, W.D. 1975. Breeding biology of the mountain plover. *Wilson Bulletin 87*: 6–31.

Knopf, F.L. 1994. Avian assemblages on altered grasslands. *Studies in Avian Biology 15*: 247–257.

Knopf, F.L. 1996. Mountain plover (*Charadrius montanus*). In *The birds of North America*, No. 211, ed. A. Poole and F. Gill. Philadelphia: Academy of Natural Sciences; Washington, DC: American Ornithologists' Union.

Knopf, F.L., and B.J. Miller. 1994. *Charadrius montanus*—montane, grassland, or bare-ground plover? *Auk 111*: 504–506.

Knopf, F.L., and J.R. Rupert. 1995. Habits and habitats of mountain plovers in California. *Condor 97*: 743–751.

Knopf, F.L., and J.R. Rupert. 1999a. Resident population of mountain plovers in Mexico? *Cotinga 11*: 17–19.

Knopf, F.L., and J.R. Rupert. 1999b. Use of crop fields by mountain plovers. *Studies in Avian Biology 19*: 81–86.

Laycock, W.A. 1988. History of grassland plowing and grass planting on the Great Plains. In *Impacts of the Conservation Reserve Program in the Great Plains*, ed. J.E. Mitchell. Fort Collins, CO: USDA Forest Service, GTR RM-185.

Miller, B., and F. Knopf. 1993. Survivability of mountain plovers on the Pawnee National Grasslands in 1992. *Journal of Field Ornithology 64*: 500–506.

NORTHERN ROCKY MOUNTAIN WOLF

Allen, D.L. 1979. Wolves of Minong: Their vital role in a wild community. Boston: Houghton Mifflin.

Bangs, E.E., and S.H. Fritts. 1996. Reintroducing the gray wolf to central Idaho and Yellowstone National Park. *Wildlife Society Bulletin 24*: 402–413.

Bergerud, A.T., W. Wyett, and J.B. Snider. 1988. The role of wolf predation in limiting a moose population. *Journal of Wildlife Management 47*: 977–988.

Carbyn, L.N., S.H. Fritts, and D.R. Seip. 1995. *Ecology and conservation of wolves in a changing world.* Edmonton, Alberta: Canadian Circumpolar Institute.

Coppinger, R., and L. Coppinger. 1995. Interactions between livestock guarding dogs and wolves. Pp. 523–526 in *Ecology and conservation of wolves in a changing world*, ed. L.N. Carbyn, S.H. Fritts, and D.R. Seip. Edmonton, Alberta: Canadian Circumpolar Institute.

Errington, P.L. 1946. Predation and vertebrate populations. *Quarterly Review of Biology 21*: 144–177, 221–245.

Fritts, S.H., E.E. Bangs, J.A. Fontaine, W.G. Brewster, and J.F. Gore. 1995. Restoring wolves to the northern Rocky Mountains of the United States. Pp. 107–125 in *Ecology and conservation of wolves in a changing world*, ed. L.N. Carbyn, S.H. Fritts, and D.R. Seip. Edmonton, Alberta: Canadian Circumpolar Institute.

Fritts, S.H., E.E. Bangs, J.A. Fontaine, M.R. Johnson, M.K. Phillips, E.D. Koch, and J.R. Gunson. 1997. Planning and implementing a reintroduction of wolves to Yellowstone National Park and Central Idaho. *Restoration Ecology 5*: 7–27.

Gasaway, W.C., R.O. Stephenson, J.L. Davis, P.K. Shepherd, and O.E. Burris. 1983. Interrelationships of wolves, prey, and man in interior Alaska. *Wildlife Monograph 84.*

Gese, E.M., and L.D. Mech. 1991. Dispersal of wolves (*Canis lupus*) in northeastern Minnesota, 1969–1989. *Canadian Journal of Zoology 69*: 2946–2955.

Gunson, J.R. 1983. Wolf predation of livestock in western Canada. Pp. 102–105 in *Wolves in Canada and Alaska: Their status, biology, and management*, ed. L.N. Carbyn. Canadian Wildlife Service Report Series, No. 45. Edmonton, Alberta: Canadian Wildlife Service.

Haber, G.C. 1996. Biological, conservation, and ethical implications of exploiting and controlling wolves. *Conservation Biology 10*: 1068–1081.

Lopez, B.H. 1978. *Of wolves and men.* New York: Charles Scribner's Sons.

McIntyre, R. 1995. *War against the wolf.* Stillwater, MN: Voyageur Press.

Mech, L.D. 1970. *The wolf: The ecology and behavior of an endangered species.* New York: Natural History Press, Doubleday.

Mech, L.D. 1995. The challenge and opportunity of recording wolf populations. *Conservation Biology 9*: 269–278.

Mech, L.D., L.G. Adams, T.J. Meier, J.W. Burch, and B.W. Dale. 1998. *The wolves of Denali.* St. Paul: University of Minnesota Press.

Mowat, F. 1963. *Never cry wolf.* New York: Dell.

Murie, A. 1944. *The wolves of Mount McKinley.* Washington, DC: U.S. Government Printing Office.

National Research Council. 1997. *Wolves, bears, and their prey in Alaska.* Washington, DC: National Academy Press.

Peterson, R.O. 1995. *The wolves of Isle Royale.* Minocqua, WI: Willow Creek Press.

Phillips, M.K., and D.W. Smith. 1996. *The wolves of Yellowstone.* Stillwater, MN: Voyageur Press.

Smith, D.W., W.G. Brewster, and E.E. Bangs. 1999. Wolves in the Greater Yellowstone ecosystem: The restoration of a top carnivore in a complex management environment. Pp. 102–125 in *Carnivores in Ecosystems: The Yellowstone Experience*, ed. T.W. Clark, A.P. Curlee, S.C. Minta, and P.M. Kareiva. New Haven, CT: Yale University Press.

Theberge, J.B. 1975. *Wolves and wilderness.* Toronto, Canada: J.M. Dent & Sons.

Thompson, I.D., and R.O. Peterson. 1988. Does wolf predation alone limit the moose population in Pulaskwa Park? A comment. *Journal of Wildlife Management 52*: 556–559.

U.S. Fish and Wildlife Service. 1987. *Restoring America's wildlife, 1937–1987.* Washington, DC: U.S. Department of the Interior, Government Printing Office.

U.S. Fish and Wildlife Service. 1994. *The reintroduction of gray wolves to Yellowstone National Park and Central Idaho. Final environmental impact statement.* Helena, MT: U.S. Fish and Wildlife Service.

Young, S.P., and E.A. Goldman. 1944. *The wolves of North America.* New York: Dover Publications.

ORINOCO RIVER DOLPHIN

Best, R.C., and V.M.F. Da Silva. 1989. Biology, status and conservation of *Inia geoffrensis* in the Amazon and Orinoco River basins. Pp. 23–34 in *Biology and conservation of the river dolphins*, ed. W.F. Perrin, R.L. Brownell, Z. Kaiya, and L. Jiankang. Gland, Switzerland: IUCN.

Best, R.C., and V.M.F. Da Silva. 1993. *Inia geoffrensis. Mammalian Species 426*: 1–8.

Boede, E.O. 1991. Delfin del Orinoco, *Inia geoffrensis*, mamífero acuático de la Orinoquía y Amazonía. Sociedad de Ciencias Naturales La Salle. *Natura 91*: 13–16.

Boede, E.O., and E. Mujica-Jorquera. 1992. *Informe respectivo de toma de muestras cutáneas en 6 ejemplares de* Inia geoffrensis, *Rio Arauca, Estado Apure, Venezuela* Report to PROFAUNA-MARNR (Wildlife Services of the Ministry of Environment). Caracas, Venezuela: Informe PROFAUNA.

Boede, E.O., and E. Mujica-Jouquera. 1999. Esfuerzos de conservacion para Tonina, *Inia geoffrensis. Revista Natura 114*: 21–25.

Boede, E.O., E. Mujica-Jorquera, and N. de Boede. 1998. Management of the Amazon River dolphin *Inia geoffrensis* at Valencia Aquarium, Venezuela. *International Zoo Yearbook 36*: 214–222.

Correa-Viana, M., and T. O'Shea. 1992. El Manati en la tradición y folklore de Venezuela. *Revista Unellez de Ciencia y Tecnología 10*(1–2): 7–13.

Da Silva, V.M.F. 1994. Aspects of the biology of the Amazonian dolphins genus *Inia* and *Sotalia fluviatilis*. Ph.D. thesis, University of Cambridge, United Kingdom.

Da Silva, V.M.F. 1995. Conservation of the fresh water dolphin with special emphasis on *Inia geoffrensis* and *Sotalia Fluviatilis*. In *Delfines y otros mamíferos acuáticos de Venezuela. Una política para su conservación*. Fundación de la Academia de Ciencias Físicas, Matemáticas y Naturales, FUDECI. Valencia, Venezuela: Clemente Editores.

Gewalt, W. 1978. Unsere Tonina *Inia geoffrensis* (Blainville 1817) Expedition 1975. *Der Zoologische Garten 48*: 323–384.

Hershkovitz, P. 1963. Notes on South American dolphins of the Genera *Inia, Sotalia*, and *Tursiops. Journal of Mammalogy 44*: 99–103.

Kendall, S., F. Trujillo, and S. Beltrán. 1995. *Dolphins of the Amazon and Orinoco*. Bogota, Colombia: Fundación Omacha.

Rodríguez, J.P., and F. Rojas-Suárez. 1995. *Libro Rojo de la Fauna Venezolana*. Caracas, Venezuela: Provita, Fundación Polar.

Trebbau, P. 1975. Measurements and some observations on the freshwater dolphin *Inia geoffrensis* in the Apure River, Venezuela. *Der Zoologische Garten 45*: 153–167.

Trebbau, P., and H.J. van Bree. 1974. Notes concerning the freshwater dolphin *Inia geoffrensis* (Blainville, 1817) in Venezuela. *Säugetierkunde Verlag Paul Parey Hamburg 39*: 50–57.

PIPING PLOVER

Bean, M.J., S.G. Fitzgerald, and M.A. O'Connell. 1991. *Reconciling conflicts under the Endangered Species Act: The habitat conservation planning experience*. Washington, DC: World Wildlife Fund.

Bent, A.C. 1929. Life histories of North American shorebirds. *U.S. Natural Museum Bulletin 146*: 236–246.

Canzanelli, L., and M. Reynolds. 1996. Negotiated rule making as a resource and visitor management tool. *Park Science 16*: 1, 16–17.

Currier, P.J., and G.R. Lingle. 1993. Habitat restoration and management for least terns and piping plovers by the Platte River Trust. P. 92 in *Proceedings, the Missouri River and its tributaries: Piping plover and least tern symposium*, ed. K.F. Higgins and M.R. Brashier. Brookings: South Dakota State University.

Cuthbert, F.J., and L.C. Wemmer. 1999. The Great Lakes Recovery Program for the piping plover: A progress report. Pp. 8–17 in *Proceedings, piping plovers and least terns of the Great Plains and nearby: A symposium/workshop, Feb. 2–5, 1998*, ed. K.F. Higgins, M.R. Brashier, and C.D. Kruse. Brookings, SD: South Dakota State University.

Defenders of Wildlife. 1996. Press release. December 4, 1996. Defenders of Wildlife, 1101 Fourteenth Street, NW, Suite 1400, Washington, DC.

Haig, S.M. 1992. Piping plover. Pp. 1–15 in *The birds of North America*, No. 2, ed. A. Poole, P. Stettenheim, and F. Gill. Philadelphia: Academy of Natural Sciences; Washington, DC: American Ornithologists' Union.

Haig, S.M., and J.H. Plissner. 1993. Distribution and abundance of piping plovers: Results and implications of the 1991 international census. *Condor 95*: 145–156.

McPhillips, N., A. Cramm, and B. Beehler. 1996. Conservation spotlight: Piping plovers plucked from perilous predicament. *Endangered Species UPDATE 13*: 14.

Palmer, R.S. 1967. Piping plover. Pp. 168–169 in *The shorebirds of North America*, ed. G.D. Stout. New York: Viking Press.

Plissner, J.H., and S.M. Haig. 1997. *1996 International piping plover census*. Corvallis, OR: U.S. Geological Survey—Biological Resources Division, Forest and Rangeland Ecosystem Science Center.

Powell, A.N., F.J. Cuthbert, L.C. Wemmer, A.W. Doolittle, and S.T. Feirer. 1997. Captive-rearing piping plovers: Developing techniques to augment wild populations. *Zoo Biology 16*: 461–477.

Rohlf, D.J. 1989. *The Endangered Species Act: A guide to its protections and implementation*. Stanford, CA: Stanford Environmental Law Society.

Russell, R.P., Jr. 1983. The piping plover in the Great Lakes region: A review of the current status and historical population of this Blue-listed species. *American Birds 37*: 951–955.

USFWS. 1988. *Recovery plan for piping plovers breeding on the Great Lakes and Northern Great Plains*. St. Paul, Minneapolis: U.S. Fish and Wildlife Service.

USFWS. 1990. *Missouri River biological opinion on Missouri River mainstem system*. Denver, CO: U.S. Fish and Wildlife Service.

USFWS. 1996. *Piping plover* (Charadrius melodus), *Atlantic Coast population, revised recovery plan*. Hadley, MA: U.S. Fish and Wildlife Service.

Wilcox, L. 1959. A twenty-year banding study of the piping plover. *Auk 76*: 129–152.

PO'O-ULI (BLACK-FACED HONEYCREEPER)

Atkinson, C.T., K.L. Woods, R.J. Dusek, L.S. Sileo, and W.M. Iko. 1995. Wildlife disease and conservation in Hawaii: Pathogenicity of avian malaria (*Plasmodium relictum*) in experimentally infected Iiwi (*Vestiaria coccinea*). *Parasitology 111*: 859–869.

Baldwin, P.H., and T.L.C. Casey. 1983. A preliminary list of foods of the po'o-uli. *'Elapio 43*: 53–56.

BRD. 1997. *Maui Critically Endangered Forest Bird Project, quarterly report, Jan.–Mar.* (Unpublished data.)

Casey, T.L.C. 1978. *Comparative ecology of four Hawaiian honeycreepers*. Master's thesis, Colorado State University, Fort Collins.

Casey, T.L.C., and J.D. Jacobi. 1974. A new genus and species of bird from the island of Maui, Hawaii (Passeriformes: Drepanididae). *Occasional Papers of the B.P. Bishop Museum 24*: 215–226.

Clark, T.W. 1997. *Averting extinction*. New Haven: Yale University Press.

Clark, T.W., R.P. Reading, and A.L. Clarke. 1994. Synthesis. In *Endangered species recovery, finding the lessons, improving the process*, ed. T.W. Clark, R.P. Reading, and A.L. Clarke. Washington, DC: Island Press.

DLNR. 1988. *Hanaw'i Natural Area Reserve Management Plan*. Honolulu: Department of Land & Natural Resources.

DLNR. 1992a. *Hawaii's extinction crisis: A call to action*. Honolulu: Department of Land & Natural Resources, U.S. Fish & Wildlife Service, and The Nature Conservancy.

DLNR. 1992b. *Limited surveys of forest birds and their habitats in the State of Hawaii*. PR Job Progress Report W-18-R-17, Job R-II-D. Honolulu: Department of Land & Natural Resources.

James, H.F., T.W. Stafford Jr., D.W. Steaman, S.L. Olson, P.S. Martin et al. 1987. Radiocarbon dates on bones of extinct birds from Hawaii. *Proceedings of the National Academy of Sciences 84*: 2350–2354.

Kepler, C.B., T.K. Pratt, A.M. Ecton, A. Englis Jr., and K.M. Fluetsch. 1996. Nesting behavior of the po'o-uli. *Wilson Bulletin 108*: 620–638.

Mosher, S.M., S.G. Fancy, G.D. Lindsey, and M.P. Moore. 1996. Predator control in a Hawaiian rainforest. In *Proceedings of the Hawaii Conservation Conference*, University of Hawaii.

Mountainspring, S., T.L.C. Casey, C.B. Kepler, and J.M. Scott, 1990. Ecology, behavior, and conservation of the po'o-uli (*Melamprosops phaeosoma*). *Wilson Bulletin 102*: 109–122.

Pratt, T.K., K.B. Kepler, and T.L.C. Casey. 1997. Po'o-uli (*Melamprosops phaeosoma*) In *The birds of North America*, No. 272. Philadelphia: Academy of Sciences.

Scott, J.M., S. Mountainspring, F.L. Ramsey, and C.B. Kepler. 1986. *Forest bird communities of the Hawaiian Islands: Their dynamics, ecology, and conservation*. Studies in Avian Biology No. 9, Cooper Ornithological Society. Lawrence, KS: Allen Press.

USFWS. 1975. *Endangered and threatened wildlife, listings of endangered and threatened fauna*. 40 FR 44151. Washington, DC: U.S. Fish & Wildlife Service.

USFWS. 1984. *The Maui-Moloka'i forest birds recovery plan*. Portland, OR: U.S. Fish & Wildlife Service.

USFWS. 1997. Initiating recovery of the po'o-uli (*Melamprosops phaeosoma*) and other endangered forest birds in East Maui. Honolulu. (Unpublished report.)

van Riper, C., III, S.G. van Riper, M.L. Goff, and M. Laird. 1986. The epizootiology and ecological significance of malaria in Hawaiian land birds. *Ecological Monographs 56*: 327–344.

RED-COCKADED WOODPECKER

Allen, D.H. 1991. An insert technique for constructing artificial red-cockaded woodpecker cavities. U.S.D.A. Forest Service, General Technical Report SE-73. Asheville, NC: Southeastern Forest Experiment Station.

Allen, D.H., K.E. Franzreb, and R.E.F. Escano. 1993. Efficacy of translocation strategies for red-cockaded woodpeckers. *Wildlife Society Bulletin 21*: 155–159.

Bonnie, R. 1996. A market-based approach to conservation of the red-cockaded woodpecker on private lands. Pp. 102–110 in *A symposium on the economics of wildlife resources on private lands*, ed. R. Johnson. Auburn, AL: Auburn University Press.

Carrie, N.R., K.R. Moore, S.A. Stephens, and E.L. Keith. 1998. Influence of cavity availability on red-cockaded woodpecker group size. *Wilson Bulletin 110*: 93–99.

Conner, R.N., D. Saenz, D.C. Rudolph, and R.N. Coulson. 1998. Southern pine beetle–induced mortality of pines with natural and artificial red-cockaded woodpecker cavities in Texas. *Wilson Bulletin 110*: 100–109.

Copeyon, C.K. 1990. A technique for constructing cavities for the red-cockaded woodpecker. *Wildlife Society Bulletin 18*: 303–311.

Costa, R. 1992. *Draft red-cockaded woodpecker procedures manual for private lands*. Atlanta: U.S. Fish and Wildlife Service, Southeast Region.

Costa, R., and R.E.F. Escano. 1989. Red-cockaded woodpecker: Status and management in the southern region in 1986. U.S.D.A. Forest Service, Technical Publication R8-TP 12, Southern Region. Atlanta.

DeFazio, J.T., Jr., M.A. Hunnicutt, M.R. Lennartz, G.L. Chapman, and J.A. Jackson. 1987. Red-cockaded woodpecker translocation experiments in South Carolina. *Proceedings of the Annual Conference of the Southeastern Association Fish and Wildlife Agencies 41*: 311–317.

Engstrom, R.T., L.A. Brennan, W.L. Neel, R.M. Farrar, S. Lindeman, W.K. Moser, and S.M. Hermann. 1996. Silvicultural practices and red-cockaded woodpecker management: A reply to Rudolph and Conner. *Wildlife Society Bulletin 24*: 334–338.

Engstrom, R.T., and G. Mikusinski. 1998. Ecological neighborhoods in red-cockaded woodpecker populations. *Auk 115*: 473–478.

Gaines, G.D., K.E. Franzreb, D.H. Allen, K.S. Laves, and W.L. Jarvis. 1995. Red-cockaded woodpecker management on the Savannah River site: A management/research success story. Pp. 81–88 in *Red-cockaded woodpecker: Recovery, ecology and management*, ed. D.L. Kulhavy, R.G. Hooper, and R. Costa. Nacogdoches, TX: Center for Applied Studies in Forestry, College of Forestry, Stephen F. Austin State University.

Jackson, J.A. 1989. The southeastern pine forest ecosystem and its birds: Past, present, and future. Pp. 119–159 in *Bird conservation 3*, ed. J.A. Jackson. Madison: University of Wisconsin Press.

Jackson, J.A. 1994. Red-cockaded woodpecker *Picoides borealis*. In *The birds of North America*, No. 85, ed. A. Poole and F. Gill. Philadelphia: Academy of Natural Sciences; Washington, DC: American Ornithologists' Union.

Jackson, J.A. 1995. The red-cockaded woodpecker: Two hundred years of knowledge, twenty years under the Endangered Species Act. Pp. 42–48 in *Red-cockaded woodpecker: Recovery, ecology and management*, ed. D.L. Kulhavy, R.G. Hooper, and R. Costa. Nacogdoches, TX: Center for Applied Studies in Forestry, College of Forestry, Stephen F. Austin State University.

Jackson, J.A. 1997. Niche concepts and habitat conservation planning. *Endangered Species Update 14*: 48–50.

Jackson, J.A., and B.J.S. Jackson. 1986. Why do red-cockaded woodpeckers need old trees? *Wildlife Society Bulletin 14*: 318–322.

Jackson, J.A., and S.D. Parris. 1995. The ecology of red-cockaded woodpeckers associated with construction and use of a multi-purpose range complex at Fort Polk, Louisiana. Pp. 277–282 in *Red-cockaded woodpecker: Recovery, ecology and management*, ed. D.L. Kulhavy, R.G. Hooper, and R. Costa. Nacogdoches, TX: Center for Applied Studies in Forestry, College of Forestry, Stephen F. Austin State University.

Lennartz, M.R., R.G. Hooper, and R.F. Harlow. 1987. Sociality and cooperative breeding of red-cockaded woodpeckers (*Picoides borealis*). *Behavioral Ecology & Sociobiology 20*: 77–88.

Ligon, J.D. 1968. Sexual differences in foraging behavior in two species of Dendrocopos woodpeckers. *Auk 85*: 203–215.

Ligon, J.D. 1970. Behavior and breeding biology of the red-cockaded woodpecker. *Auk 87*: 255–278.

Ligon, J.D., W.W. Baker, R.N. Conner, J.A. Jackson, F.C. James, D.C. Rudolph, P.B. Stacey, and J.R. Walters. 1991. The red-cockaded woodpecker: On the road to oblivion? *Auk 108*: 200–201.

Ligon, J.D., P.B. Stacey, R.N. Conner, C.E. Bock, and C.S. Adkisson. 1986. Report of the American Ornithologists' Union Committee for the Conservation of the Red-cockaded Woodpecker. *Auk 103*: 848–855.

McFarlane, R.W. 1992. *A stillness in the pines: The ecology of the red-cockaded woodpecker*. New York: W.W. Norton.

Ramey, P. 1980. Seasonal, sexual, and geographical variation in the foraging ecology of red-cockaded woodpeckers (*Picoides borealis*). Master's thesis, Mississippi State University, Mississippi State.

Richardson, D., and R. Costa. 1998. *Draft strategy and guidelines for the recovery and management of the red-cockaded woodpecker and its habitats on National Wildlife Refuges*. Atlanta: U.S. Fish and Wildlife Service.

Richardson, D.M., and J.M. Stockie. 1995. Response of a small red-cockaded woodpecker population to intensive management at Noxubee National Wildlife Refuge. Pp. 98–105 in *Wilderness and natural areas in the eastern United States: A management challenge*, ed. D.L. Kulhavy and R.N. Conner. Nacogdoches, TX: Center for Applied Studies, School of Forestry, Stephen F. Austin State University.

Rudolph, D.C., and R.N. Conner. 1996. Red-cockaded woodpeckers and silvicultural practice: Is uneven-aged silviculture preferable to even-aged? *Wildlife Society Bulletin 24*: 330–333.

U.S. Department of the Interior. 1968. *Rare and endangered fish and wildlife of the United States.* U.S. Bureau of Sport Fisheries and Wildlife, Research Publication 34. Washington, DC: U.S. Department of the Interior.

Watson, J.C., R.G. Hooper, D.L. Carlson, W.E. Taylor, and T.E. Milling. 1995. Restoration of the red-cockaded woodpecker population on the Francis Marion National Forest: Three years post Hugo. Pp. 172–182 in *Wilderness and natural areas in the eastern United States: A management challenge*, ed. D.L. Kulhavy and R.N. Conner. Nacogdoches, TX: Center for Applied Studies, School of Forestry, Stephen F. Austin State University.

RED WOLF

Dowling, T.E., W.L. Minckley, M.E. Douglas, P.C. Marsh, and B.D. Demarais. 1992. Response to Wayne, Nowak, and Phillips and Henry: Use of molecular characteristics in conservation biology. *Conservation Biology* 6:600–603.

Henry, V.G. 1992. Finding on a petition to delist the red wolf (*Canis rufus*). *Federal Register* 57: 1246–1250.

Henry, V.G. 1998. Finding on a petition to delist the red wolf (*Canis rufus*). *Federal Register* 63(236): 64799–64800.

Mangun, W.R., J.N. Lucas, J.C. Whitehead, and J.C. Mangun. 1996. Valuing red wolf recovery efforts at Alligator River NWR: Measuring citizen support. Pp. 165–171 in *Proceedings Defenders of Wildlife's Wolves of America Conference.* Washington, DC: Defenders of Wildlife.

Mech, L.D. 1996. A new era of carnivore conservation. *Wildlife Society Bulletin* 24: 397–401.

Nowak, R., M.K. Phillips, V.G. Henry, W.C. Hunter, and R. Smith. 1995. The origin and fate of the red wolf. Pp. 409–415 in *Ecology and conservation of wolves in a changing world*, ed. L.N. Carbyn, S.H. Fritts, and D.R. Seip. Canadian Circumpolar Institute, Occasional Publication 35: 1–642.

Parker, W.K. 1987. A plan for reestablishing the red wolf on Alligator River National Wildlife Refuge, North Carolina. U.S. Fish and Wildlife Service Red Wolf Management Series, Technical Report 1: 1–21.

Parker, W.T., and M.K. Phillips. 1991. Application of the experimental population designation to recovery of endangered red wolves. *Wildlife Society Bulletin* 19: 73–79.

Phillips, M.K., and D.W. Smith. 1998. Gray wolves and private landowners in the Greater Yellowstone Area. In *Transactions of the 63rd North American Wildlife and Natural Resource Conference*, Orlando, FL.

Quintal, P.K.M. 1995. *Public attitudes and beliefs about the red wolf and its recovery in North Carolina.* Master's thesis, North Carolina State University, Raleigh.

Riley, G.A., and R.T. McBride. 1972. A survey of the red wolf *Canis rufus*. U.S. Department of the Interior Special Science Report Wildlife No. 162. Washington, DC.

Rosen, W.E. 1997. *Recovery of the red wolf in northeastern North Carolina and the Great Smoky Mountains National Park: Public attitudes and economic impacts. Final report.* Ithaca: Cornell University.

Shaw, J.H. 1975. Ecology, behavior, and systematics of the red wolf *Canis rufus*. Ph.D. thesis, Yale University.

Theberge, J. 1998. Wolf country: Eleven years tracking the Algonquin wolf. Toronto: McClelland & Stewart.

U.S. Fish and Wildlife Service. 1989. *Red Wolf recovery/species survival plan.* Atlanta, GA: USFWS.

Wayne, R.K., N. Lehman, and T.K. Fuller. 1995. Conservation genetics of the gray wolf. Pp. 399–407 in *Ecology and conservation of wolves in a changing world*, ed. L.N. Carbyn,

S.H. Fritts, and D.R. Seip. Canadian Circumpolar Institute, Occasional Publication 35: 1–642.

SCARLET MACAW

Beissinger, S.R., and E.H. Bucher. 1992. Sustainable harvesting of parrots for conservation. Pp. 73–117 in *New world parrots in crisis: Solutions from conservation biology*, ed. S.R. Beissinger and N.F.R. Snyder. Washington, DC: Smithsonian Institution Press.

Boo, E. 1990. *Ecotourism: The potentials and pitfalls*. World Wildlife Fund #72. Baltimore, MD: WWF.

Butler, P.J. 1992. Parrots, people, pressure, and pride. Pp. 25–46 in *New world parrots in crisis: Solutions from conservation biology*, ed. S.R. Beissinger and N.F.R. Synder. Washington, DC: Smithsonian Institution Press.

Coc, R., L. Marsh, and E. Platt. 1998. The Belize Zoo: Grassroots efforts in education and outreach. Pp. 389–396 in *Timber, tourists, and temples: Conservation and development in the Maya Forest of Belize, Guatemala, and Mexico*, ed. R.B. Primack, D. Bray, H.A. Galletti, and I. Ponciano. Washington, DC: Island Press.

Forshaw, J.M. 1989. *Parrots of the world*, 3rd ed. Melbourne: Lansdowne Editions.

Iñigo-Elias, E.E. 1996. *Ecology and breeding biology of the scarlet macaw* (Ara macao) *in the Usumacinta drainage basin of Mexico and Guatemala*. Ph.D. thesis, University of Florida, Gainesville.

Iñigo-Elias, E.E., and M.A. Ramos. 1991. The Psittacine trade in Mexico. Pp. 380–392 in *Neotropical wildlife use and conservation*, ed. J.G. Robinson and K.H. Redford. Chicago: University of Chicago Press.

Marineros, L., and C. Vaughan. 1995. Scarlet macaws of Carara: Perspectives for management. Pp. 445–467 in *The large macaws: Their care, breeding and conservation*, ed. J. Abramson, B.L. Speer, and J.B. Thomsen. Fort Bragg, CA: Raintree Publications.

Munn, C.A. 1988. Macaw biology in Manu National Park. *Parrotletter 1*: 821–823.

Munn, C.A. 1992. Macaw biology and ecotourism, or "When a bird in the bush is worth two in the hand." Pp. 47–72 in *New world parrots in crisis: Solutions from conservation biology*, ed. S.R. Beissinger and N.F.R. Synder. Washington, DC: Smithsonian Institution Press.

Norris, R., J.S. Wilber, and L.O. Morales-Marin. 1998. Community-based ecotourism in the Maya Forest: Problems and potentials. Pp. 327–343 in *Timber, tourists, and temples: Conservation and development in the Maya Forest of Belize, Guatemala, and Mexico*, ed. R.B. Primack, D. Bray, H.A. Galletti, and I. Ponciano. Washington, DC: Island Press.

Sanz, V., and A. Grajal. 1998. Successful reintroduction of captive-raised yellow-shouldered Amazon parrots on Margarita Island, Venezuela. *Conservation Biology 12*: 430–441.

Snyder, N.F.R., S.R. Derrickson, S.R. Beissinger, J.W. Wiley, T.B. Smith, W.D. Toone, and B. Miller. 1996. Limitations of captive breeding in endangered species recovery. *Conservation Biology 10*: 338–348.

Snyder, N.F.R., S.E. Koenig, J. Koschmann, H.A. Snyder, and T.B. Johnson. 1994. Thick-billed parrot releases in Arizona. *Condor 96*: 844–862.

Swanson, T.M. 1992. Economics and animal welfare: The case of the live bird trade. Pp. 43–57 in *Perceptions, conservation, and management of wild birds in trade*, ed. J.B. Thomsen, S.R. Edwards, and T.A. Mulliken. Cambridge: TRAFFIC International.

Wiedenfeld, D.A. 1994. A new subspecies of scarlet macaw and its status and conservation. *Ornithologia Neotropical 5*: 99–104.

SNOW LEOPARD

Ahlborn, G.G., and R.M. Jackson. 1988. Marking in free-ranging snow leopards in west Nepal: A preliminary assessment. Pp. 25–49 in *Proceedings of the fifth international snow leopard symposium*, ed. H. Freeman. Seattle, WA: Wildlife Institute of India and International Snow Leopard Trust.

Barnes, L. 1989. Cat skin trade in Kathmandu, Nepal. *Traffic Bulletin 11*(4): 65.

Chundawat, R.S., and G.S. Rawat. 1994. Food habits of snow leopard in Ladakh, India. Pp. 127–132 in *Proceedings of the seventh international snow leopard symposium*, ed. J.L. Fox and Du Jizeng. Seattle, WA: International Snow Leopard Trust.

Fox, J.L. 1994. Snow leopard conservation in the wild—a comprehensive perspective on a low density and highly fragmented population. Pp. 3–15 in *Proceedings of the seventh international snow leopard symposium*, ed. J.L. Fox and Du Jizeng. Seattle, WA: International Snow Leopard Trust.

Heinen, J.T., and B. Leisure. 1993. A new look at the Himalayan fur trade. *Oryx 27*: 231–238.

Hemmer, H. 1972. *Uncia uncia. Mammalian Species 20*: 1–5.

Jackson, R., and G. Ahlborn. 1989. Snow leopards (*Panthera uncia*) in Nepal: Home range and movements. *National Geographic Research 5*: 161–175.

Jackson, R. and J.L. Fox. 1997. Snow leopard conservation: Accomplishments and research priorities. Pp. 128–145 in *Proceedings of the eighth international snow leopard symposium*, ed. R. Jackson and A. Ahmad. Lahore: International Snow Leopard Trust and WWF-Pakistan.

Jackson, R.M., G. Ahlborn, M. Gurung, and S. Ale. 1996. Reducing livestock depredation losses in the Nepalese Himalaya. *Proceedings 17th Vertebrate Pest Conference 17*: 241–247.

Koshkarev, E. 1994. Poaching in the former USSR. *Snow Line 12*(2): 6–7.

Nowell, K., and P. Jackson, eds. 1996. Wild cats: Status survey and action plan. Cambridge, UK: IUCN/SSC Cat Specialist Group.

O'Gara, B.W. 1988. Snow leopards and sport hunting in the Mongolian People's Republic. Pp. 215–225 in *Proceedings of the fifth international snow leopard symposium*, ed. H. Freeman. Seattle, WA: Wildlife Institute of India and International Snow Leopard Trust.

Oli, M.K., I.R. Taylor, and M.E. Rogers. 1993. Diet of the snow leopard (*Panthera uncia*) in the Annapurna Conservation Area, Nepal. *Journal Zoology (London) 231*: 365–370.

Oli, M.K., I.R. Taylor, and M.E. Rogers. 1994. Snow leopard *Panthera uncia* predation of livestock—an assessment of local perceptions in the Annapurna Conservation Area, Nepal. *Biological Conservation 68*: 63–68.

PEER (Public Employees for Environmental Responsibility). 1996. Tarnished trophies: The Department of Interior's wild sheep loophole. *White Paper No. 7*: 1–29.

Schaller, G.B., H. Li, Talipu, H. Lu, J. Ren, M. Qiu, and H. Wang. 1987. Status of large mammals in the Taxkorgan Reserve, Xinjiang, China. *Biological Conservation 42*: 53–71.

SOUTHERN SEA OTTER

Benz, C. 1996. Evaluating attempts to reintroduce sea otters along the California coastline. *Endangered Species Update 13*(12): 31–35.

Demaster, D.P., C. Marzin, and R.J. Jameson. 1996. Estimating the historical abundance of sea otters in California. *Endangered Species Update 13*(12): 79–81.

Estes, J.A. 1991. Catastrophes and conservation: Lessons from sea otters and the Exxon *Valdez*. *Science 254*: 1596.

Estes, J.A. 1996. The influence of large, mobile predators in aquatic food webs: Examples from sea otters and kelp forests. Pp. 65–72 in *Aquatic predators and their prey*, ed. S.P.R. Greenstreet and M.L. Tasker. Oxford: Blackwell Scientific.

Estes, J.A., D.F. Doak, J.R. Bodkin, R.J. Jameson, D. Monson, J. Watt, and M.T. Tinker. 1996. Comparative demography of sea otter populations. *Endangered Species Update 13*(12): 11–13.

Jarman, W.M., C.E. Bacon, J.A. Estes, M. Simon, and R.J. Norstrom. 1996. Organochlorine contaminants in sea otters: The sea otter as a bio-indicator. *Endangered Species Update* 13(12): 20–22.

Ralls, K., D.P. Demaster, and J.A. Estes. 1996. Developing a criterion for delisting the southern sea otter under the U.S. Endangered Species Act. *Conservation Biology 10*: 1528–1537.

Ralls, K., T.C. Eagle, and D.B. Siniff. 1996. Movement and spatial use patterns of California sea otters. *Canadian Journal of Zoology 74*: 1841–1849.

Riedman, M.L., and J.A. Estes. 1990. *The sea otter* (Enhydra lutris): *Behavior, ecology, and natural history*. Biological Report 90. Washington, DC: U.S. Fish and Wildlife Service.

Thomas, N.J., and R.A. Cole. 1996. The risk of disease and threats to the wild population. *Endangered Species Update 13*(12): 23–27.

VanBlaricom, G.R. 1996. Saving the sea otter population in California: Contemporary problems and future pitfalls. *Endangered Species Update 13*(12): 85–91.

STURGEONS AND PADDLEFISHES (ACIPENSERIFORMES)

Anonymous. 1997a. *About sturgeon and CITES.* Office of Management Authority, CITES Fact Sheet Series (08/97): 1–5.

Anonymous. 1997b. The mafia has won fishing guards. Will the border guards beat the mafia? *Izvestiya*, 15 September 1997. (In Russian.)

Anonymous. 1997c. Violations of the Russian Federation laws on the protection of marine biological resources. *Zeleonyi Mir [Green World] 15*: 8–9. (In Russian.)

Bemis, W.E., E.K. Findeis, and L. Grande. 1997. An overview of Acipenseriformes. Pp. 25–71 in *Sturgeon biodiversity and conservation*, ed. V.J. Birstein, J.R. Waldman, and W.E. Bemis. Dordrecht, Netherlands: Kluwer Academic Publishers.

Birstein, V.J. 1993. Sturgeons and paddlefishes: Threatened fishes in need of conservation. *Conservation Biology 7*: 773–787.

Birstein, V.J. 1997. Concluding remarks: The current status of sturgeons, threats to their survival, the caviar trade and international efforts needed to save them. Pp. 71–88 in *Sturgeon stocks and caviar trade workshop*, ed. V.J. Birstein, A. Bauer, and A. Kaiser-Pohlmann. Gland, Switzerland: IUCN.

Birstein, V.J., W.E. Bemis, and J.R. Waldman. 1997. The threatened status of acipenseriform fishes: A summary. Pp. 427–435 in *Sturgeon biodiversity and conservation*, ed. V.J. Birstein, J.R. Waldman, and W.E. Bemis. Dordrecht, Netherlands: Kluwer Academic Publishers.

Birstein, V.J., J. Betts, and R. DeSalle. 1998. Molecular identification of *Acipenser sturio* specimens: A warning note for recovery plans. *Biological Conservation 84*: 97–101.

Birstein, V.J., and R. DeSalle. 1998. Molecular phylogeny of Acipenserinae. *Molecular Phylogenetics and Evolution 9*: 141–155.

Birstein, V.J., P. Doukakis, B. Sorkin, and R. DeSalle. 1998. Population aggregation analysis of three caviar-producing species of sturgeons and implications for the species identification of black caviar. *Conservation Biology 12*: 766–775.

Birstein, V.J., R. Hanner, and R. DeSalle. 1997. Phylogeny of the Acipenseriformes: Cytogenetic and molecular approaches. Pp. 127–155 in *Sturgeon biodiversity and conservation*, ed. V.J. Birstein, J.R. Waldman, and W.E. Bemis. Dordrecht, Netherlands: Kluwer Academic Publishers.

De Meulenaer, T., and C. Raymakers. 1996. *Sturgeons of the Caspian Sea and the international trade in caviar*. Cambridge, UK: TRAFFIC International.

DeSalle, R., and V.J. Birstein. 1996. PCR identification of black caviar. *Nature 377*: 673–674.

Dumont, H. 1995. Ecocide in the Caspian Sea. *Nature 377*: 673–674.

Gardiner, B.G. 1984. Sturgeons as living fossils. Pp. 148–152 in *Living fossils*, ed. N. Eldredge and S.M. Stanley. New York: Springer-Verlag.

IUCN. 1996. *1996 IUCN red list of threatened animals*. Gland, Switzerland: IUCN.

Khodorevskaya, R.P., G.F. Dovgopol, O.L. Zhuravleva, and A.D. Vlasenko. 1997. Present status of commercial stocks of sturgeons in the Caspian Sea basin. Pp. 209–219 in *Sturgeon biodiversity and conservation*, ed. V.J. Birstein, J.R. Waldman, and W.E. Bemis. Dordrecht, The Netherlands: Kluwer Academic Publishers.

Khodorevskaya, R.P., and A.S. Novikova. 1995. Status of beluga sturgeon, *Huso huso*, in the Caspian Sea. *Journal of Ichthyology 35*: 59–68.

Mrosovsky, N. 1997. IUCN's credibility critically endangered. *Nature 389*: 436.

Ryder, J.A. 1888/1890. The sturgeons and sturgeon industries of the eastern coast of the United States, with an account of experiments bearing upon sturgeon culture. *Bulletin of the United States Fish Commission 8*: 231–326.

Sabeau, L. 1997. Sturgeon aquaculture in France. Pp. 43–49 in *Caspian Environment Program. Proceedings from the first bio-network workshop, Bordeaux, November 1997*, ed. H. Dumont, S. Wilson, and B. Wazniewicz. Washington, DC: World Bank.

Taylor, S. 1997. The historical development of the caviar trade and the caviar industry. Pp. 45–54 in *Sturgeon stocks and caviar trade workshop*, ed. V.J. Birstein, A. Bauer, and A. Kaiser-Pohlmann. Gland, Switzerland: IUCN.

Waldman, J.R., and I.I. Wirgin. 1998. Status and restoration options for Atlantic sturgeon in North America. *Conservation Biology 12*: 631–638.

WCMC (World Conservation Monitoring Centre). 1996. Checklist of CITES species: A reference to the appendices to the Convention on International Trade in Endangered Species of Wild Fauna and Flora. Geneva, Switzerland: CITES Secretariat/World Conservation Monitoring Centre.

Wilczynski, P. 1998. Subject: Caspian Environment Program—Bio-resources Network Proceeding. A letter to Caspian Bio-resources Network members dated June 24, 1998.

Zholdasova, I. 1997. Sturgeons and the Aral Sea ecological catastrophe. Pp. 373–380 in *Sturgeon biodiversity and conservation*, ed. V.J. Birstein, J.R. Waldman, and W.E. Bemis. Dordrecht, The Netherlands: Kluwer Academic Publishers.

YUMA CLAPPER RAIL

Anderson, B.W., and R.D. Ohmart. 1985. Habitat use by clapper rails in the lower Colorado River valley. *Condor 87*: 116–126.

Bennett, W.W., and R.D. Ohmart. 1978. *Habitat requirements and population characteristics of the clapper rail* (Rallus longirostris yumanensis) *in the Imperial Valley of California*. Livermore, CA: University of California, Lawrence Livermore Lab.

Conway, C.J. 1990. Seasonal changes in movements and habitat use by three sympatric species of rails. Master's thesis, University of Wyoming, Laramie.

Conway, C.J., W.R. Eddleman, S.H. Anderson, and L.R. Hanebury. 1993. Seasonal changes in Yuma clapper rail vocalization rate and habitat use. *Journal of Wildlife Management 56*: 282–290.

Eddleman, W.R. 1989. *Biology of the Yuma clapper rail in the southwestern U.S. and northwestern Mexico*. Final Report, Intra-Agency Agreement No. 4-AA-30–02060. Yuma, AZ: U.S. Bureau of Reclamation, Yuma Projects Office.

Eddleman, W.R. and C.J. Conway. 1994. Clapper rail. Pp. 167–179 in *Management of migratory shore and upland game birds in North America*, ed. T.C. Tacha and C.E. Braun. Washington, DC: International Association of Fish and Wildlife Agencies.

Eddleman, W.R., and C.J. Conway. 1998. Clapper rail (*Rallus longirostris*) No. 340 in *The birds of North America*, ed. A. Poole and F. Gill. Philadelphia: The Birds of North America.

Eddleman, W.R., F.L. Knopf, B. Meanley, F.A. Reid, and R. Zembal. 1988. Conservation of North American rallids. *Wilson Bulletin 100*: 458–475.

Glenn, E.P., C. Lee, R. Felger, and S. Zengel. 1996. Effects of water management on the wetlands of the Colorado River Delta, Mexico. *Conservation Biology 10*: 1175–1186.

Gould, G.I., Jr. 1975. Yuma clapper rail study—censuses and habitat distribution, 1973-74.

Administrative Report No. 75–2. Sacramento: Wildlife Management Branch, California Department of Fish and Game.

Grinnell, J. 1914. An account of the mammals and birds of the lower Colorado Valley with special reference to the distributional problems presented. *University of California Publications in Zoology 12*: 51–294.

Jarman, W.M. 1991. *Identification and levels of organochlorine compounds in birds and marine mammals* (Falco peregrinus anatum, Falco mexicanus, Rallus longirostris levipes). Ph.D. thesis, University of California, Santa Cruz.

Klaas, E.E., H.M. Ohlendorf, and E. Cromartie. 1980. Organochlorine residues and shell thicknesses in eggs of the Clapper Rail, Common Gallinule, Purple Gallinule, and Limpkin (Class Aves), eastern and southern United States, 1972–74. *Pesticide Monitoring Journal 14*: 90–94.

Krebs, C.J. 1991. The experimental paradigm and long-term population studies. *Ibis 133* (Suppl. 1): 3–8.

Lonzarich, D.G., T.E. Harvey, and J.E. Takekawa. 1992. Trace element and organochlorine concentrations in California clapper rail (*Rallus longirostris obsoletus*) eggs. *Archives of Environmental Contaminants and Toxicology 23*: 147–153.

Nichols, J.D. 1991. Extensive monitoring programs viewed as long-term population studies: The case of North American waterfowl. *Ibis 133* (Suppl. 1): 89–98.

Nichols, J.D. 1999. Monitoring is not enough: On the need for a model-based approach to migratory bird management. In *Strategies for bird conservation: The Partners-in-Flight planning process*, ed. R. Bonney, D. Pashley, L. Niles, and R. Cooper. Ithaca, NY: Cornell Lab of Ornithology.

Odom, R.R. 1975. Mercury contamination in Georgia rails. *Proceedings of the Annual Conference of the Southeastern Association of Game and Fish Commissions 28*: 649–658.

Ohmart, R.D., W.O. Deason, and S.J. Freeland. 1975. Dynamics of marsh land formation and succession along the lower Colorado River and their importance and management problems as related to wildlife in the arid Southwest. *Transactions of the North American Wildlife and Natural Resource Conference 40*: 240–251.

Ohmart, R.D., and R.E. Tomlinson. 1977. Foods of western clapper rails. *Wilson Bulletin 89*: 332–336.

Piest, L., and J. Campoy. 1998. *Report of Yuma clapper rail surveys at Ciénega de Santa Clara, Sonora*. Unpublished report. Yuma: Arizona Game and Fish Department.

Roth, R.R. 1972. The daily and seasonal behavior patterns of the clapper rail (*Rallus longirostris*) with notes on food habits and environmental pollutant contamination. Master's thesis, Louisiana State University, Baton Rouge.

Rusk, M.K. 1991. Selenium risk to Yuma clapper rails and other marsh birds of the lower Colorado River. Master's thesis, University of Arizona, Tucson.

Shuford, W.D. 1993. *The Marin County breeding bird atlas*. Bolinas, CA: Bushtit Books.

Small, A. 1994. *California birds: Their status and distribution*. Vista, CA: Ibis Publishing.

Smith, P.M. 1975. *Yuma clapper rail study, Mohave County, Arizona, 1973*. Progress Report, Project W-54-R-6, Job II-5.9. Sacramento: California Department of Fish and Game.

Sykes, G. 1937. *The Colorado delta*. American Geographic Society Special Publication No. 19. Washington, DC: Carnegie Institute; New York: American Geographical Society.

Todd, R.L. 1986. *A saltwater marsh hen in Arizona: A history of the Yuma clapper rail* (Rallus longirostris yumanensis). Completion Report, Federal Aid Project W-95-R. Phoenix: Arizona Game and Fish Department.

Tomlinson, R.E., and R.L. Todd. 1973. Distribution of two western clapper rail races as determined by responses to taped calls. *Condor 75*: 177–183.

U.S. Fish and Wildlife Service. 1983. Yuma clapper rail recovery plan. Albuquerque, NM: U.S. Fish and Wildlife Service.

INTERDISCIPLINARY PROBLEM SOLVING IN ENDANGERED SPECIES CONSERVATION: THE YELLOWSTONE GRIZZLY BEAR CASE

Ascher, W., and R. Healy. 1990. *Natural resource policymaking in developing countries.* Durham, NC: Duke University Press.

Brewer, G.D. 1992. Business and environment: A time for creative and constructive coexistence. The 25th Annual William K. McInally Memorial Lecture, School of Business Administration, University of Michigan. Ann Arbor, MI: University of Michigan.

Brewer, G.D., and T.W. Clark. 1994. A policy sciences perspective: Improving implementation. Pp. 391–413 in *Endangered species recovery: Finding the lessons, improving the process,* ed. T.W. Clark, P. Reading, and A.C. Clarke. Washington, DC: Island Press.

Brunner, R.D. 1991. Policy movement as a policy problem. *Policy Sciences 24*: 65–98.

Brunner, R.D. 1993a. Myths, scientific and political. Pp. 4–12 in *The objectivity crises: Rethinking the role of science in society. Chairman's report to the Committee on Science, Space, and Technology,* ed. G.E. Brown. House of Representatives, U.S. Congress. Washington, DC: U.S. GPO, 68–054.

Brunner, R.D. 1993b. *Policy and global change research: A modest proposal.* Paper presented at the Fourteenth Annual Research Conference, Association for Public Policy Analysis and Management, October 30, 1993, Washington, DC.

Brunner, R.D. 1995. *Notes on basic concepts of the policy sciences.* Unpublished course notes. Boulder: Department of Political Science, University of Colorado.

Brunner, R.D., and W. Ascher. 1992. Science and social responsibility. *Policy Sciences 25*: 295–331.

Clark, T.W. 1992. Practicing natural resource management with a policy orientation. *Environmental Management 16*: 423–433.

Clark, T.W. 1993. Creating and using knowledge for species and ecosystem conservation: Science, organizations, and policy. *Perspectives in Biology and Medicine 36*: 497–525, appendices.

Clark, T.W. 1996. Appraising threatened species recovery processes: Some pragmatic recommendations for improvements. Pp. 1–22 in *Back from the brink: Refining the threatened species recovery process,* ed. S. Stephens and S. Maxwell. Transactions of the Royal Zoological Society, Sydney, New South Wales. Canberra: Australian Nature Conservation Agency.

Clark, T.W. 1997. *Averting extinction: Reconstructing endangered species recovery.* New Haven, CT: Yale University Press.

Clark, T.W. In press. Interdisciplinary problem-solving: Next steps in the Greater Yellowstone Ecosystem. *Policy Sciences.*

Clark, T.W., and R.D. Brunner. 1996. Making partnerships work: An introduction to decision process. *Endangered Species Update 13*(9): 1–4.

Clark, T.W., and D.E. Casey. 1992. *Tales of the grizzly: Thirty-nine stories of grizzly bear encounters in the wilderness.* Moose, WY: Homestead Press.

Clark, T.W., T. Donnay, P. Schuyler, P. Curlee, P. Cymerys, T. Sullivan, L. Sheeline, R. Reading, R. Wallace, A. Marcer-Batlle, Y. DeFretes, and T.K. Kennedy Jr. 1992. Conserving biodiversity in the real world: Professional practice using a policy orientation. *Endangered Species Update 9*(5, 6): 5–8.

Clark, T.W., and S.C. Minta. 1994. *Greater Yellowstone's future: Prospects for ecosystem science, management, and policy.* Moose, WY: Homestead Press.

Clark, T.W., and R.L. Wallace. 1998. Understanding the human factor in endangered species recovery: An introduction to human social process. *Endangered Species Update 15*(1): 2–9.

Clark, T.W., and R.L. Wallace. 1999. The professional in endangered species conservation: An introduction to standpoint clarification. *Endangered Species Update 16*: 9–13.

Clark, T.W., A.R. Willard, and C.R. Cromley, eds. In press. *Foundations of natural resources policy and management.* New Haven, CT: Yale University Press.

Crick, B. 1992. In defense of politics. Chicago: University of Chicago Press.

Culhane, R.J. 1981. Public lands politics: Resources for the future. Baltimore: Johns Hopkins University Press.

Dery, D. 1984. *Problem definition in policy analysis.* Lawrence, KS: University of Kansas Press.

Dewey, J. 1910. How we think. Pp. 177–356 in *The middle works 1899–1924.* Vol. 6 (1910–1911), ed. J. Boydston and J. Dewey. Carbondale: Southern Illinois University Press.

Dryzek, J.S. 1995. Toward an ecological modernity. *Policy Sciences 28:* 231–242.

Flannery, T.F. 1994. *The future eaters: An ecological history of Australasian lands and people.* Chatswood, New South Wales: Reed Books.

Grumbine, E. 1994. What is ecosystem management? *Conservation Biology 8:* 27–38.

Knight, R.P., B.M. Blanchared, and P. Schullery. 1999. Yellowstone bears. In *Carnivores in ecosystems: The Yellowstone experience,* ed. T.W. Clark, A.P. Curlee, S.C. Minta, and P.M. Kareiva. New Haven, CT: Yale University Press.

Lasswell, H.D. 1970. From fragmentation to configuration. *Policy Sciences 2:* 439–446.

Lasswell, H.D. 1971. *A pre-view of the policy sciences.* New York: American Elsevier Publishing.

Lasswell, H.D., and A. Kaplan. 1950. *Power and society: A framework for political inquiry.* New Haven, CT: Yale University Press.

Lasswell, H.D., and M.S. McDougal. 1992. *Jurisprudence for a free society: Studies in law, science, and policy.* New Haven, CT: New Haven Press.

Latour, B. 1987. *Science in action.* Cambridge, MA: Harvard University Press.

Leopold, A. 1949. *A sand county almanac.* New York: Ballantine Books.

Lipmann, W. 1965. *Public opinion.* New York: The Free Press.

Mattson, D.J., and J.J. Craighead. 1994. The Yellowstone grizzly bear recovery program: Uncertain information, uncertain policy. Pp. 101–129 in *Endangered species recovery: Finding the lessons, improving the process,* ed. T.W. Clark, R.P. Reading, and A.C. Clarke. Washington, DC: Island Press.

May, R. 1991. *The cry for myth.* New York: W.W. Norton.

McDougal, M.S., W.M. Reisman, and A.R. Willard. 1989. The world community: A planetary social process. *University of California, Davis Law Review 21:* 807–972.

Norgaard, R.B., and J.A. Dixon. 1986. Pluralistic project design: An argument for combining economic and coevolutionary methodologies. *Policy Sciences 19:* 297–317.

Pool, R. 1990. Struggling to do science for society. *Science 248:* 672–673.

Primm, S.A. 1996. A pragmatic approach to grizzly bear conservation. *Conservation Biology 10:* 1026–1035.

Primm, S.A., and T.W. Clark. 1996. Making sense of the policy process for carnivore conservation. *Conservation Biology 10:* 1036–1045.

Reading, R.P., and B.J. Miller. 1994. The black-footed ferret recovery program: Unmasking professional and organizational weaknesses. Pp. 101–129 in *Endangered species recovery: Finding the lessons, improving the process,* ed. T.W. Clark, R.P. Reading, and A.C. Clarke. Washington, DC: Island Press.

Simon, H. 1985. Human nature in politics: The dialogue of psychology with political science. *American Political Science Review 79:* 293–304.

U.S. Fish and Wildlife Service. 1993. *Grizzly bear recovery plan.* Denver: U.S. Fish and Wildlife Service.

Viederman, S., G.K. Meffe, and C.R. Carroll. 1997. The role of institutions and policymaking in conservation. Pp. 545–574 in *Introduction to conservation biology,* ed. G. Meffe and C.R. Carroll. Sunderland, MA: Sinauer Associates.

Wallace, R.L., and T.W. Clark. 1999. Understanding and solving problems in endangered species conservation: An introduction to problem orientation. *Endangered Species Update 16:* 28–34.

Watson, J.D. 1968. *The double helix.* New York: New American Library.

Weiss, J.A. 1989. The power of problem-definition: The case of government paperwork. *Policy Sciences 22*: 97–121.

Willard, A.R., and C. Norchi. 1993. The decision seminar as an instrument of power and enlightenment. *Political Psychology 14*: 575–606.

CONCLUSIONS: CAUSES OF ENDANGERMENT AND CONFLICTS IN RECOVERY

Bacow, L.S., and M. Wheeler, 1984. *The politics of environmental mediation.* New York: Columbia University Press.

Bingham, G.A. 1986. *Resolving environmental disputes: A decade of experience.* Washington, DC: Conservation Foundation.

Clark, T.W. 1996. Learning as a strategy for improving endangered species conservation. *Endangered Species Update 13*(1,2): 5–6, 22–24.

Clark, T.W. 1997. *Averting extinction: Reconstructing endangered species recovery.* New Haven, CT: Yale University Press.

Clark, T.W., R.P. Reading, and A.L. Clarke, eds. 1994. *Endangered species recovery: Finding the lessons, improving the process.* Washington, DC: Island Press.

Clark, T.W., and R.L. Wallace. 1998. Understanding the human factor in endangered species recovery: An introduction to human social process. *Endangered Species Update 15*(1): 2–9.

Clark, T.W., A.R. Willard, and C.R. Cromley, eds. In press. *Foundations of natural resources policy and management.* New Haven, CT: Yale University Press.

Costanza, R., R. d'Arge, R. de Groot, S. Farber, M. Grasso, B. Hannon, S. Naeem, K. Limberg, J. Paruelo, R.V. O'Neill, R. Rasking, P. Sutton, and M. van den Belt. 1997. The value of the world's ecosystem services and natural capital. *Nature 387*: 253–260.

Crowfoot, J.E., and J.M. Wondolleck. 1990. *Environmental disputes: Community involvement in conflict resolution.* Washington, DC: Island Press.

Hollings, C.S. 1978. *Adaptive environmental assessment and management.* New York: John Wiley & Sons.

Kellert, S.R. 1985. Social and perceptual factors in endangered species management. *Journal of Wildlife Management 49*: 528–536.

Kellert, S.R. 1996. *The value of life: Biological diversity and human society.* Washington, DC: Island Press.

Pimentel, D., C. Wilson, C. McCullum, R. Huang, T. Dwen, J. Flack, Q. Tren, T. Saltman, and B. Cliff. 1997. Economic and environmental benefits of diversity. *BioScience 47*: 747–757.

Reaka-Kudla, M.L., D.E. Wilson, and E.O. Wilson, eds. 1996. *Biodiversity II: Understanding and protecting our biological resources.* Washington, DC: National Academy Press.

Rolston, H., III. 1994. *Conserving natural value.* New York: Columbia University Press.

Wallace, R.L., and T.W. Clark. 1999. Solving problems in endangered species conservation: An introduction to problem orientation. *Endangered Species Update 16*(2): 28–34.

Wilcove, D.S., D. Rothstein, J. Dubow, A. Phillips, and E. Losos. 1998. Quantifying threats to imperiled species in the United States. *BioScience 48*: 607–615.

Wilson, E.O., ed. 1988. *Biodiversity.* Washington, DC: National Academy Press.

Wilson, E.O. 1992. *The diversity of life.* Cambridge, MA: Belknap Press of Harvard University Press.

Wondolleck, J.M., S.L. Yaffee, and J.E. Crowfoot. 1994. A conflict management perspective: Applying the principles of alternative dispute resolution. Pp. 305–326 in *Endangered species recovery: Finding the lessons, improving the process*, ed. T.W. Clark, R.P. Reading, and A.L. Clarke. Washington, DC: Island Press.

Woodroffe, R., and J.R. Ginsberg. 1998. Edge effects and the extinction of populations inside protected areas. *Science 280*: 2126–2128.

Index

Contributors

Sukhiin Amgalanbaatar, GTZ—Nature Conservation & Buffer Zone Development Project, Khudaldaany gudamj-5, Ulaanbaatar–11, MONGOLIA

Paul C. Banko, Hawaii National Park, P.O. Box 44, Hawaii National Park, HI 96718 USA

Robert E. Beck Jr., Aquatic & Wildlife Resources Division, Department of Agriculture, Government of Guam, 192 Dairy Road, Mangilao, Guam 96923 USA

John L. Behler, Department of Herpetology, Wildlife Conservation Society, 2300 Southern Blvd., Bronx, NY 10460 USA

Steven R. Beissinger, Department of Environmental Science, Policy and Management, Hillgard Hall #3110, University of California, Berkeley, CA 94720–3110 USA

José F. Bernal-Stoopen, Unidad de Zoológicos de la Cuidad de México, Primera Sección del Bosque de Chapultepec, 11850 México, D.F., MEXICO

Vadim J. Birstein, Visiting Scientist, American Museum of Natural History, Central Park West at 79th Street, New York, NY 10024 USA

Ernesto O. Boede, Centro Veterinario "Los Colorados," Avenida Guzman Blanco No. 103, Urbanizacion Los Colorados, Valencia, VENEZUELA

Axel Bräunlich, Nature Conservation International & BirdLife International, Brusseler Str. 46, 13353 Berlin, GERMANY

David Chiszar, Dept. Psychology, Muenzinger Bldg., Campus Box 345, University of Colorado, Boulder, CO 80309–0345 USA

Tim W. Clark, Yale University and Northern Rockies Conservation Cooperative, P.O. Box 2705, Jackson, WY 83001 USA

Mark Collins, Maui Forest Bird Recovery Project, 2464 Olinda Road, Makawao, HI 96768 USA

Paul Conry, Department of Land and Natural Resources, Division of Forestry and Wildlife, 1151 Punchbowl Street, Rm. 325, Honolulu, HI 96813 USA

Courtney J. Conway, Department of Natural Resource Sciences, Washington State University, 2710 University Drive, Richland, WA 99352 USA

Paul Stephen Corn, USGS-BRD and Aldo Leopold Wilderness Research Institute, Box 8089, Missoula, MT 59807 USA

Blair Csuti, Oregon Zoo, 4001 SW Canyon Road, Portland, OR 97221–2799 USA

María Rosa Cuesta, Bioandina Foundation, Av. 4 Boívar, Torre General Golfredo Masinii, Apartado Postal 131, Mérida, Estado Mérida, VENEZUELA

Francesca J. Cuthbert, Department of Fisheries and Wildlife, University of Minnesota, St. Paul, MN 55108 USA

James M. Dietz, Department of Zoology, University of Maryland, College Park, MD 20742 USA

Stephen J. Dinsmore, Dept. of Fishery & Wildlife Biology, Colorado State University, Fort Collins, CO 80523–1474 USA

William R. Eddleman, Department of Biology, Southeast Missouri State University, Cape Girardeau, MO 63701 USA

Nina Fascione, Defenders of Wildlife, 1101 14th St., NW, Suite 1400, Washington, D.C. 20005 USA

Anna T.C. Feistner, Research Department, Durrell Wildlife Conservation Trust, Les Augres Manor, Trinity, Jersey JE3 5BP, Channel Islands, UNITED KINGDOM

Sarah L. Fowler, Nature Conservation Bureau, 36 Kingfisher Court, Hambridge Road, Newbury, Berkshire RG14 5SJ UNITED KINGDOM

Jack Frazier, CINVESTAV, Apartado Postal 73 "Cordemex," Mérida, Yucatán, C.P. 97310 MEXICO. *Present mailing address:* Conservation and Research Center, National Zoological Park, Smithsonian Institution, 1500 Remount Road, Front Royal, VA 22630 USA

Dean Gibson, Duke University Primate Center, 3705 Erwin Road, Durham, NC 27705 USA

Luis Mariano González, Ministerio de Medio Ambiente, Dirección General de Conservación de la Naturaleza, Gran Vía de San Francisco, 4, 28005 Madrid, SPAIN

Rick Haeffner, Denver Zoo, 2900 East 23rd Ave., Denver, CO 80205 USA

Jerome A. Jackson, Whitaker Center, College of Arts & Sciences, Florida Gulf Coast University, 10501 FGCU Blvd. South, Ft. Myers, FL 33965–6565 USA

Rodney M. Jackson, International Snow Leopard Trust, 4649 Sunnyside Avenue North, Suite 325, Seattle, WA 98103 USA

Brian T. Kelly, Red Wolf Recovery Program, Alligator River National Wildlife Refuge, P.O. Box 1969, Manteo, NC 27954 USA

Solomon N. Kyalo, East African Wild Life Society, P.O. Box 20110, Nairobi, KENYA

Carlos López-González, University of California–Davis, 2400 Pole Line Rd., Apt. 2, Davis, CO 95616 USA

Tiit Maran, Foundation "Lutreola," Tallinn Zoo, Paldiski Road 145, EE0035 Tallinn, ESTONIA

David J. Mattson, USGS Forest & Rangeland Ecosystem Science Center, Department of Fish & Wildlife Resources, University of Idaho, Moscow, ID 83844–1136 USA

Brian Miller, Denver Zoological Foundation, 2900 East 23rd Ave., Denver, CO 80205 USA

Gary S. Miller, 13520 SE Wiese Road, Boring, OR 97009 USA

M.G.L. Mills, South African National Parks, Endangered Wildlife Trust and University of Pretoria, Private Bag X402, Skukuza 1350 SOUTH AFRICA

Numi Mitchell, The Conservation Agency, 67 Howland Avenue, Jamestown, RI 02835 USA

Henry M. Mix, Nature Conservation International, Mainzer Str. 21, 10247 Berlin, GERMANY

Esmeralda Mujica-Jorquera, Zoo Consult, C.A., Apartado Postal 1567, Valencia 2001, Estado Carabobo, VENEZUELA

Erin Muths, USGS-BRD, Midcontinent Ecological Science Center, 4512 McMurray Ave., Fort Collins, CO 80525 USA

Peter Myroniuk, Zoological Parks & Gardens Board, Royal Melbourne Zoological Gardens, P.O. Box 74, Parkville, Victoria 3052 AUSTRALIA

Rodrigo Núñez, Fundación Cuixmala, Apartado Postal 161, San Patricio, Jalisco, C.P. 48980 MEXICO

R. Andrew Odum, Toledo Zoological Society, P.O. Box 4010, Toledo, OH 43609 USA

Jane M. Packard, Department of Wildlife & Fisheries Science, Texas A&M University, Room 210, Nagle Hall, College Station, TX 77843 USA

Michael K. Phillips, Turner Endangered Species Fund, P.O. Box 190, Galletin Gateway, MT 59730 USA

Katherine Ralls, National Zoological Park, Smithsonian Institution, Washington, D.C. 20008 USA

Galen B. Rathbun, Piedras Blancas Field Station, Western Ecological Research Center, U.S. Geological Survey, San Simeon, CA 93452–0070 USA. *Current address*: Department of Ornithology & Mammalogy, California Academy of Sciences, Golden Gate Park, San Francisco, CA 94118 USA

Ravi Chellam, Wildlife Institute of India, P.B. No. 18 (Chandrabani), Dehra Dun 248 001 INDIA

Richard P. Reading, Denver Zoological Foundation, 2900 East 23rd Ave., Denver, CO 80205 USA

Katherine Renton, Durrell Institute of Conservation & Ecology, University of Kent, Canterbury, Kent, CT2 7NJ UNITED KINGDOM

Mary Rowen, U.S. AID, 3133 Connecticut Avenue #519, Washington, D.C. 20008 USA

Vasant K. Saberwal, Institute for Social and Economic Change, Nagarbhavi P.O., Bangalore 560 072 INDIA

Julie A. Savidge, University of Nebraska, Lincoln, Dept. of Forestry, Fisheries and Wildlife, 202 Natural Resources Hall, East Campus, Lincoln, NE 68583–0819 USA

Peter Schuyler, The Santa Catalina Island Conservancy, P.O. Box 752, Avalon, CA 90704–0752 USA

J. Michael Scott, Department of Fish & Wildlife, University of Idaho, Moscow, ID 83844–1141 USA

Michelle Pellissier Scott, University of New Hampshire, Department of Zoology, Rudman Hall, Durham, NH 03824–2617 USA

John Seebeck, Natural Resources & Environment, 240 Victoria Parade, P.O. Box 500, East Melbourne, Victoria 3003 AUSTRALIA

Claudio Sillero-Zubiri, Department of Zoology, Oxford University, South Parks Road, Oxford, OX1 3PS UNITED KINGDOM

Douglas W. Smith, Yellowstone Center for Resources, P.O. Box 168, Yellowstone National Park, WY 82190 USA

Eleanor J. Sterling, American Museum of Natural History, Central Park West at 79th St., New York, NY 10024–5192 USA

Raman Sukumar, Centre for Ecological Sciences, Indian Institute of Science, Bangalore-560012 INDIA

Elides Aquiles Sulbarán, El Campito. Residencia Doña Chepa, Piso 4, Apartamento B-4, Merida, VENEZUELA

John Thorbjarnarson, International Programs, Wildlife Conservation Society, 2300 Southern Blvd., Bronx, NY 10460 USA

Julián Treviño-Villarreal, Instituto de Ecología y Alimentos, Universidad Autónoma de Tamaulipas, Blvd. A. López Mateos #928, 7040 Cd. Victoria, Tam. MEXICO

Amy L. Vedder, Wildlife Conservation Society, 185th Street & Southern Ave., Bronx, NY 10460 USA

Richard L. Wallace, School of Forestry & Environmental Studies, Yale University, 205 Prospect Street, New Haven, CT 06511 USA. *Current address:* Eckerd College/BES, 4200 54th Avenue South, St. Petersburg, FL 33701 USA

Xiaoming Wang, Department of Biology, East China Normal University, 3663 Zhong Shan Road, Shanghai 200062 P.R. CHINA

William Weber, Wildlife Conservation Society, 185th Street & Southern Ave., Bronx, NY 10460 USA

Lauren C. Wemmer, Department of Fisheries and Wildlife, University of Minnesota, St. Paul, MN 55108 USA

Stuart Williams, Institute of Zoology, Zoological Society of London, Regent's Park, London NW1 4RY UNITED KINGDOM